DATE DUE

JUN 02 2008			
SEP 02 2009			

Günther Seliger (Ed.)

Sustainability in Manufacturing

Recovery of Resources in Product and Material Cycles

With 216 Figures

Springer

Editor

Professor Dr.-Ing. Günther Seliger

Technische Universität Berlin
Institut für Werkzeugmaschinen und Fabrikbetrieb
Pascalstr. 8–9
10587 Berlin Germany

seliger@mf.tu-berlin.de

Library of Congress Control Number: 2006939126

ISBN 978-3-540-49870-4 Springer Berlin Heidelberg New York

Springer is a part of Springer Science+Business Media

springer.com

Typesetting and image editing: by Steffen Schneider and Yetvart Ficiciyan, Berlin
Production: LE-TEX Jelonek, Schmidt & Vöckler GbR, Leipzig
Cover: Frido Steinen-Broo, eStudio Calamar, Spain

Printed on acid-free paper 68/3100YL - 5 4 3 2 1 0

Preface and Acknowledgements

Product life in a cycle economy environment has become a leading idea around manufacturing in the 21st century. Global population increase must be stopped by peaceful means, if we want to sustain ecological, social and economical conditions of human life. Therefore increasing the standard of living in hardly developed regions and societies becomes a central challenge. To achieve this goal in global scale and in ecological compatibility, energy and resource consumption per capita must be considerably reduced from the too high level of hitherto technology shaping developed industrial societies. Australia, Canada, Japan, South Korea, USA, Western Europe all together count for less than one fifth of the 2006 global population of around 6.5 billion but consume disproportionate global resources. According to http://www.footprintnetwork.org, since 1985, humanity consumes more resources than can be regenerated by our planet – in 2001 about "1.2 globes". Consequently the use productivity of resources must be increased. Utilizing the dynamic forces of market competition in a framework of legal incentives for developing ecologically friendly technology can support achieving this ecological goal.

If the emerging societies of China, India and Latin America counting for more than one half of global population on their advancement remain within the existing patterns of resource consumption as coined by the current technological products and processes of the developed countries, globe cannot sustain. This is not a long term challenge of the coming century but rather a task for sustainable engineering within the coming two or three decades. This book describes approaches of development in life cycle management around manufacturing engineering addressing issues of sustainability. The contents stem from 12 years of university research within a cooperative project which in a great measure was financed by the German National Science Foundation (DFG).

DFG has granted support for the interdisciplinary Collaborative Research Center (Sonderforschungsbereich, Sfb 281) "Disassembly Factories for Regaining Resources in Product and Material Cycles" from 1995 from 2006. Setting up economical and ecological criteria in life cycle engineering and management and, related to these criteria, developing technology in design for ease of disassembly and recycling, in processes for disassembly, in planning tools for reuse, remanufacturing and recycling, in logistics and urban development have been essential paths of research within Sfb 281. Also, forming a global network of research institutes on life cycle engineering and sustainability in design and manufacturing (www.sustainablemanufacturing.org) was initiated by this activity. Yearly, conferences so far 2004 in Berlin, Germany (ISBN 3-937151-21-4), 2005 in Shanghai, China, and 2006 in Sao Paulo, Brazil, and continuous exchange of graduate and PhD students in joint projects of teaching and research (http://www.gpd.tu-berlin.de; http://www.global-engineering-teams.org) coin this global initiative.

I thank professors Leo Alting from Copenhagen, Denmark, Jay Lee from Cincinnatti and Jun Ni from Ann Arbor, USA, Joao Oliveira and Henrique Rozenfeld from Sao Paulo, Brazil, Shozo Takata from Tokyo, Japan, Wang Chengtao from Shanghai, China, and Prof. Hesselbach from Braunschweig, Germany, who as representatives of this network have contributed to this book. 55 Ph.D. theses and more than 400 scientific publications have been and are published in the framework of Sfb 281. 13 transfer projects in cooperation with 18 industrial partners prove the impact of research results in practical application. I thank my colleagues in Berlin, professors Helmut Baumgarten, Peter Bayerer, Lucienne Blessing, Wolfgang Beitz (†), Hans-Jürgen Ewers (†), Günter Fleischer, Frank-Lothar Krause, Herbert Reichl, Günter Spur and Eckart Uhlmann, who together with their PhD and master students carried out these research activities. Also, I want to thank my former PhD students Ansgar Kriwet, Claudia Hentschel, Martin Wagner, Katrin Müller, Alexander Stenzel, Bahadir Basdere, Carsten Franke, and Sebastian Kernbaum for their strong engagement as managers of Sfb 281. I thank DFG for their large scale financial support of our joint effort, namely Dr. Funk, Dr. Hollmann and Dr. Kunze for their continuous cooperation. I thank all the members of the academic evaluation teams who in four intense inspection procedures 1994, 1997, 2000 and 2003 recommended supporting our activities, namely Prof. Lechner (†), Prof. Renz, Prof. Spath and Prof. Scholz-Reiter as reporters to the senate of DFG. Finally, I thank all the authors contributing to this book and especially my Ph.D. students Sebastian Kernbaum and Marco Zettl who helped me in putting everything together.

Berlin, January 28th 2007
Günther Seliger
Spokesman Sfb 281

Table of Contents

Table of Authors

Ackermann, Robert, Dr.-Ing.
(Chapters 3.2, 3.3)
Institute for Environmental Engineering,
Technical University Berlin

Alting, Leo, Prof., Ph.D.
(Chapter 3.1)
Department of Manufacturing Engineering and Management,
Technical University of Denmark, Kongens Lyngby, Denmark

Baumgarten, Helmut, Prof. Dr.-Ing.
(Chapter 3.6)
Institute for Technology and Management,
Technical University Berlin

Bayat, Nayim, Dr.-Ing.
(Chapter 5.1)
Institute for Machine Tools and Factory Management,
Technical University Berlin

Blessing, Lucienne, Prof. Dr.-Ing.
(Chapter 4.1, 4.2)
Institute of Engineering Design, Micro and Medical Technology,
Technical University Berlin

Butz, Christian, Dr.-Ing.
(Chapter 3.6)
Institute for Technology and Management,
Technical University Berlin

Chengtao, Wang, Prof., Ph.D.
(Chapter 2.4)
School of Mechanical and Power Engineering,
Shanghai Jiao Tong University, Shanghai, China

Consiglio, Stefano, Dipl.-Ing.
(Chapter 5.2)
Institute for Machine Tools and Factory Management,
Technical University Berlin

Djurdjanovic, Dragan, Dr.
(Chapter 7.2)
Cooperative Research Center on Intelligent Maintenance Systems,
University of Michigan, Ann Arbor, USA

Dose, Julia, Dr.-Ing.
(Chapters 3.2, 3.3)
Institute for Environmental Engineering,
Technical University Berlin

El Mernissi, Adil, Dipl.-Ing.
(Chapter 5.4)
Institute for Machine Tools and Factory Management,
Technical University Berlin

Elmer, Carl-Friedrich, Dipl.-Volkswirt
(Chapter 3.5)
School of Economics & Management,
Technical University Berlin

Fleischer, Günter, Prof. Dr.-Ing.
(Chapters 3.2, 3.3)
Institute for Environmental Engineering,
Technical University Berlin

Friedrich, Thomas, Dipl.-Ing.
(Chapter 5.5)
Institute for Machine Tools and Factory Management,
Technical University Berlin

Früsch, Ingo, Dipl.-Ing.
(Chapter 5.3)
Institute for Machine Tools and Factory Management,
Technical University Berlin

Gegusch, René, Dipl.-Ing.
(Chapter 7.3)
Institute for Machine Tools and Factory Management,
Technical University Berlin

Gries, Bruno, Dipl.-Ing.
(Chapter 4.1)
Institute of Engineering Design, Micro and Medical Technology,
Technical University Berlin

Guelere Filho, Américo, Eng. MSc.
(Chapter 4.6)
Department of Production Engineering at the Engineering School of São Carlos, University of São Paulo, São Carlos, Brazil

Harms, Robert, Dipl.-Ing.
(Chapter 5.5)
Institute for Machine Tools and Factory Management,
Technical University Berlin

Hauschild, Michael Z., Ass. Prof., Ph.D.
(Chapter 3.1)
Department of Manufacturing Engineering and Management,
Technical University of Denmark, Kongens Lyngby, Denmark

Herrmann, Christoph, Dr.-Ing.
(Chapter 3.4)
Institute for Machine Tools and Production Technology,
Technical University Braunschweig

Hesselbach, Jürgen, Prof. Dr.-Ing. Dr. h.c.
(Chapter 3.4)
Institute for Machine Tools and Production Technology,
Technical University Braunschweig

Hollan, Robert, Dipl.-Ing.
(Chapter 5.4)
Institute for Machine Tools and Factory Management,
Technical University Berlin

Jungk, Holger, Dipl.-Ing.
(Chapter 4.4)
Institute for Machine Tools and Factory Management,
Technical University Berlin

Kernbaum, Sebastian, Dipl.-Ing.
(Chapters 6.1, 6.2)
Institute for Machine Tools and Factory Management,
Technical University Berlin

Kind, Christian, Dipl.-Ing.
(Chapter 4.4)
Institute for Machine Tools and Factory Management,
Technical University Berlin

Klett, Jan, Dipl.-Ing.
(Chapter 4.2)
Institute of Engineering Design, Micro and Medical Technology,
Technical University Berlin

Krause, Frank-Lothar, Prof. Dr.-Ing.
(Chapters 4.3, 4.4)
Institute for Machine Tools and Factory Management,
Technical University Berlin

Lee, Jay, Prof. Dr.
(Chapter 7.2)
Cooperative Research Center on Intelligent Maintenance Systems,
University of Cincinnati, Cincinnati, USA

Luger, Tobias, Dipl.-Wirtsch.-Ing.
(Chapter 3.4)
Institute for Machine Tools and Production Technology,
Technical University Braunschweig

Middendorf, Andreas, Dipl.-Ing.
(Chapter 7.4)
Research Center for Microperipheric Technologies,
Technical University Berlin

Ming, Chen, Prof., Ph.D.
(Chapter 2.4)
School of Mechanical and Power Engineering
Shanghai Jiao Tong University, Shanghai, China

Ni, Jun, Prof. Dr.
(Chapter 7.2)
Cooperative Research Center on Intelligent Maintenance Systems
University of Michigan, Ann Arbor, USA

Odry, Daniel, Dipl.-Ing.
(Chapter 7.3)
Institute for Machine Tools and Factory Management,
Technical University Berlin

Oliveira, João Fernando Gomes, Prof. Dr.
(Chapter 4.6)
Department of Production Engineering at the Engineering School of São Carlos,
University of São Paulo, São Carlos, Brazil

Ometto, Aldo Roberto, Prof. Dr.
(Chapter 4.6)
Department of Production Engineering at the Engineering School of São Carlos,
University of São Paulo, São Carlos, Brazil

Pietschmann, Nils, Dipl.-Ing.
(Chapter 3.6)
Institute for Technology and Management,
Technical University Berlin

Reichl, Herbert, Prof. Dr.-Ing. Dr.-Ing. E.h.
(Chapter 7.4)
Research Center for Microperipheric Technologies,
Technical University Berlin

Reise, Carsten, Dipl.-Wirtsch.-Ing.
(Chapter 7.6)
Institute for Machine Tools and Factory Management,
Technical University Berlin

Romahn, Andreas, Dipl.-Inform.
(Chapter 4.3)
Institute for Machine Tools and Factory Management,
Technical University Berlin

Rothenburg, Uwe, Dipl.-Ing.
(Chapter 4.3)
Institute for Machine Tools and Factory Management,
Technical University Berlin

Rozenfeld, Henrique, Prof. Dr.-Ing.
(Chapter 4.5)
Department of Production Engineering at the Engineering School of São Carlos,
University of São Paulo, São Carlos, Brazil

Seliger, Günther, Prof. Dr.-Ing.
(Chapters 1, 2.1, 2.2, 2.3, 5.2, 5.5, 6.1, 6.2, 7.3, 7.5, 7.6, 8)
Institute for Machine Tools and Factory Management,
Technical University Berlin

Sönnichsen, Christian, Dipl.-Ing., Dipl.-Inf.
(Chapter 5.5)
Institute for Machine Tools and Factory Management,
Technical University Berlin

Takata, Shozo, Prof. Dr.
(Chapter 7.1)
Faculty of Science and Engineering,
WASEDA University, Tokyo, Japan

Uhlmann, Eckart, Prof. Dr. h. c. Dr.-Ing.
(Chapters 5.1, 5.3, 5.4, 5.5)
Institute for Machine Tools and Factory Management,
Technical University Berlin

Wang, Haixia, Dr.
(Chapter 7.2)
Cooperative Research Center on Intelligent Maintenance Systems,
University of Cincinnati, Cincinnati, USA

Wenzel, Henrik, Ass. Prof.
(Chapter 3.1)
Department of Manufacturing Engineering and Management,
Technical University of Denmark, Kongens Lyngby, Denmark

Zettl, Marco, Dipl.-Ing.
(Chapter 7.5)
Institute for Machine Tools and Factory Management,
Technical University Berlin

1 Introduction

Günther Seliger, Berlin, Germany

Humankind is confronted with global challenges related to economy, ecology and socio-policy, e.g. movement of labour, global warming, and population increase. Since besides the traditional Western world, also emerging countries like China and India with their 1.3 and 1.1 billion residents counting for about two fifths of present global population are demanding the life-style of industrialized countries, the resources and the ecosystem of the globe come under increased pressure. A big challenge of humankind is the global warming effect, implicating accidents in nature. It is estimated that with the end of the 21st century the global average temperature will have increased between 1.4 to 5.8 degrees centigrade compared to the actual values. Reasons are human consumption of fossil fuels, the cutting of forests, and the practice of certain farming methods [1]. Nowadays global energy consumption is around seven times as much as in 1960 [2].

It becomes apparent that industrialized as well as emerging countries have to face these challenges together in order to conserve the resources and the ecosystem of the planet for future generations. On international political level first steps are global treaties as well as regional regulations and incentive systems. The Kyoto Protocol to the United Nations Framework Convention on Climate Change negotiated in 1997 is an example for such a global treaty. Since September 2005, a total of 156 countries agreed on the reduction of their carbon dioxide emissions and five other green house gases [1]. Regional legislations are for example the European Union directives on Waste of Electric and Electronic Equipment (WEEE) [3] and End-of-Live Vehicle (ELV) [4]. Often these activities are addressing only ecological aspects without taking economics and competition into account.

The principle of sustainability as a mission statement for development moves in the spot of attention. Sustainability comes from forestry where it means that only so much wood shall be gained as can be regained by reforestation. The World Commission on Environment and Development defined sustainability in the so called Brundtland-Report "Our Common Future" in 1987 as meeting the needs of the present without compromising the ability of future generations to meet their own needs [5]. The keyword sustainability is widely used by companies, institutions and governments, often synonymous for environmental activities. Global players like e.g., Daimler Chrysler AG and Bayer AG, are publishing their sustainability-related activities in so called sustainability reports. A Global Reporting Initiative (GRI) is aiming on the standardization of such reports along the so called triple-bottom-line of economy, ecology and sociopolicy [6].

Being proactive and innovative with respect to sustainable development is often regarded as successful rather in the long term than in short term. Companies considering sustainability issues are perhaps better acknowledged because of their strategic risk management. An indicator for this might be the stronger growth of the Dow-Jones-Sustainability-Index (DJSI) compared to the Dow-Jones-Index (DJI). Companies implementing aspects of sustainability are listed in the DJSI, which is a subset of the DJI [7].

It is within this framework of ecological, economical and social challenge that aspects of design and manufacturing with special reference to disassembly and recycling in life cycle management and engineering are addressed in this book. Design and manufacturing considerably determine how wealth is created and what the relation is between useful functionality and exploited resources.

DFG supported Collaborative Research Center on Disassembly Factories for the Recovery of Resources in Product and Material Cycles (Sfb 281) from 1995 to 2006 has connected engineering disciplines of manufacturing and information technology, design, environmental sciences, and architecture and also economists for the design of a political framework for ease of sustainable development.

Throughout the whole product life cycle, products are adapted to different usage phases by disassembly and reassembly in maintenance and remanufacturing, thus reaching the limit of the overall wear. A new paradigm in production industries, allowing maximal use with minimal resource consumption, competes with traditional economical models like bigger physical quantities for piece cost decrease. New built-operate-transfer concepts, facility management, business areas in service and maintenance or leasing are indicators, showing that ecological strategies for long-life cycles, increasing the length of usage phases and the intensity of use become more and more economically competitive. If functionality, according to type, location, time and date, quality and quantity, can be provided at competitive prices, purchasing the use instead of a comparable product becomes more attractive.

For the adaptation of products and components to new usage phases, information about condition and usability can be provided continuously. The concept of the Life Cycle Unit (LCU) proposes to provide adaptation relevant data on the product design and condition for all the different life cycle phases. Universal tools and processes contribute to a flexible and cost effective disassembly. In interlinked disassembly systems with automated and manual workplaces, disassembly processes are integrated. The flexibility and productivity of the processes are experimentally investigated. With concepts for product modularisation, the ability to use products in multiple usage phases is increased. Applying design for recycling concepts decrease the effort for the adaptation of old products at the end of the respective usage phase. Integrating distribution and reverse logistics contributes to ease of access on remanufactured products. Modular equipment and facilities are objects of effective and efficient cycle economy.

So are planning tools for the fast adaptation of factories to continuously changing programs of products to be remanufactured. Respective information technological architectures have been implemented. Thus, the gap can be bridged between temporally separated functional areas of product planning and development, manufacturing, use and reuse, and recycle in a cycle economy. Computational simulation helps with fast analysis of a large number of alternatives and to exclude recognized errors in real processes.

Complex ecological valuation gets a function of guidance as a manageable tool for a continuous ecological improvement of product and process design. Economical potentials of life cycle engineering have been assessed by the development of a framework for efficient environmental policies.

During the first funding period (1995-1997), disassembly was addressed as an end-of-life option. Emphasis was put on increasing the efficiency of disassembly processes. Equipment and tools supporting disassembly were developed. In the second funding period (1998-2000), disassembly and reassembly have been considered with respect to the various adjustment processes, which components undergo during respective product life cycles. Approaches for the recovery of components and material were designed under economical, ecological and technological criteria. A prototypical hybrid disassembly cell including automated and manual operations for washing machines was set-up. The third funding period (2001-2003) involved the identification of business cases where disassembly technology can provide competitive advantages. The collection, disassembly, remanufacturing and reassembly of used electronic consumer goods and automotive components were analyzed by scenarios and identified as promising areas. Based on the example of mobile telephones, ecological and economic evaluations were carried out and logistic take-back concepts implemented; a hybrid disassembly system for disassembly-friendly mobile telephones was designed and built up. A considerable part of these activities was conducted in cooperation with both industry and research partners from the global network of the Sfb 281. In the fourth and final funding period (2004-2006), the Sfb 281 could achieve the following results of research:

- Guarantee a fast and cost-efficient disassembly of components, for recycling or re-use purposes, by modular and highly flexible processes, equipment and tools.
- Increase the operational availability of the disassembly system by making it flexible in regard to different products and their respective different conditions. For this purpose, a macro-oriented cell control and a protocol for data exchange were developed.
- A conceptual tool for module configuration according to the economic and ecologic evaluation of products and processes has been developed. Reuse-friendly, modular product structures are considered in this tool.
- Procedures for planning disassembly on demand have been implemented, also for setting-up and adapting disassembly networks dependent on part programs to be disassembled.
- Obstacles for global cycle economy and methods to overcome them have been identified.
- A holistic method for valuating different products according to Life Cycle Assessment (LCA) criteria was generated and implemented for practical application in Product Development Processes (PDP).
- Information technological tools and infrastructure for design methodology, product condition monitoring and adaptation were developed.
- A disassembly knowledge platform (http://demonty.ipk.fraunhofer.de) accessible by internet has been set-up to communicate the results of the Sfb 281 to interested public domain. The platform is designed as a tool for both learning and knowledge management in development processes.

This book is kind of final report about the research activities within Sfb 281 supplemented by contributions given by outstanding partners from the developed global network on sustainability in engineering. The chapter on global framework presents changes in distribution of wealth in global scale and increase in use productivity of resources as goals to be set. Directions are described on how innovation especially in globally networked university education might be managed. Paths of technology development for increasing the use productivity of resources along the product life cycles are identified. An insight is given on how China as the population wise biggest country on globe is dealing with sustainable recycling of automotive vehicles.

The chapter on life cycle engineering and management is introduced by presenting an intense analysis of international state of the art and future perspectives. Complex tools for Life Cycle Assessment (LCA) are adapted to feasible Product Development Processes (PDP). Reasonable environmental targets and appropriate incentives to reach these targets in a legal and institutional framework are introduced. Reverse logistics help implementing cycle economy.

Product and process development according to sustainability are contents of the following three chapters on product development, processes and tools for disassembly, planning for reuse, remanufacturing and recycling. Concrete developments of Sfb 281 on technical equipment for disassembly integrated in a hybrid disassembly cell for automated and manual operations are described. Software tools for planning and design with respective information and communication technological infrastructures [8] are introduced. Brazilian perspectives on product development and cleaner production are given in the chapter on product development.

The chapter on enabling for sustainability in engineering demonstrates how new strategies on maintenance combined with micro system technology for supervision and communication might change the paradigm of selling products to customers into the paradigm of selling functionality to users. Thus price wise competitivity enabled by technology contributes to environmental performance in the sense of increased use productivity of resources. Modularity contributes to ease of remanufacturing. Competence management in life long education helps implementing sustainable processes and products.

Finally a sketch of a roadmap is given for future research and development on sustainability in engineering.

References

1 http://unfccc.int (Access: October 2006).

2 Wackernagel, M.; Ecological Footprint Accounting – Comparing Earth's Biological Capacity with an Economy's Resource Demand. In: Keiner, M. (ed.): The Future of Sustainability. Springer-Verlag, Berlin, Heidelberg, New York 2006, pp. 193-209.

3 The European Parliament, the Council: Directive of the European Parliament and of the Council on Waste from Electrical and Electronic Equipment (WEEE). Brussels, 2002.

4 The European Parliament, the Council: Directive of the European Parliament and of the Council on End-of-Life Vehicles. Brussels, 2002.

5 N.N.: Our Common Future. The World Commission on Environment and Development. University Press, New York, Oxford 1987.

6 Global Reporting Initiative: www.globalreporting.org (Access: October 2006).

7 Sustainability Dow Jones Index: www.sustainability-indexes.com (Access: October 2006).

8 http://demonty.ipk.fraunhofer.de (Access: November 2006).

2 Global Framework

This chapter is addressing aspects of how human living can be sustained on our planet earth. Challenges of sustainability are in overcoming the unequal distribution of wealth in different regions of the globe. Also the global population growth must be stopped by better standards of living without increasing resource consumption. A reference model of management and technology, innovative processes and products by modern research and education in different societal frames is presented. University education in student project teams from different continents developing and realizing innovative products and processes according to criteria of sustainability are identified as new means to accelerate creative imagination and implementation. The power of student engineering can help overcoming the barriers of management in established institutions often bound to traditional thinking habits and short term daily requirements. Approaches towards sustainability in engineering with focus on manufacturing are presented and illustrated by actual developments on vehicle recycling in China.

2.1 Wealth Distribution and Use Productivity

Günther Seliger, Berlin, Germany

Sustainability is directed to enhancing human living standards while improving the availability of natural resources and ecosystems for future generations. More than half of global value creation of today is achieved by less than one tenth of the global population. A sustainable political, economical and social stability can only be achieved if mankind is able to create jobs and living conditions of human dignity worldwide and not only in the technologically developed regions of East Asia, North America and Western Europe.

Human creative imagination, knowledge, experience, skills and initiative by entrepreneurial action coin our living conditions. Overcoming poverty requires a minimum of resources, abilities and qualification provided for everyone in the world to take initiative for value adding activities. On the other hand equal level of wealth for everyone with hardly any differences in people's living conditions takes away incentive for useful engagement. The course of accumulate income earned by a community population from the poorest to the richest member is described by the Lorenz curves. The equity factor relates individual welfare to overall equity (Figure 1).

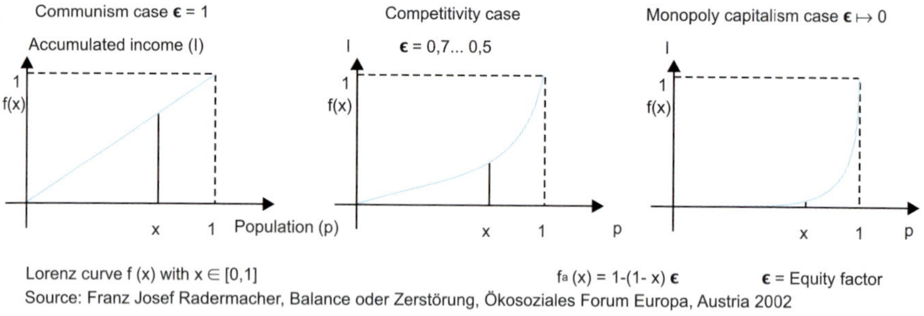

Fig. 1: Lorenz curve and equity factor: quantitative instrument for describing the inequality in a society [1]

Empirical statistical data show significant differences in equity factors within different countries and regions all over the globe (Figure 2). In reaching sustainability within the present globalisation process the crucial rule of achieving a higher level of equity worldwide is considered to be essential. More equity is a key concept discussed in the European Union to find consensus between regions of different levels of wealth. It is based on strictly limiting environmental burdens and co-funding development [2]. It is integrated in the idea of tenfold growth with tenfold eco-efficiency [3].

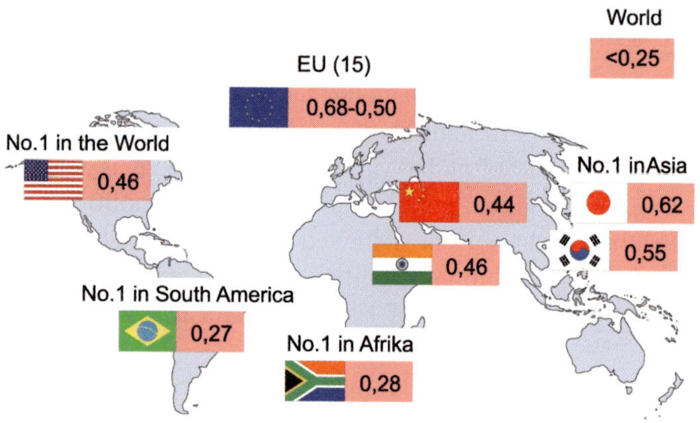

Fig. 2: Differences in equity factors [1]

The exponential growth of population on the globe must be stopped. Apart from unacceptable developments as war or disease a higher standard of living is the only means to reduce population increase. And if this shall happen without exceeding ecological limits the use productivity of resources must be raised (Figure 3).

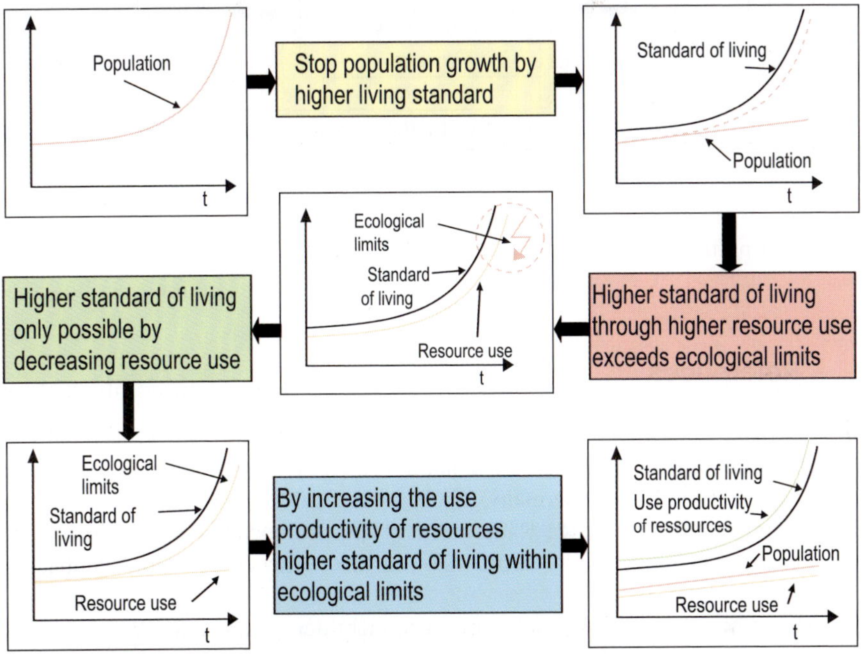

Fig. 3: Logical path for increasing use productivity of resources [1]

Engineering challenges derived from this requirement are design of processes and products with improved usefulness and less environmental harm. Technology interpreted as science systematically exploited for useful purposes offers huge potentials to contribute. Technology enables for processes transforming natural resources into products to meet human needs. The interaction between research and education imposes dynamics on how creative solutions are developed for relevant tasks. Due to new means of transport and communication knowledge and value creation is no longer limited to niches of wealth but more and more accessible by everyone from everywhere at anytime. These dynamics must be mastered by management considering chances for cooperation and risks of competition. Different societal frames with different value systems considering economical, ecological and sociopolitical issues in different regions of the globe have to be taken into account (Figure 4).

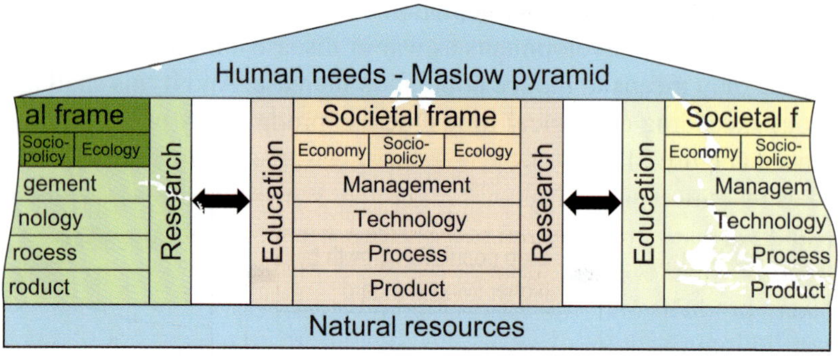

Fig. 4: Engineering challenges by sustainability [1]

References

1 Seliger, G.: Global Sustainability – A Future Scenario. In: Proceedings of the Global Conference on Sustainable Product Development and Life Cycle Engineering, Berlin, Germany, 29.09. – 01.10.2004, pp. 29-35.

2 Kämpke, T.; Pestel, R.; Radermacher, F.J.: European Journal of Law and Economics, vol. 15, iss. 2, pp. 129-163 (35), Kluwer Academic Publishers, March 2003.

3 Schmidt-Bleek, F.: Wieviel Umwelt braucht der Mensch? MIPS – Das Maß für ökologisches Wirtschaften, Birkhaeuser, 1993; English translation: The Fossil Makers – Factor 10 and More, available at www.Factor10 Institute.org (Access: November 2006).

2.2 Education by Projects

Günther Seliger, Berlin, Germany

Management is challenged by innovation in sustainability. Innovation is defined as invention plus diffusion i.e. success on the market [1]. Both education and business are essential factors for the success of sustainability in a competitive market environment.

Today products and processes emerge by cooperation of generalists and experts in interdisciplinary teams. Project success depends considerably on the ability to communicate, present and document information. Another element of success is the ability to exploit creative potentials for useful applications (Figure 1). The balance between cooperation and competition among partners in development networks has to be diligently kept [2]. The global orientation of more and more big and small companies pushes international cooperation and calls for team members whose social as well as intercultural competencies are well developed.

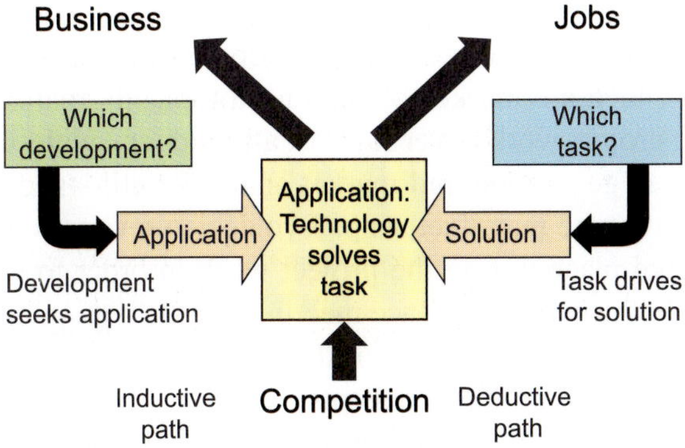

Fig. 1: Inductive and deductive innovation path [1]

Compared to ex-cathedra teaching work on projects in study courses enables students to achieve competencies and skills in addition to technical knowledge as well as to develop their personal attitudes.

In global experimental teaching courses [4, 5] jointly organized by universities from different continents the objective is on integrated transfer of methodical and social competencies utilizing expertise gained by means of case studies [3]. The few months long courses consist of internet-lectures organized by the teaching staff of the cooperating universities and a project-oriented work in continentally mixed student teams. During the project work, the students develop and prototype globally marketable technical products. The modular design of the products offers chances of adaptation to various global regions. Two meetings, early and late in the respective term, supplement the virtual meetings via internet communication by face to face communication of all participants.

Lectures are transmitted via modern multi-point and broadband video-conference systems, which allow location independent dialogues between students as well as between students and teaching staff. Moreover, international experts and industrial partners are able to communicate and supervise the students. The lectures refer to the topics marketing, conceptual design, product engineering, manufacturing and distribution.

The project work is based on the initiative of the students, who are assigned into multi-national teams of six students with two from each of the respective countries. The students choose the product to be developed by themselves based on a generic presetting. E.g. an “internet-ready product”, a so-called “multiple-use-product” or a so called “learnstrument” has to be developed by the student teams. The idea of application in different markets of developed and emerging countries is considered. Also projects to be realized by student teams are given by industrial cooperation partners.

Student project teams are directed to developing technical products for the global market, which are marketable and prototypically realizable by their own skills. The project work covers almost all product development phases, from product idea generation and market analysis until standard assembly drawings and prototype manufacturing. Two design reviews combined with presentations take place during the project work in order to determine the team progress. In the two face to face meetings, at the beginning and the end of the course, the participants meet for the direct exchange of ideas and the implementation of their product concepts.

Experiences gained from education project course work indicate deficits concerning methodological and knowledge competencies in the cooperative development of product concepts, in the development of assessment criteria, in the documentation and communication of solutions and experiences, in the cooperative development of initiative, in the short and clear formulation using synchronous communication media. Deficits with respect to personal and social competencies have been identified in the cooperation before the first personal meeting, in the mutual respect for personal and cultural peculiarities, in handling of own expertise limits, in recognition of and intervention for solving technical and interpersonal problems i.e. said different from heard different from understood different from accepted different from own action. Deficits in competencies of decision-making and taking over responsibility could be discovered in the team decision process, in the sharing of work between the students under consideration of their different qualifications and preferences, in the balance of details and range of development, in the balance of work load and development quality. Organizational issues could be improved in the balance between management and control and moderation of teaching staff, in the coordination of lecture contents and knowledge necessary for project work, and in the resources availability due to regressive budgets of universities.

Continuous education becomes a crucial task of management in an industrial and societal environment which, due to technological developments and their accessibility by everyone from anywhere at anytime, is more and more coined by innovation. Success in running paths of sustainability is considerably dependent on how people can be convinced. Leadership in modern management requires abilities in teaching.

References

1 Seliger, G.: Product Innovation – Industrial Approach. Annals of the CIRP, Vol. 50/2/, 2001.

2 Seliger, G.; Karl, H.; Weber, H.: Co-operative Design, Manufacturing, and Assembly of Complex Products. Annals of the CIRP, Vol. 46/1/, 1997.

3 Kim, J.-W.; Kim, D.-M.; Consiglio, S.; Severengiz, S.; Seliger, G.; Patil, L.; Dutta, D.: A Global Collaboration to Teach Global Product Development: Faculty Perspectives. In: Proceedings of the 2006 ASEE Annual Conference and Exposition, CDRom, Chicago, IL, June 18-21, 2006.

4 http://www.gpd.tu-berlin.de (Access: November 2006).

5 http://www.global-engineering-teams.org (Access: November 2006).

2.3 Processes and Products

Günther Seliger, Berlin, Germany

Another challenge for management on their way to sustainability is how to apply chances of economic competitivity for ecological goals. Today, the remanufacturing of expensive, long living investment goods, e.g. machine tools, jet fans, military equipment or automobile engines, is state of the art. Remanufacturing has to be extended to a large number of consumer goods with short life cycles and relatively low values. Reuse is an alternative to material recycling to comply with recovery rates and quantities as well as special treatment requirements as prescribed by European legislation with the directive on Waste of Electrical and Electronic Equipment (WEEE) [1].

Some remanufacturing cases are widely known, e.g. the remanufacturing of single use cameras (Eastman Kodak and Fuji Film), toner cartridges (Xerox), photocopiers (Fuji Xerox, Australia, Netherlands and UK), commercial cleaning equipment (Electrolux) and brand name computers (IBM, France, Germany, USA; HP, Australia). Remanufacturers are OEMs themselves who have integrated new distribution models such as leasing or "pay per use" with remanufacturing strategies. Other remanufacturing practices, e.g. for washing machines (ENVIE, France), personal computers (ReUse network, Germany), accumulators (teldeon, Germany), cordless phones, car stereos, FM radios (Topp Companies, USA) and mobile phones (ReCellular, USA; Greener Solutions, UK) are less popular, due to the fact that OEMs are not involved. Products are not sold through regular retail channels established by OEMs [2].

Specific economical chances for remanufacturing can be identified in the selling use approach. Only the use of the product and not the product itself is the object of the trade. Product manufacturers or third party companies can act as use providers. The use provider offers the functionality of the product to the customer without passing the product out of his possession. He is responsible for the accessibility of the required utilization and the treatment of equipment over the entire lifetime. The use provider manages the costs of investment, transport, operation, maintenance and disposal. Consequently, the customers only pay for the use that they obtain by the product and not the product itself.

Selling use becomes more competitive, if the product providing the use can be adapted to multiple usage phases. Adaptation processes are necessary due to the change of technical features and user needs. Kinds of adaptation are maintenance, repair, remanufacturing, up- and downgrading, enlargement and reduction as well as, rearrangement and modernization. Adaptation often requires disassembly and reassembly. Additional processes are cleaning, treatment, component supply and removal, inspection, and sorting. A balanced strategy of preventive maintenance and repair preserves or increases the residual value of the products.

Fig. 1: Exemplary products enabled for sustainability [3]

Requirements on products for the selling use approach are modularity, integrability, customization, convertibility as well as diagnosability to support customer driven adaptability. The implementation of these properties increases the applicability and the availability of a product in multiple usage phases. In the selling use approach the use provider is responsible for the availability of the use at the right place and time in adequate quality. Therefore the use provider needs system-accompanying quality management, information and communication systems to guarantee product pursuit and product access. Leasing-, rent- and service-contracts regulate the responsibilities between customer and use provider. Selling use becomes competitive to selling products, once the idle capacity costs are higher than the extra costs to be paid for logistics, information and communication. Logistics include all necessary processes to provide the demanded use at the right place and time in adequate quality. Thus the effort for adaptation between two usage phases and the transport of the product are included as well. Modern information and communication give real time access for suppliers and customers on demand and availability of the specified function of use to be met by respectively adapted products. The Figure 1 depicts examples of products from Sfb 281 that were enabled for sustainability by means of process and product design and changes for new business models [3].

References

1 Directive 2002/96/EC of the European Parliament and of the Council on Waste Electrical and Electronic Equipment (WEEE) of 27 January, 2003.
2 Franke, C.; Seliger, G.; Yakut, E.: Mobile Phone Remanufacturing Program Planning in Networks. In: Proceedings of the Global Conference on Sustainable Product Development and Life Cycle Engineering, Berlin, Germany 2004.
3 Seliger, G.: Global Sustainability – A Future Scenario. In: Proceedings of the Global Conference on Sustainable Product Development and Life Cycle Engineering, Berlin, Germany 2004, pp. 29-35.

2.4 Sustainable Recycling of End-of-Life Vehicle in China

Wang Chengtao, Chen Ming, Shanghai, China

By the end of year 2005, the volume of in-use vehicle was 35 million in China. It was estimated that the volume of End-of Life Vehicle (ELV) was 2.1 million and nearly 60 % of which were dismantled while others were still running. With the rapid development of automobile industry in China, it is estimated that, by the year 2010, the volume of in-use vehicle will be 55 million, and the end-of life vehicle will be 3.3 million; by the year 2020, the volumes will be 100 million and 8 million, respectively (Figure 1). At the year of 2001, Decree #307 about ELV disposal was issued by State Department of P. R. China. This is the highest law concerning about the disposal of ELV in China today. The ELV recycling industry is full of opportunities and challenges in China.

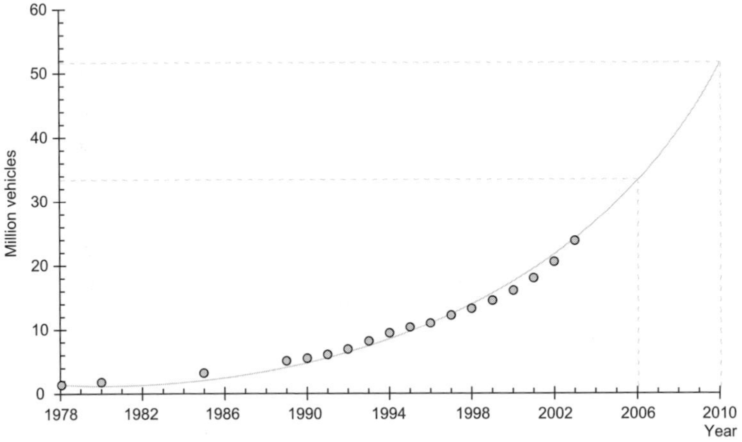

Fig. 1: Volume of in use vehicle in China

By the end of year 2005, Chinese government had implemented her 10th Five Year Plan (2001~2005) and set up 11th Five Year Plan (2006~2010) for the development of national economy. But along with the economic rapid growth, the issues of environmental protection and social inequity and inequality have also become increasingly tangible, and choice a sustainable way for China become stringent and urgent. On May of 2004, the National Development and Reform Commission (NDRC) issued the "Automobile Industry Development Policy". This was a policy for economic development of Chinese automobile industry mainly aiming at "making the automobile industry a pillar industry in the national economy by the year 2010" and neglecting systematically addressing sustainability of automobile industry.

A new regulation about "The Motor Vehicle Product Recovery Technology Policy" was issued on February of 2006 by the NDRC, the Ministry of Science and Technology and the State Environmental Protection Administration (SEPA). According to the regulation, Chinese automobile manufacturers and importers will be responsible for collecting and recycling vehicles from 2010 or designating other authorized end-of life vehicle dismantlers to handle the business. The new regulation was designed to encourage domestic automobile producers to use more environmental benign materials. According to the regulation, Chinese automobile manufacturers are banned to using hazard and difficult-to-recycle materials in manufacturing. The new policy will surely raise the costs to automobile manufacturers and importers and poses a higher demand on technology innovation. But over the long term, it will help to increase the sustainability of the Chinese automobile industry.

2.4.1 Recycling Industry

Without professional equipments and management specifications, and, with small production scale, low dismantling efficient, low recycling rate, and environment pollution, it is that Chinese ELV dismantling industry is now facing. Almost all ELV dismantling plants have little idea about environmental protection and scientific, reasonable disposal technique. The waste fluids are not removed before dismantling process. Today, most of ELVs are dismantled manually in China with pneumatic and electric tools and oxyacetylene cutting because of the low cost of labour. The dismantling processing is random and unreasonable, for example, in order to dismantle a part for sale, the connection or other parts around it will be destroyed possibly. The secondary-parts are sold without any inspection and processing, partly because customs would like to see the condition of ELV from which the part is dismantled and determine its quality and price, and partly because the plants have no capability to processing and inspection. Shredding process is simplified to cut metal into big pieces for convenient transportation other than to grind into fist-like pieces.

Scopes of ELV

Depending on Chinese ELV standard (Issued by formal National Committee of Economy and Trade) the civil vehicles which is registered in China, should be discarded as ELV depending the following technical specifications:

- mini size commercial vehicles, including cross-country type, total mileage up to 300,000 km,
- light commercial vehicles, including cross-country type, total mileage up to 400,000 km,
- heavy, medium commercial vehicles, including cross country type, total mileage up to 400,000 km,
- passenger vehicles, including cross-country type, total mileage up to 500,000 km,
- others, total mileage up to 450,000 km,
- mini size commercial vehicles, including cross-country type, commercial vehicles with trailer and taxicab, total service period up to 8 years,
- light commercial vehicles and others, total service period up to 10 years;
- the vehicles whose total service period over the fixed number of year as mentioned above, except the taxicab with less than 19 passengers, light and mini size commercial vehicle including cross-country type, could prolong their service period up to half of the fixed number of year as mentioned above on the condition of inspection up to grade by vehicle administration according to national vehicle exhausts standard (GB7258—2004).

Characteristics of ELV in China

The characteristics of ELV in China are mentioned as following:

- There is difference in condition of ELV of eastern, central and western regions of China. The ELV of eastern China is in better condition than that of western China because of the advancing in economy of eastern China and consequently the higher requirement of vehicle users;
- There is a big number of end of life commercial vehicles in China, and the proportion of personal car is small at present. It is a tendency of increasing in proportion of end-of-life personal car from now on;
- There is a tendency of increasing in obsolete imported vehicles because of the big imports of vehicle in early 1990s. Passenger cars accounted for the biggest proportion in imported vehicles. For instance, in 2000, the proportion of passenger cars was 51 % (21,614 vehicles) and the proportion of cross-country vehicles was 20 % (8,668 vehicles). The proportion

of imported trucks and buses added up to less than 30 %. The recycling of end of life imported vehicles is mainly in metal materials because the utilities of vehicle are limited to business and official, and therefore, there is no market for reused parts in this stratum. On the other hand, some domestic auto parts remanufacturing plants imported cores from US and Japan, and sale abroad after remanufactured. So the gaps between the recycling of end of life imported vehicle and the outbound remanufacturing of auto parts and assemblies should be considered.

Patterns of ELV Recycling Industry in China

There are four scopes of industry in ELV recycling and remanufacturing in China:

- reverse logistics and dismantling of ELV,
- remanufacturing of typical parts and assemblies,
- recycling of materials;
- waste disposal and environmental protection in ELV recycling industry.

There are six kinds of ways to reuse and recycle materials after ELV dismantling in China:

- *Parts reuse:* Some reusable parts come into vehicle after-sales market or return to vehicle manufacturer for assembling;
- *Reuse after remanufacture:* The remanufactured parts mainly come into vehicle after-sales market for maintenance;
- Reuse in debasement: Some parts are not qualified in vehicles but could be applied in other products or services, such as the obsolete tires in bank of river;
- *Material recycle:* Most of recyclable materials, such as metals and plastics, are reclaimed as recycled materials;
- *Energy recovery:* Some materials, such as tires, lubricating oils, are recycled by the way of incineration, and the recovered energy could be used for recycling industry itself or other applications;
- *Landfill:* The shredder residues have to be land filled in junk yards.

Vehicles Recycling and Reverse Logistic Infrastructure in China

There is difference in economic development of eastern, central and western regions of China. It comes into being the characteristic Chinese economic development pattern. The regional difference in economic requirements and consumptions leads to the obvious difference in vehicle consumptions. The new vehicles begin their life cycle in eastern and end of life in western. The development and layout of Chinese vehicle recycling industry should consider this social fact. Therefore, we propose a model of vehicles recycling and reverse logistic infrastructure in China (Figure 2).

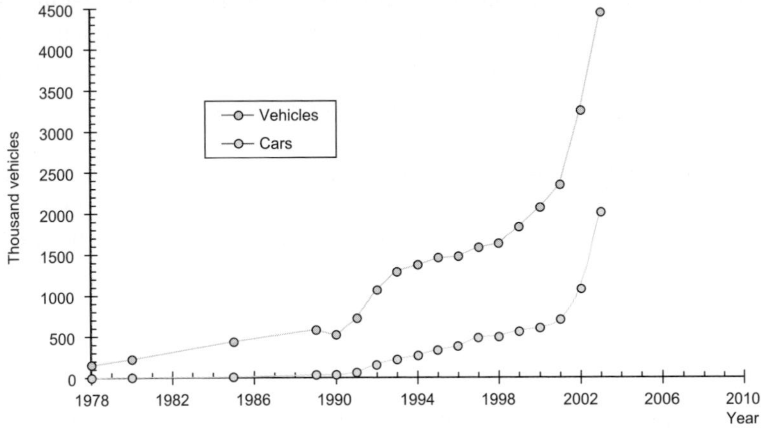

Fig. 2: Production of Vehicles in China

In central and western regions of China, a lot of ELV dismantlers should be established and accept obsolete vehicles. After being dismantled, the auto bodies, tires, plastics and used oils should be treated locally. For instance, bodies are recycled as scrap steel; tires and plastics are recycled as crushed powder. The effective and low-cost process and equipment should be developed and applied in dismantlers of this regions.

The repairable parts and assemblies which are collected in dismantlers or service stations should be sent to remanufacture industry after gathering of a certain amount. The remanufacturing enterprises could be specialized or comprehensive, and could be independent or belonged to vehicle manufacturers. The remanufacturing parts came into spare parts market. The failure parts in service stations are replaced and treated as the same as those in dismantlers.

Dismantlers collect direct reusable parts and send to Original Equipment Manufacturers (OEM) or vehicle manufacturers respectively via logistics between service stations and vehicle manufacturers. The automotive suppliers should establish their specialized workshops or departments for dealing with those parts which could be reassembled after strictly quality checking and testing.

2.4.2 Remanufacturing Activities

A Shanghai-based end-of-life vehicle recycling pilot demonstration system, which could disassemble obsolete commercial vehicles for spare parts, and recycle the rubber, plastic and metal material out of which vehicles are made. The project is the first ELV recycling system in Shanghai after the promulgation of Decree #307, and may lead to the establishment of other similar systems in regions at prefecture level and above with qualification cognizance all over China.

The dismantling plant, which is the heart of the system, includes equipment for disassembling big size commercial vehicles (vehicle mass >5 ton), including pre-treatment for the safe, environment-friendly removal of liquids, and also capability for separating the different materials for recycling. Combined, the rubber, plastic, metal and reused parts and assemblies are weighed more than 12,500 ton/a, and annual profit is about RMB ¥ 2,280,000 (i.e. US$ 274,368). Preliminary indications are that the project could be profitable and sustainable in China. Figure 3 shows the layout of pilot demonstration ELV dismantling plant.

However, the most significant advantages are the low cost of labour, which might allow the low cost facility to process vehicles, as well as the large potential source of vehicles and the significant numbers of consumers of reusable and remanufactured parts.

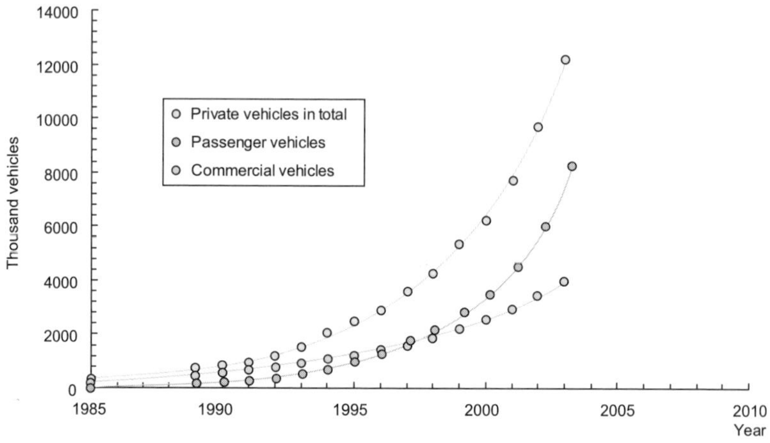

Fig. 3: Volume of private vehicles in China

The dismantling plant is housed in a facility of at least 20,000 m^2, much of which is used for storage. However, there will be space to expand the facility to incorporate some of the material specific recycling and parts remanufacturing modules. For example, today, an auto engine parts remanufacturing workshop is on establishing. It will result in an expansion of the secondary-parts market in China, which would compete with new, OEM-compliant and -guaranteed parts. And, the remanufacturing and sales of parts are likely to be important parts of the pilot system's success.

Shanghai Jiao Tong University, with financial support from the Science and Technology Commit of Shanghai Municipal Government, is collaborating with joint venture entrepreneurs who manage the facility. The university is responsible for environmental issues, and for the co-ordination of the factory layout and remanufacturing issues.

In fact, the level of development on ELV recycling industry in China is still low, and Shanghai should be responsible for establishing an pilot demonstration engineering system on ELV reduce, reuse and recycle, prompting the sustainable and orderly development of national-wide ELV recycling. Therefore, the goal of the pilot demonstration system is to establish an ELV recycling engineering system which matches to the reputation of international metropolis of Shanghai, and thoroughly remould the ELV recycling industry from extensive to intensive and environmentally benign industry and thereby promoting the sustainable utilization of natural resource and advancing the orderly development of ELV recycling. Depending on the policies and laws about ELV disposal including the Decree #307 issued by State Department in 2001, Shanghai municipal government consolidated the ELV recycling industry by means of administrative command, and the former 14 official designated ELV recycling enterprises were cut and merged to present 4 enterprises.

Technical Innovation

To achieving the goal mentioned above, the ELV pilot system project is focused on the following key finds and technical innovations:

- *A scientific and practical administration system for ELV recycling in Shanghai is established.* An municipal-wide information platform for ELV recycling administration is constructed. It is made up of "Shanghai ELV Administration System" and "Operation Flow Management System Software for ELV Dismantler", and it carries out the multi-layered and multi-noded functional management of monitoring, coordinating and administration among municipal governmental departments, ELV dismantler managing departments and ELV dismantling workshops. The recommendations for ELV related local policy and administrative regulation are proposed as well.

- *A pilot demonstration ELV dismantling plant is established.* The project focused on following 6 tasks:
  - demonstration of visual image and corporation image (VI/CI),
  - demonstration of recycling process of ELV,
  - demonstration of facilities for dismantling,
  - demonstration of clean production of ELV dismantling,
  - demonstration of logistic and storage of ELV dismantler,
  - demonstration of reused parts and assemblies examination and remanufacturing.

The technique innovations of the project focused on following:

Based on the abundance of Chinese labour and the increasingly strict requirement of environmental protection, a systematic green production specification and technological process is developed for ELV dismantling plant, and it results the improving rate of ELV recycling. The proportion of reused parts is raised from 22.5 % up to present 31.7 % in weight and the landfill is reduced from 10 % to 4 %.

Based on the requirement of residual fatigue life prediction for reused parts, an evaluation measurement and patented apparatus are developed for residual fatigue life prediction of used auto crank, which is taken as experimental objective. The dangerous sections of used auto crank are evaluated for cracks and stress concentration via eddy current testing and metal magnetic memory inspecting respectively, and therefore, it comprehensively evaluated whether the used parts are worthy of being remanufactured.

- *A primary pilot demonstration network of reused auto parts is established* (Figure 4). It includes some downstream designated reuse and remanufacture industrial enterprises which is specialized in reclaiming and disposal of used auto parts. And the aspects of corporation image, qualification, technological process, equipments and environmental protection are of comparative sense of demonstration.

Effects in Practise

Economic feasibility – The general investment of Shanghai pilot demonstration system of ELV recycling is about RMB ¥ 10 million (i.e. US$ 1.2 million), and its annual profit and revenue duty in total is about RMB ¥ 4,100,000 (i.e. US$ 496,368). The rate of return on investment is 22.8 %.

Achievement in recycling – The proportion of reused parts is raised from 22.5 % up to present 31.7 % in weight and the landfill is reduced from 10 % to 4 %.

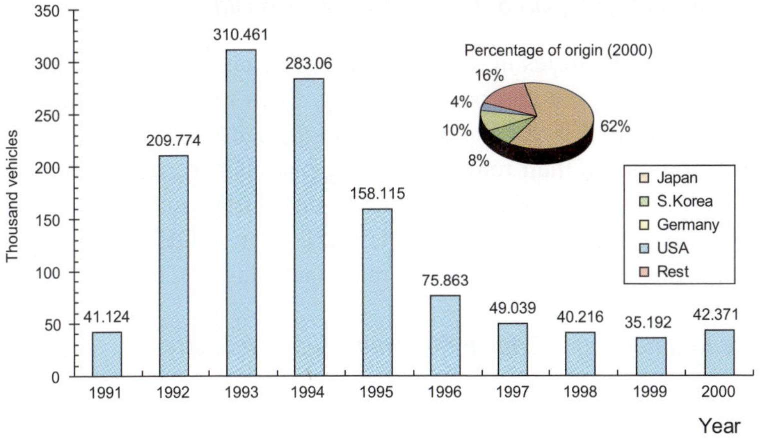

Fig. 4: Imported vehicles in China

Engine Remanufacture

There are, and only, two engine remanufacturing factories in China. One is for *Volkswagen* engine remanufacturing in *Shanghai*, (*Shanghai Volkswagen* Allied Developing Co., LTD.) and the other is for *Styr* engine remanufacturing in *Jinan*, *Shandong* Province (*Jinan Fuqiang* Power Co., LTD.), which is a Sino-Britain joint venture enterprise and a member of Production Engine Remanufacturers Association (PERA, www.pera.org).

Because of the potential market in remanufacturing, plenty of experienced technicians and lower cost of labour, nowadays, China is becoming a hot area for remanufacturing operations and investments. For example, an US based engine remanufacturing equipment provider, Sunnen Mechanical Co., LTD., has landed in China for two years. Sunnen says it will provide not only the remanufacture equipments but also the technologies and services for Chinese engine remanufacture industry. Caterpillar (Caterpillar China Investment Co., LTD.) has also established its Asia & Pacific Remanufacturing Services Office in Shanghai, today. However the engine remanufacturing is still restricted and controlled in China, because of worrying about the illegal piecing together the vehicles as mentioned before.

Dynamotor, Actuating Motor Remanufacture and Others

There are some factories in remanufacturing dynamotors, actuating motors, constant velocity joint and power steering. Such remanufacturing factories are small and flexible, and, some of them are automotive parts OEM supplier. The quality of their remanufacturing parts is perfect, and they are provided with the same quality guarantee as new OEM automotive parts. The price of those secondary-parts are 1/3~1/2 of new OEM one. It seems a brighter future of secondary-parts remanufacturing in China.

Surface Engineering in Engine Remanufacturing Industry

The normal procedure of remanufacture, for example, main bearing houses, is to enlarge the holes and replace thick bearings. It is difficult for user to repair here further, because the thick bearing is hard to get in market. With surface engineering, it is easy to reach the dimension and surface condition as same as original parts required, besides lower cost, The surface engineering technique is applied in practice for *Styr* engine remanufacturing (Figure 5). The practiced surface engineering techniques include:

- high-speed electric arc spray technique for repairing main bearing houses in engine body,
- electric brush plating technique for repairing cam journals, crank journals, rod and other bearing houses,
- micro pulse cold weld technique for repairing flaw in engine body.

A State National Key Lab for Remanufacture was established in Beijing, year 2003, aiming at fundamental research and practical applications of surface engineering technology for rapidly developing of remanufacturing industry in China. The applications of newly developed nano-scale surface engineering, such as nano-materials brush electroplating, nano-materials thermal spraying and nano-materials self-repairing antifriction additive technology, show the superiority in remanufacturing industry to the traditional way.

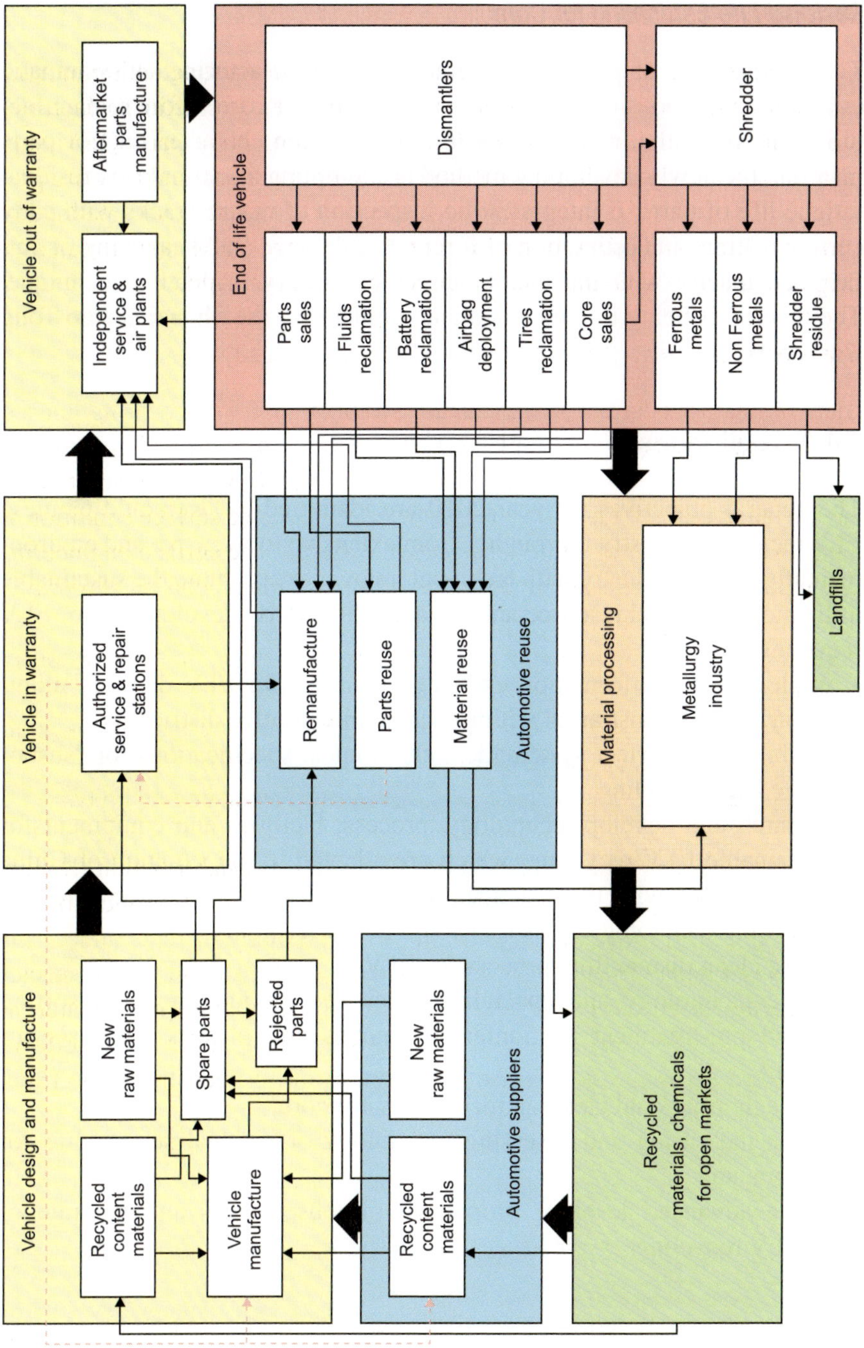

Fig. 5: Vehicles recycling & reverse logistic infrastructure in China

Residual Life Estimating for Core

The residual life of parts is important for remanufacturing. A systematic estimating method is established for residual life of core before remanufacturing. Base on the measuring of flaw and broken characteristics of parts material, the newly developing method is attempting to estimate the residual fatigue life of parts. It integrates the inspection of macro cracks with eddy current testing, and estimation of forepart of damage and weakening of anti fatigue capacity with the metal magnetic memory inspecting technique. The method is being implemented and validated at the Shanghai Jiao Tong University.

2.4.3 Challenges

The general objectives till year 2010 are identified as "Remould Chinese ELV recycling industry thoroughly from extensive to intensive and environmentally benign industry with technical innovation, promote the sustainable utilization of natural resource and advance the orderly development of ELV recycling":

- implement the information exchange platform for ELV disposal monitoring and administration within local administrative district;
- establish and perfect systematic technological specifications for "sustainable ELV recycling";
- research and develop technology, process, facilities and equipmentsfor sustainable ELV recycling, which are adapted to the situation of China and accorded with the practicalities of Chinese ELV industry, and the following technological issues should be focused on:
  - the clean dismantling process for ELV,
  - the technology and specialized facilities on identification, sorting and pre-treatment of nonferrous metals and plastics materials for ELV,
  - the thermal/cold cleaning technology for core,
  - the measuring and assessing technology of residual service life for core, and
  - the advanced level of proportion of reused parts and the rate of ELV recycling.

Outline for the State-of-the-Art

- *The investigation on the information system for monitoring, management and administration of whole life cycle of vehicles (design, manufacture, sale, aftersales service, dismantling, scraping, reuse and recycle) and its technique specification system.* Subjected to the Decree #307, the information system for ELV monitoring and administration and the systematic local administrative specifications should be developed within local administrative district. The auto manufacturers should be involved in collaborative researches on systematic specifications, legislations and sustainable ELV recycling activities. It is lacking of extended manufacturers' responsibility for ELV disposal till now in China.
- *The investigation on the state-of-the-art in the clean dismantling process for ELV.* New technologies, such as water jet cutting, should be applied in practice of dismantling in order to control and reduce volume of gas, liquid and solid waste in process.
- *The investigation on the state-of-the-art in the identification, sorting and pre-treatment of materials for ELV.* The technology and facilities specialized on identification, sorting and pre-treatment of nonferrous metals and plastics materials for ELV should be developed in order to upgrade the efficiency and quality of material recycling. A portable infrared based plastics identification technology and apparatus could be applied in non metal recycling of ELV. The recycling of nonferrous metals such as aluminium, magnesium, copperalloys would be implemented depending on casting alloy or forging alloy. The identification and sorting of nonferrous parts would be mainly manual operation and it is adaptable to the situation of China. The thermal/cold cleaning technology and equipment could be applied in cleaning of core.
- *The investigation on the state-of-the-art in the measuring and assessing of residual service life for ELV core.* The model, criteria and facilities for residue service life measuring and assessing should be developed based on non-destructive testing techniques such as, metal magnetic memory and infrared thermal imaging, and the mechanism of damaging and degradation of reused parts on complexservicing circumstance as well.
- *The investigation on the state-of-the-art in the surface engineering for ELV core.*
- *The investigation on the "3E – Environment Energy Economy" assessment for sustainable ELV recycling industry in China.* Environmental management of the total materials cycle must evaluate the effect of changes in processing or material substitutions, where materials use, reuse,

component remanufacture, and materials recycling can be considered and the overall costs and impacts assessed.
- *The approaches on education and dissemination for sustainable ELV recycling in China.*

Roadmap

The approaches for development of ELV recycling industry in China should be considering followings:
- establish the information system of monitoring, management and administration for whole life cycle of vehicles and its technological specification system;
- establish the big scaled "Recycling Centre for ELV" and its well organized reverse logistics. The centres could be either the state owned and designated ELV recycling corporations or the steel industry invested and steel scrap oriented ELV recycling enterprise. The centres should be responsible for the research and development collaboration of ELV recycling among government, industry and university;
- establish the "Remanufacture Centre for auto parts" which relies on the technologies and specifications of auto manufacturer and supplier, and it should be collaborating with insurance agent for the potential auto repairing market;
- establish the broad international communication and cooperation with, for example, US, EU and Japan on technique, management experience and industrialization of ELV recycling.

The following points should be considered on approaches for development of sustainable ELV recycling in China:
- R&D for easy-to-recycle materials and parts;
- diminishing the category of materials used in vehicle manufacture;
- design for dismantling and design for recycling;
- extending the category of reused and remanufactured parts, and raising its ratio to recycling in materials;
- reducing and disposing shred residues;
- R&D for clean energy recovery;
- Life Cycle Assessment for ELV recycling techniques.

3 Life Cycle Engineering and Management

Engineering and social sciences have developed methods and tools to evaluate and implement technological and societal processes in order to cope with the challenges of sustainability. Alting, Hauschild, and Wenzel give an overview of the international development of life cycle engineering and its future perspectives. Fleischer, Dose, and Ackermann describe how to compare different options of Reuse, Recycling and Recovery (RRR) in product life cycle with respect to environmental impact. Disassembly proves to be a relevant but not the only determining factor. For reducing the complexity of Life Cycle Assessment (LCA) integrated in Product Development Processes (PDP), they propose Life Cycle Inventory (LCI) modules and demonstrate the viability of this Life Cycle Approach (LCAp) on the example of a washing machines lye pump. Hesselbach, Herrmann, and Luger present a procedure and a software for the assessment of a product's recyclability to be integrated into product development and process planning. von Hirschhausen and Elmer, coming from a perspective of institutional economics, analyze a legal and institutional framework that provides reasonable environmental targets and appropriate incentives to reach these targets in an efficient way. Baumgarten, and Butz specify the challenges of the cycle economy for reverse logistics, present paths of development for redistribution and how to create a network of disassembly factories in Europe thus illustrating the importance of the logistics methodology for the implementation of a cycle economy.

3.1 Status and Perspectives

Leo Alting, Michael Hauschild, Henrik Wenzel, Lyngby, Denmark

Life cycle assessment was developed as an analytical tool to help assess the environmental impacts of products or services. For a product to perform its function it must be developed, manufactured, distributed to its users and maintained during use. The production of the product requires a number of supportive to the manufacturing functions. Resources must be extracted and converted into materials, components or ancillary substances from which the product is made. Infrastructure must provide its function to the plant and its employees. When the product no longer serves its purpose, it must be refitted, recycled or disposed of in other ways. Between these activities, transportation processes form a physical link. All the activities consume resources and cause environmental impacts. In order to get an impression of the total environmental impacts caused by the product or service, the analysis must thus focus on the product system or the life cycle of the product which is shown in a general form in Figure 1.

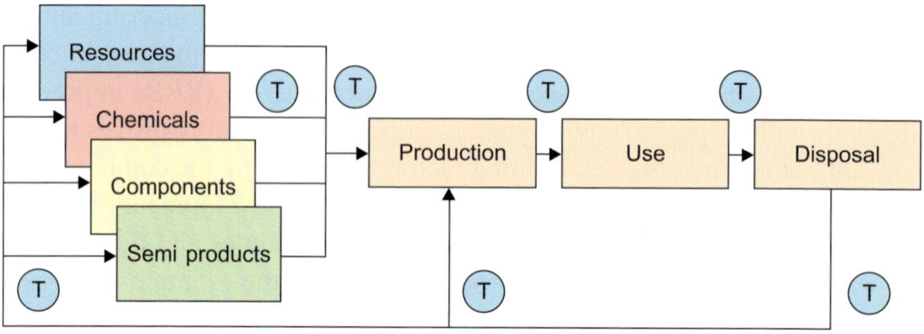

Fig. 1: The product system or life cycle of a product. Transportation processes (circumscribed T) form a physical link between processes and life cycle stages (modified after [1]).

Decisions made by a company influence a number of actors along the life cycle of its products. These actors are providing the needs of the company, using and servicing its products or taking care of the products when they are discarded. With the quest for sustainability, a company's responsibility is extended to cover not only its own processes but also the other activities caused by the company's demands to its suppliers and their suppliers. The responsibility thus extends up-stream in the product chain, but also down-stream to include the impact that the company has on its products' behaviour during their end-of-life treatment. A company which has the aim to operate in a sustainable manner needs to think in the whole product chain, and not just on those links which belong to its own sphere of legal responsibility. Furthermore, the company needs not only to consider the environmental and economic aspects but also the social impacts from its activities. The holistic system's perspective which is applied in life cycle assessment enables the company to disclose the problem shifting which occurs when solutions to (environmental) problems at one place in a product's life cycle create new problems elsewhere in the life cycle. This characteristic makes LCA a valuable decision support tool in companies, which aim at developing their activities in a sustainable direction.

3.1.1 History

The development of life cycle assessment methodology has its roots back in the late 1960's and early 1970's. The first studies applying a life cycle perspective on a process system took place in the USA, focusing on the environmental impacts of different types of beverage containers [2]. The studies have been inspired from substance flow analysis. Environmental awareness at that time was characterised by a particular concern for resource depletion as illustrated a few years earlier by the report to the Club of Rome: The Limits to growth [3], and inspired by the experience of the first global oil crisis in 1973. Many of the environmental impacts of our industrial activities, our energy systems and our use of chemicals were still to be discovered at that time. A quantitative assessment of the environmental impacts, caused by the emissions from the product system, was thus not performed in these early studies. The focus was mainly on the consumption of energy and other resources, and the name given to the assessment technique was 'Resource and Environmental Profile Analysis' (REPA).

The idea of assessing the environmental impacts of products in a life cycle perspective lead a quiet life throughout most of the 1970's but, in the early 1980's, it experienced a renaissance initiated by a fierce public debate. In Europe, the extensive use of resources for packaging of products received a lot of public attention, and governments in a number of European countries commissioned analyses of the resource consumption and environmental emissions for different beverage container systems, such as beer cans and milk containers [4, 5, 6, 7, 8]. Although the studies tried to answer the same question, and although they compared very similar packaging technologies (returnable bottles made from glass or PC, and milk cartons), they reached quite different conclusions as to which system had the lowest environmental impact. It became clear from this type of comparative studies that if LCA should retain its credibility, and function as support tool for decision makers in government and industry, a development of central parts of the methodology accompanied by international consensus building was indispensable. The Society of Environmental Toxicology and Chemistry (SETAC) became the international organisation to host the global community of LCA researchers, and throughout the 1990's, a number of consecutive international SETAC working groups moved the methodology development and international consensus building a good step forward [9, 10, 11, 12, 13].

In parallel to the scientific process taking place in SETAC, the International Standards Organization (ISO) initiated a global standardisation process for LCA. Four standards were developed for LCA and its main phases and issued in the ISO 14000 series of standards for Environmental Management, (ISO 14040-14043 [14, 15, 16, 17]). These standards describe the minimum requirements for the performance of LCA and define its framework as shown in Figure 2.

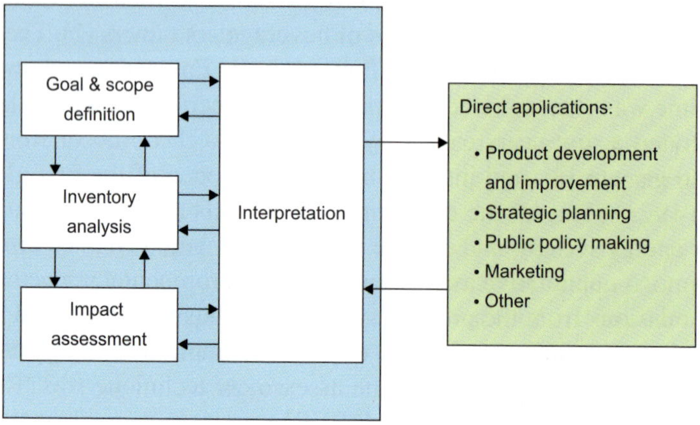

Fig. 2: The framework of life cycle assessment according to the ISO 14040 standard.

The standardisation harmonised the use of the methodology and increased the credibility of the results, and at the turn of the millennium, the use of LCA was widespread among industries and governments in most industrialised countries in Europe, North America and Asia. Today, industries use LCA for comparing and focusing on alternatives in product development, for documentation of environmental performance in marketing, and for decision support in environmental management, while governments use the tool for analysis of societal system choices, e.g. analysis of waste management systems [18, 19, 20], and as the analytical backbone of the Integrated Product Policy (IPP) in ecolabelling schemes and for green public purchasing.

In contrast, the use of LCA in developing countries is quite rare, and in an attempt to support a global dissemination of the interest in the environmental performance of products, the United Nations Environment Program in 2002, in collaboration with SETAC, launched the Life Cycle Initiative under the UNEP Program for Sustainable Consumption. The initiative has the mission: 'To develop and disseminate practical tools for evaluating the opportunities, risks, and trade-offs associated with products and services over their whole life cycle'. A global user survey performed for guiding the definition of the Life Cycle Initiative revealed that for the impact assessment phase of LCA, the highest priorities of the users are [21]:

- transparency in the methodology,
- scientific confidence and co-operation,
- development of recommended factors and methodologies with uncertainty described as well as
- development of a methodology for the various impact categories with a specific interest in developing countries (salinisation, erosion, soil depletion).

The Life Cycle Initiative has an emphasis on the dissemination of life cycle approaches throughout the world, and in recognition of the current needs, a specific attention to the implementation in developing economies is given. The identification and development of recommended methodological practices takes place in ad hoc task forces, the recommendations of which must be endorsed by the International Life Cycle Panel (ILCP), a body of independent experts and practitioners of LCA appointed by UNEP and SETAC in order to become official recommendations from the UNEP/SETAC Life Cycle Initiative [22]. The UNEP/SETAC Life Cycle Initiative has largely replaced SETAC as the forum for international collaboration and consensus building on life cycle assessment methodology.

3.1.2 Methodology

As illustrated in Figure 2, a life cycle assessment in accordance with the ISO 14040 standard proceeds iteratively through four phases. The use of LCA results in different applications is perceived as lying outside the framework, and the standards make no attempts to standardise these – LCA per se is perceived as a decision support tool, not a decision tool. The four phases of LCA are described in the following sections.

Goal and Scope Definition

In this first phase, the goal and intended use of the LCA is defined, and the assessment is scoped in terms of boundaries of the product system (Figure 1), temporal and technological scope of the processes in the product system, and assessment parameters to be considered in the assessment. The function to be provided by the system is meticulously described in qualitative terms and quantified in the functional unit which helps define the reference flow of products for the LCA, i.e. the number of product units for which the collection of data is done. For a packaging study, the functional unit may thus be 'packaging of 1000 litres of milk in 1 litre containers' identifying that the relevant comparison may be between 1000 carton boxes and 40 returnable PC-bottles (which on average can be used 25 times). For a comparison of outdoor wall paints, the functional unit might be: "coverage and protection of 10 m^2 of wall surface for five years (with further specification of the colour fastness and of the geographically determined climatic conditions and exposure to sunlight). Based on this definition, the quantity of paint and the number of treatments needed for keeping the wall coated for five years can be determined for the different paints which are to be compared.

It is a fundamental characteristic of life cycle assessment that its object is defined by the function or service that must be provided. This is in accordance with the comparative nature of most applications of LCA. For a fair comparison, it is essential that the systems that are compared actually provide the same function to the user.

Life Cycle Inventory

After scoping the product system, the inventory analysis collects information on the input and output (environmental exchanges) of all the processes within the boundaries of the product system (Figure 3). The compilation of inventory data for each individual process quantifies the input and output associated with the reference flow of products as derived from the functional unit. The data is typically presented in an aggregated form for the whole product system, as total emissions of substance X or total use of resource Y per functional unit. This function-specificity is a fundamental characteristic of the Life Cycle Inventory (LCI) and the ensuing impact assessment, and consistent with the purpose of LCA to evaluate the environmental impacts associated with providing the service which is specified by the functional unit.

Life Cycle Impact Assessment

The purpose of the life cycle impact assessment phase is to interpret the inventory results in terms of their potential impacts on the areas of protection of the LCA, i.e. those aspects of our world which the use of the LCA shall help protect:

Areas of protection for Life Cycle Inpact Assessment (LCIA) are:

- human health,
- natural environment,
- natural resources and
- man-made environment.

Life cycle impact assessment applies a holistic perspective on environmental impacts. In principle it attempts to model any impact from the product system which can be expected to damage one or more areas of protection. This means that LCIA addresses not only the toxic impacts from chemical emissions (as environmental risk assessment does) but also the other impacts associated with emissions of air pollutants (global warming, stratospheric ozone depletion, acidification, photochemical ozone and smog formation) or waterborne pollutants (eutrophication and oxygen depletion), as well as the environmental impacts from different forms of land use, from noise and from radiation, as well as the loss of renewable and non-renewable resources. Some LCIA methods also include the human health impacts from the occupational exposure from operating the processes in the life cycle [24].

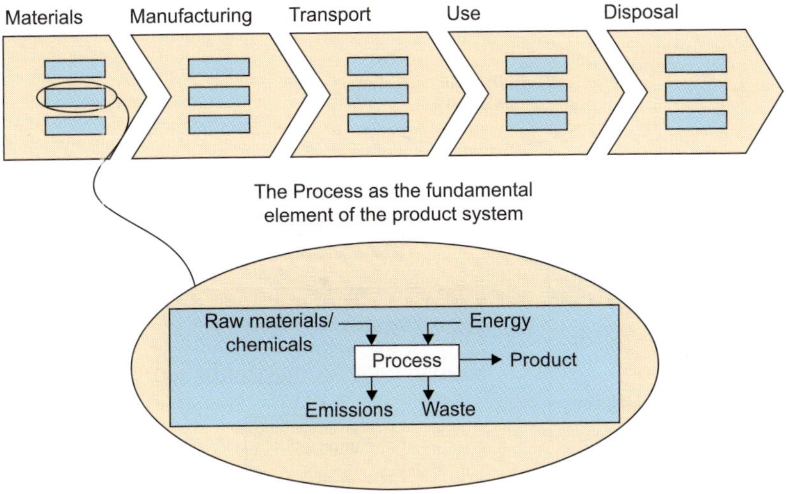

Fig. 3: The life cycle inventory analysis collects information about environ mental exchanges of all processes in the product system, comprising input of raw materials, components, chemicals, and energy and output of air- and waterborne emissions, waste, and products delivered to other processes (based on [23])

If the inventory analysis for the product system has been thorough, the inventory will contain a multitude of substance emissions and input of different resources. Some of these exchanges are environmentally significant and even small amounts can be of importance. Others are of no significance whatsoever. Hence regarding environmental exchanges, the target of the impact assessment is to translate the emissions into their potential impacts on the areas of protection by applying the best available knowledge about causal relations between emissions and their effects in the environment as illustrated in Figure 4.

For greenhouse gases like CO_2 and CH_4, an impact that appear early in the causality chain in Figure 4 would be an increment in the atmosphere's ability to absorb infrared radiation.

Among later impacts are an increase in the atmospheric heat content, propagation of the global marine and soil compartments causing changes in regional and global climates and sea-level rise, eventually damaging several of the areas of protection: human health, natural environment and man-made environment. For greenhouse gases, the fate processes in Figure 4 would be the degradation and transport of the gases in the troposphere, the stratosphere, and the global water and soil compartments, and they would be integrated in the chain of impacts all the way from emission to the areas of protection.

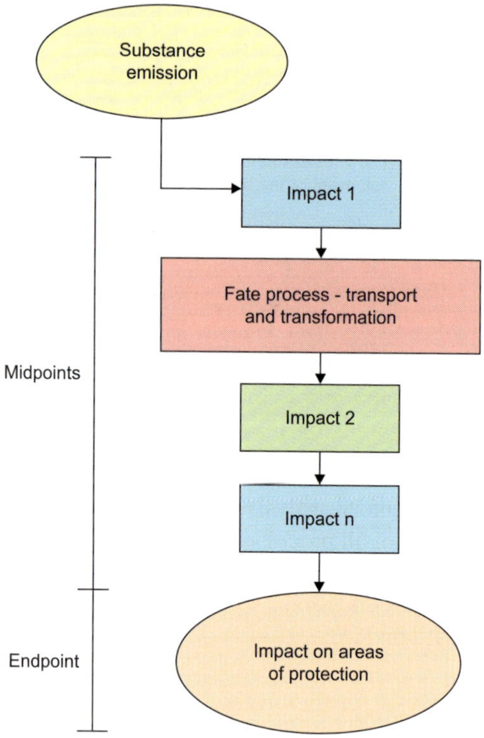

Fig. 4: Modelling of the impacts, which emissions cause on areas of protection, are based on causal relationships known from the environmental sciences (modified from [23])

For the consumption of resources, the severity which is applied in the impact assessment is typically derived from the scarcity of the resource – the relationship between the economically feasible known reserve and the current consumption.

The life cycle impact assessment proceeds through four steps [16]:

The first step is the Selection of impact categories and their classification. Here, categories of environmental impacts of relevance to the study are defined. In most LCA studies, existing impact categories can just be adopted. Next, the substance emissions from the inventory are assigned to the impact categories according to their ability to contribute to different environmental problems. Figure 5 illustrates environmental impact categories which are often modelled in LCIA.

The second step is the Characterisation where the impact from each emission is modelled according to the environmental mechanism (Figure 4) and expressed as an impact score in a unit common to all contributions within the impact category, e.g. kg CO_2-equivalents for all greenhouse gases. Following Characterisation, the contributions from different substance emissions can be summed within each impact category, and the inventory data translated into a profile of environmental impact scores and resource consumptions.

The third step is the Normalisation where the impact scores and resource consumptions from the Characterisation are related to a common reference in order to facilitate comparisons across impact categories. Life cycle assessment is often used for comparative studies ('is alternative 'A' preferable to alternative 'B'?'), and comparison across impact categories is necessary when there are trade-offs between the categories, i.e. when improvements in one impact category are obtained at the expense of another impact category. Normalisation expresses the magnitude of the impact scores on a scale which is common to all the categories of impact (typically the background impact from society's total activities). In Figure 5 the impact scores are expressed in person equivalents, PE. The unit PE represents the annual impact from an average person and is useful for bringing the rather diverse environmental impacts on a common scale [23].

The fourth and final step of the impact assessment is the Valuation where a ranking or weighting is performed of the different environmental impact categories and resource consumptions reflecting the relative importance they are assigned in the study. The Valuation is needed when trade-off situations occur as described under Normalisation. Where Normalisation expresses the relative magnitudes of the impact scores and resource consumptions, Valuation expresses their relative significance considering the goal of the study.

According to the ISO standard, the first two steps of the impact assessment are mandatory while Normalisation and Valuation are optional. The valuation step is the most normative part of the methodology. Here, preferences and stakeholder values are applied. There is no objective way to perform the valuation, and hence no 'correct' set of ranks or weighting factors. The ISO standard for LCIA does not permit valuation in studies supporting comparative assertions disclosed to the public [16].

The ISO standard for LCIA refrains from a standardisation of detailed methodological choices. Over the last decade, several well-documented methodologies for life cycle assessment have been developed filling this gap [24, 25, 26, 27, 28, 29, 30, 31].

Figure 5 shows an output from one of these methodologies, the EDIP97 methodology (Environmental Design of Industrial Products) [24, 26] applied to two refrigerator designs which employ HFC134a and pentane/isobutane as refrigerant and insulation foaming agent respectively.

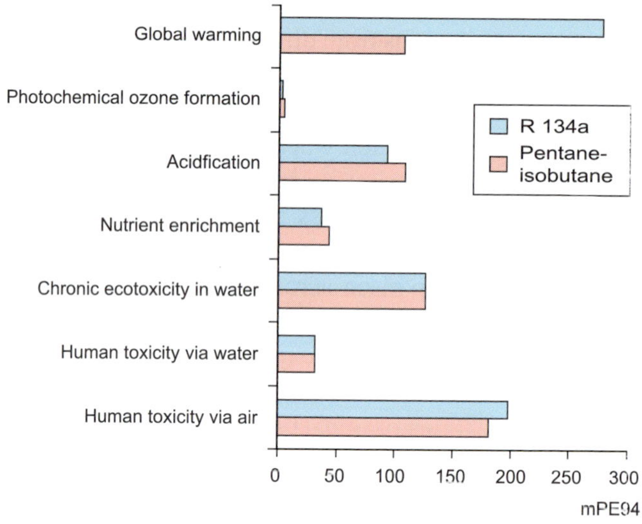

Fig. 5: Impact profiles for two refrigerator designs. All impacts are normalised and expressed in a common unit – the person equivalent, PE (or rather milli PE) [32]

The results show the impact scores for a number of environmental impact categories. All impact scores are normalised and expressed in a common unit – the person equivalent, PE (or rather milli PE), representing the annual impact from an average person in a reference year [32].

Decisions based on a comparison between impact categories require some sort of valuation as not all impact categories can be assumed to have the same importance (modified from [23]).

Life cycle impact assessment is still under development, and apart from the global impact categories, global warming (leading to climate change) and stratospheric ozone depletion, no consensus has yet been reached on how to model the impacts. Two schools can be identified [23]:

Midpoint modelling, where the impacts are modelled at some midpoint in the environmental mechanism (Figure 4). The midpoint is typically chosen as far as possible towards the areas of protection in the causality chain, i.e. at the point where further modelling is deemed too uncertain. The relation of the midpoint to the area of protection is then considered in the weighting. Midpoint modelling is the traditional approach to life cycle impact assessment, and Figure 5 shows a typical output from a midpoint oriented method.

Endpoint modelling or damage modelling where the impacts are modelled all the way to the effects they cause on the areas of protection using the best available environmental models. This school thinks that the increased uncertainty in the impact modelling is warranted by the improved interpretation of the results. The only weighting needed here is thus the weighting of the areas of protection.

As more and better environmental models become available, the optimal indicator point in a midpoint approach will move further towards the endpoint, and as larger parts of the environmental mechanism are included in the characterisation modelling, the midpoint approach will move towards the endpoint approach. Until they converge, the two approaches will compliment each other [23, 32]. Current work seeks to combine the two approaches to allow the user to apply a mixture of them [34].

Interpretation

Interpretation is the phase of the LCA where the results of the other phases are interpreted according to the goal of the study using sensitivity and uncertainty analysis. The outcome of the interpretation may be a conclusion serving as a recommendation to the decision makers, who will normally consider the environmental and resource impacts together with other decision criteria like economic and social aspects. The interpretation may also lead to the recommendation of a further iteration, reviewing and possibly revising the scope of the study, the collection of data for the inventory or the impact assessment. As illustrated in Figure 6, LCA is performed iteratively, and each phase may be revisited several times. With each iteration the uncertainty is reduced, and the assessment is completed when the results are sufficiently certain to adequately answer the questions which were posed in the goal and scope definition at the beginning.

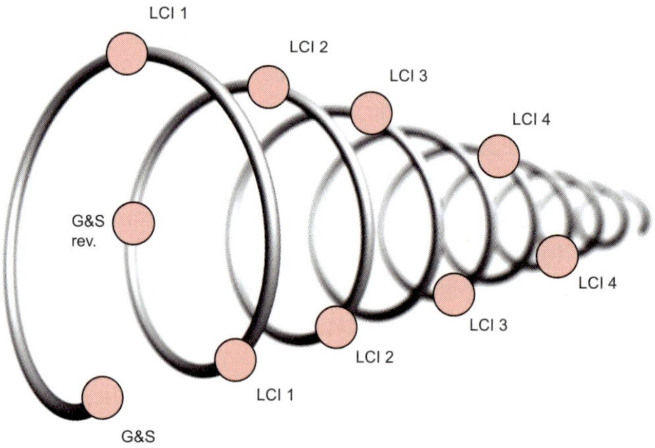

Fig. 6: The phases of life cycle assessment are performed in an iterative manner

3.1.3 Economic and Social Aspects

Traditionally, LCA considers environmental and resource impacts, and to some extent working environment impacts. LCA in its traditional form is thus prevented from addressing trade-offs between environmental concerns and social and economic concerns in the product life cycle. Since sustainability assessment includes both environmental performance, and social and economic performance, LCA's ability to support actual decision-making in companies which aim for sustainability may be questioned. There is thus a pressure to develop the methodology for inclusion of both economic/cost impacts and social impacts in life cycle assessment to make LCA a sustainability assessment tool. Both of these new disciplines are, however, quite young and little developed at present.

Life Cycle Costs and Economic Life Cycle Impact Assessment

Life Cycle Costing (LCC), which considers the economic implications on the manufacturer throughout the life cycle of a product, has had a relatively long history outside the scientific LCA community. LCC and LCA have been designed to provide answers to different questions, and focus has been narrower in LCC than the full life cycle focus in LCA. Life cycle costing has been applied to compare cost-effectiveness of different business decisions or investments from the point of view of a decision maker in a company or of a customer, and only those parts of the material life cycle of the product, where direct costs or benefits arise, are included in the product system. [35].

Nevertheless, life cycle management activities integrating the environmental LCA results in business decision making have in recent years motivated ambitions of integrating cost assessments with the environmental assessment along the product chain. This would make them more consistent by basing them on the same information about the material and energy flows of the product system. It would also allow decision makers to weigh environmental and economic impacts against each other along the product chain [35, 36, 37]. One of the strong issues is if – and how – external costs should be included in the life cycle cost assessment. For proactive companies aiming for sustainability, this would be a relevant option, and it would make LCC and environmental LCA results more compatible with most products [38]. There is, however, still no consensus on how to monetarize environmental damages in a consistent way [39].

Social Life Cycle Assessment

In our globalised economy, important stakeholder groups nowadays hold companies responsible for their social impacts through activities like child labour, corruption, discrimination of employees, and deprivation of employees of their right to organize and demand fair working conditions. Often these impacts occur far from the company headquarters, typically upstream in the product chain, but there are numerous examples where such cases have reached the media, and where globalised corporations have been held responsible for poor working conditions, not only in their own facilities, but also at their suppliers. The damage to their brand can be substantial, and for companies who claim to be sustainable, it can be devastating. Many companies thus see themselves in need of a tool which can help them make informed decisions about their social impacts throughout the life cycle of their products.

The omission of social impacts from LCIA is also, to some degree, inconsistent with the defined areas of protection since social impacts will often lead to impacts on human health, and indirectly on the sustainable use of ecosystems. Nonetheless, very little work has so far been performed in Social LCA, e.g. [40, 41], but attempts are on-going to develop LCIA for social impacts, and under the UNEP/SETAC Life Cycle Initiative, a task force has been dedicated to this topic [22].

3.1.4 Data and Tools

Life cycles easily comprise hundreds of processes and to assist the modelling of the product system, numerous software tools have been developed over the years supporting the different phases of the ISO framework (Figure 2), the system modelling, the inventory analysis, the impact assessment, and the interpretation, e.g. [42, 43, 44, 45, 46]. Collection of data for the environmental exchanges between processes in the product system and the environment is normally the most labour-intensive part of doing an LCA. To assist the inventory analysis, data has been collected in life cycle databases in a unit process form which allows them to be used as building blocks in different life cycle models. The data bases contain data on the most important processes, e.g. manufacturing, transportation, disposal, electricity and thermal energy generation, and materials, e.g. plastics, metals, biological materials, but typically only for the traditional LCA countries (North America, Europe and Japan, e.g. [47, 48, 49, 50, 51]). Some of these databases are supplied together with the software tools mentioned above.

The practical use of environmental LCA methods and software tools in the industry has revealed the need for simplifications of many applications. Hence, streamlined life cycle assessment methods have been derived from experience with the complex full methods [13, 52].

Figure 7 shows an example of the outcome from a simplified life cycle assessment, a Life Cycle Check, of the refrigerator comparison for which the environmental profile from a full LCA are shown in Table 1, highlighting the main differences between the two products in their life cycle.

Although method development work still goes on, the life cycle assessment methodology has reached the level of maturity, stability and versatility required for industrial use. Furthermore, environmental data for most of the important materials and processes are available, at least for industrialised countries, in databases accessible through the Internet or as part of commercial software.

In summary, the necessary tools exist to support a widespread use of life cycle assessment by the industry.

Tab. 1: Comparison of two refrigerator designs applying HFC134a or pentane/isobutane as refrigerant and insulation foaming agent. The results were obtained using a life cycle check (a MECO matrix for analysis of the main differences in the sources of impacts over the life cycle stages of the product [24])

	Life cycle stage			
Agents	Extraction of raw materials	Manufactu-ring stage	Use stage	Disposal stage
Materials	Fluorine for HFC134a?			
Energy				
Chemicals		Emission of HFC134a and precursors	Emission of HFC134a	Emission of rest content of HFC134a
		Emission of pentane and i-butane	Emission of pentane and i-butane	Emission of rest content of pentane and i-butane
Other			Risk of explosion for pentane and i-butane	

3.1.5 Integrated Product Policy

A product-oriented environmental policy has developed in several European countries, e.g. [53, 54, 55, 56, 57], and a coordinated Integrated Product Policy (IPP) is currently under development at the EU level [58, 59].

The tools, which are applied in the national environmental product policies and in EU's Integrated Product Policy, are typically a combination of soft regulation and support to industries who want to apply Life Cycle Engineering. The soft regulation aims on one hand at informing consumers and other downstream stakeholders about the environmental consequences of consumption and motivate them to buy the more environmentally benign products. Green public purchase policies, leading authorities to buy the most environmentally benign products, is also a way of greening the market.

On the other hand, the soft regulation aims at forming a level playing ground for those companies who wish to compete on the environmental performance of their products. It is thus important that the downstream stakeholders can get credible information helping them to identify the most environmentally benign products. To this aim, different information systems have been developed:

- official ecolabelling systems [60, 61] that develop criteria for environmentally friendly products and manage the assignment of ecolabels to products (see [62] for an overview of different national ecolabelling systems),
- official environmental product declaration system [60, 63, 64],
- standard for self claims on environmental performance on products [65], and
- standards for performing and using LCA [14, 15, 16, 17].

A green product tax is also discussed as a potential tool of the IPP. A product tax set according to the environmental impact of the product would represent a way of internalizing some of the externalities (costs due to environmental damages not paid by producers or consumers and hence the responsibility of society) associated with the product's life cycle. Products causing large environmental impacts would thus have a high tax while environmentally benign products would have a low tax. Such an Environmental Impact Tax might replace the Sales Tax or Value Added Tax (VAT), and if it were sufficiently high, consumers' wish to save money would tend to favour the green products. Table 1 lists the main tools of the Integrated Product Policy.

In addition to the attempts to push the market towards a greener consumption pattern, direct support to industries is applied in the form of subsidies to pioneers within a certain area or type of application, and public funding is given to the development of methods, tools and databases needed by LCA and design-for-environment. A number of national research projects have thus been instrumental in the international development of LCA methodology and collection of life cycle unit process data [27, 29, 30, 31, 25, 24, 28].

Another way of supporting the implementation of Life Cycle Engineering in the industry is through providing information on the opportunities it opens in terms of capacity building.

To assist the product-oriented activity, the Danish Environmental Protection Agency has organized product panels. Within prioritized product fields (at first textiles, electronics and transport of goods), selected stakeholders are invited to form an independent panel with the task of generating ideas and activities to improve the environmental aspects of their product area. The use of product panels is seen as a way of involving central stakeholders and creative minds all along the product chains and widening out the public participation in the Danish Integrated Product Policy, and today, product panels have been established within a number of other product areas [66].

Tab. 2: Tools of an integrated product policy (adapted from [1])

Tool	Purpose
Ecolabelling system	Help the consumer identify those products which have the best performance on the most important environmental parameters. Examples: EU Flower, Nordic Swan, German Blue Angel, ISO 14024 standard
Environmental Product Declaration system	Environmental profiles for products based on a full life cycle assessment. For communication to business and end consumers. Examples: Swedish EPD system, ISO 14025 report
Standards for LCA	Common standards for transparent and reproducible life cycle assessments are the backbone of an IPP. Example: ISO 14040-14043 standards
Guidelines on Design for environment	Disseminate good approaches to integration of environmental aspects in the development of new products. Example ISO/TR 14062 Guideline
Green public purchase guidelines	Guide (public) purchasers on which environmental aspects to consider when purchasing different products. Example: Green public purchase guidelines from several countries on a number of product groups (Norway: http://www.grip.no, Japan: http://www.gpn.jp, Denmark: http://www.miljoevejledning.dk)
Green product taxes	Let the product's price reflect its environmental impact to avoid the current situation where the cheapest products often have the worst environmental impacts. Example: Some attempts at CO_2 taxes

In support of the dissemination of life cycle approaches within the industry, several countries have established national LCA Centres, e.g. [67, 68, 69, 70, 71, 72]. As the first European country Denmark established a publicly funded LCA Center in 2002 in a collaboration among the three leading Danish consultants in the field. The Danish LCA Center is a knowledge centre for life cycle assessments and life cycle approaches, supporting product-oriented environmental strategies in private and public companies by assisting them in implementing life cycle thinking [73]. The activities are mainly aimed at the industry, particularly of SMEs. The highly dynamic sector of SMEs constitutes a very important part of the national industry but hitherto, attempts of introducing life cycle approaches to this sector have had rather limited success. The aims of the Danish LCA Center are:

- To assist companies with a need for environmental assessment of products in a life cycle perspective.
- To secure that the development of tools and methods for the life cycle approach in Denmark builds on a solid and scientific basis.
- To promote product-oriented environmental work in companies (Life Cycle Assessments and other Environmental Management Systems).
- To maintain the existing cooperation between Danish LCA stakeholders.

The activities of the Center build mainly on the developments of methods and tools described in previous sections, and can thus be seen as a targeted dissemination action, based on already developed material.

The main learning from the first years' operation of the Center has been that in the absence of legal requirements on LCA, DFE or other forms of life cycle thinking, the biggest challenge in the dissemination of life cycle approaches lies in motivating the small and medium-sized enterprises to take the idea onboard. Environmental arguments are not sufficient; the focus in the communication must be on the economic part of the triple bottom line. In other words, the companies must be convinced that they will actually make a profit or realise economic savings through working with their product's life cycle, even in the short term, since many SMEs operate with a very short time horizon.

Tab. 3: Tasks and activities of the Danish LCA Center (adapted from [1])

Task or activity	Purpose
Establish networks of stakeholders	Bring together key stakeholders for exchange of experience, good examples, data and collaboration.
Act as help desk for Danish stakeholders with interest in life cycle thinking	Quick answers to questions regarding LCA and related topics.
Propagate information (homepage, newsletter, organise meetings among Danish key LCA actors)	Provide overview of activities and stakeholders in the field.
Provide training and courses on LCA and related topics	Introduction and training of stakeholders who are new to LCA as well as more experienced users with special needs.
Provide software and databases on LCA	Adapt existing software to Danish LCA methodology, secure that the necessary software is available to Danish users to support their use of LCA and related disciplines.
Coordination of and participation in national and international LCA activities	Information to Danish users on the international developments, presentation of Danish experience in international collaborations like ISO, UNEP and SETAC.

3.1.6 Eco-Efficiency of Industrial Activities

In short, Life Cycle Engineering covers the engineering activities which address the industry's environmental impacts in a product life cycle perspective. An aim of Life Cycle Engineering is to improve the eco-efficiency of industrial activities, defined as the ratio between the service which is provided by the activities and the environmental impacts associated with providing this service. According to Wenzel and Alting [84], there are exactly four levels on which Life Cycle Engineering can address the eco-efficiency:

- *The product* of the company: the eco-efficiency of the product must be viewed in a life cycle perspective, and the tools for optimising the eco-efficiency of the product are thus the analysis tools reviewed in this paper (Life Cycle Assessment tools) as well as the synthesis tools (Design for Environment tools).
- *The production* or manufacturing system used to produce the product, whether located at the main manufacturer or outsourced to sub contractors. Once the eco-efficiency of the product life cycle has been optimised using LCA and DFE tools, it is time to address the eco-efficiency of the production system(s) involved in the product life cycle through re-engineering the production with attention on its environmental impacts and resource use. A central tool for this is the Process integration tool [84].
- *The process(es)* involved in the production system (or elsewhere in the life cycle) of the product. When the eco-efficiencies of the product life cycle and the production have been optimised, the attention should be directed towards the individual processes involved in the production. The tools for optimising the eco-efficiency of industrial processes are commonly denominated Cleaner Production or Process Optimisation [84].
- *The discharges* from the processes. When the other levels have been optimised, traditional emission analysis tools (Environmental Risk Assessment) and technologies for emission avoidance and emission treatment are the tools to address discharges from the processes, reducing or converting them to less harmful substances.

Design for Environment

Design for Environment (DFE) covers approaches used by product developers to reduce the environmental impacts from products in a life cycle perspective while maintaining – or preferably enhancing – the competitivity of the product in the market. Hauschild et al. point out that the environmental considerations must find their place among the many other priorities considered in the development of a new product (Figure 8) [75]. Even from an environmental point of view, the weight given to the environmental performance of the product should not be higher than that which gives the strongest competitive edge to the product. If the product does not perform well in the marketplace and supersedes other less environmentally sound products, no reduction of the load on the environment is obtained. Often, however, environmental improvement can easily be attained without impairing other important performance parameters of the product. On the contrary, the work on optimisation of the environmental performance often spurs creativity and brings innovative design solutions into the product development process which allows improvement of the overall performance together with the environmental performance [76].

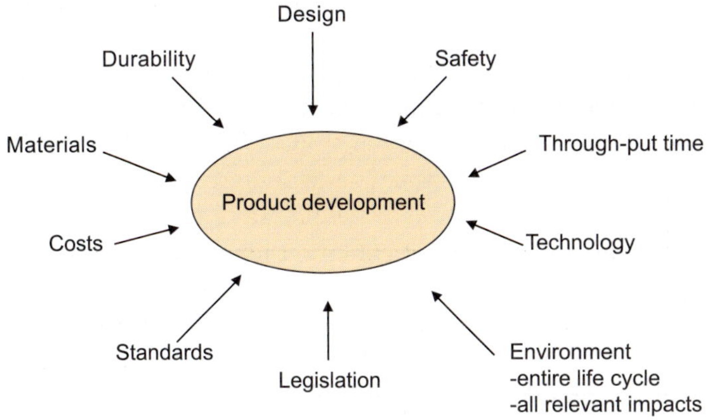

Fig. 8: Product development is the art of finding the best compromise among the many requirements for the new product set by different stakeholders (from [75])

The life of a product starts with the initial design concept. As shown by Boothroyd et al., getting the design right, at the beginning of a product's life, is the most important factor for the eventual cost of the product [77]. In addition, the impact of a product upon the environment is determined at the design phase [78], and often in the very early design phase. Hauschild et al. point at that as design proceeds from the idea stage over conceptual and detailed design towards production, the knowledge about the product, and hence its environmental properties increases strongly (full curve in Figure 9). Simultaneously, as more and more properties are fixed by the choices made in the course of the development process, the possibility to influence the environmental performance of the product (dashed curve) is reduced. It is thus largest at the early stages of the product development process where the knowledge about the product (full curve) is least [75].

Figure 9 substantiates the need for both analysis (LCA) and synthesis tools (DFE) which are applicable at the early stages of product development.

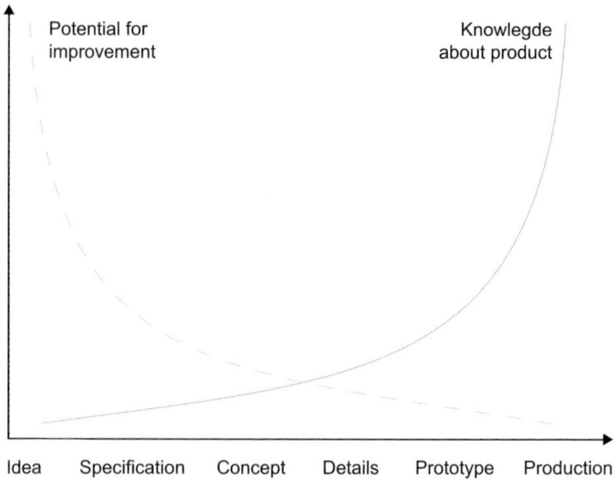

Fig. 9: Schematic representation of the development in knowledge about the product (full curve) and the possibility to influence its environmental performance (dashed curve) throughout the stages of product develop ment from Idea to Production [75]

EcoDesign also has a set of common sense criteria that a designer should review at the initial design stages. Luttropp has put these together and called them the 10 Golden Rules of EcoDesign [80]. Kaldjian [81] has a similar list. The EcoDesign rules can be summarised as:

- Do not use toxic substances, but use closed loops when necessary to do so.
- Minimize energy and resource consumption in production, use, packaging and transportation.
- Promote long life
  - especially for passive products,
  - promote maintenance,
  - use better materials, surfaces, and structural strength.
- Consider end-of-life
  - organize for upgrading, repair and recycling,
  - use few, simple, recycled, unblended materials,
  - assemble for disassembly.

These Golden rules for EcoDesign can be seen to be very similar to the foregoing list of axioms for DFE [74].

The rules can conflict with each other:

- Low weight may shorten product life.

- Renewable materials may cost more energy.
- Use of toxic materials may reduce energy consumption.

Sometimes the design rules even lead to a worse environmental impact:

- Composite materials using scarce resources may be the best choice for vehicles.
- Long lifetime may be unwanted if the product uses energy and the technology evolves quickly towards lower energy consumption.

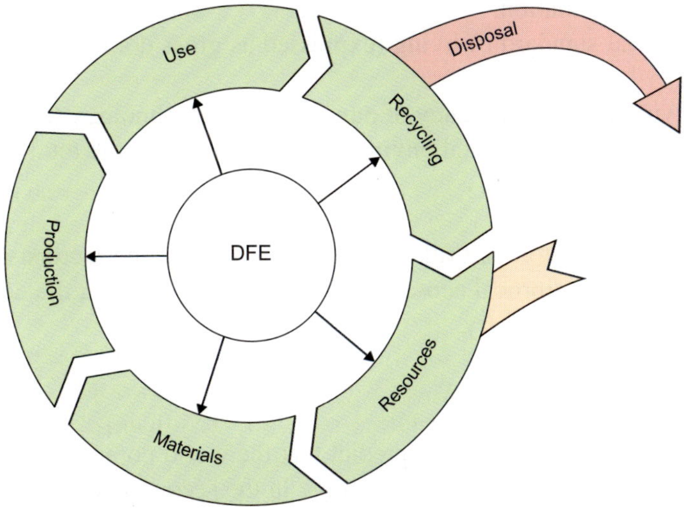

Fig. 10: The right DFE tools must be applied to give the new product an optimal environmental performance, trimming the largest environmental impacts of the product's life cycle. The corona of the life cycle represents the magnitude of the environmental impact from the different parts of the product's life cycle [82]

Among the multitude of existing DFE tools, it is thus important to choose the right one for the intended purpose. Hauschild et al. [82] present a hierarchy of focusing in DFE proceeding over:

- Determining whether the intended product is the right one to be produced also in the long term.
- Identification of the 'environmental hot spots' in the life cycle of the product.
- Selection of the DFE tool which supports optimisation of the product by reducing these hot spots (Figure 10).

Sustainable Production

A sustainable development was defined by the United Nations' World Commission on Environment and Development as: "... development that meets the needs of the present without compromising the ability of future generations to meet their own needs" [85].

In an attempt to make this definition more operational, Graedel and Allenby [86] in their 'master equation' express our total impact on the environment as a product of – what can be seen – as the causing agents:

- the global population,
- the material standard of living (expressed as gross domestic product per capita),
- the environmental efficiency of our society and technology (expressed as the ratio between the total environmental impact and the total economic activity).

The first two factors are ruled by social and economic factors while the third factor, the reciprocal eco-efficiency, is to a large extent the domain of the manufacturing industry.

The Eco-Efficiency Challenge

Optimistic population forecasts predict that the world population will stabilize somewhere around ten billion people in the second half of this century [87]. The majority will live in developing countries where the current standard of living is often far below the acceptable level by any standard. From an ethical point of view this is unacceptable and recent events have shown that the resulting social instability is a permanent threat to democratic societies, also in the industrialised countries. A future increase in the material standard of living must thus be anticipated, particularly in the developing countries, in order to reduce current regional differences and create acceptable conditions everywhere. At least a doubling in the global average material standard of living must therefore be expected towards the middle of the century.

According to the master equation, our eco-efficiency in the same period must increase by more than a factor of four in order just to stay at the current level of environmental impact. It is well known that although the state of the environment has improved locally in some areas, the general picture is still bleak and getting worse in regard to more regional and global impacts. This is clearly not sustainable, and in order to allow a reduction in total environmental impacts and accommodate uncertainties in the growth of the other two parameters of the master equation, the challenge of the eco-efficiency of our society and our manufacturing industry in particular, is often presented to require a factor 10 improvement [88, 89, 90, 91]. Particularly the technologies introduced in the developing countries must be environmentally friendly, since this is where the growth is going to be more intense.

All our activities must be addressed in order to obtain such radical improvements: Power generation, treatment of solid and liquid waste streams, infrastructure and our general material standard of living as founded on our production and consumption of products. The industry has a key role to play in several of these and particularly in the environmental improvement of our products. The industry's efforts must be based on a combination of analytical tools for analysing the impacts of products and focusing on the most important impacts, and synthesis tools for designing and developing new products with an improved environmental performance.

The Other Dimensions of Sustainability

Sustainability is not just about the environment. Elkington [90] introduced the concept of the Triple Bottom Line simultaneously addressing the social impacts, the environmental impacts, and the economic impacts of the activities of a company.

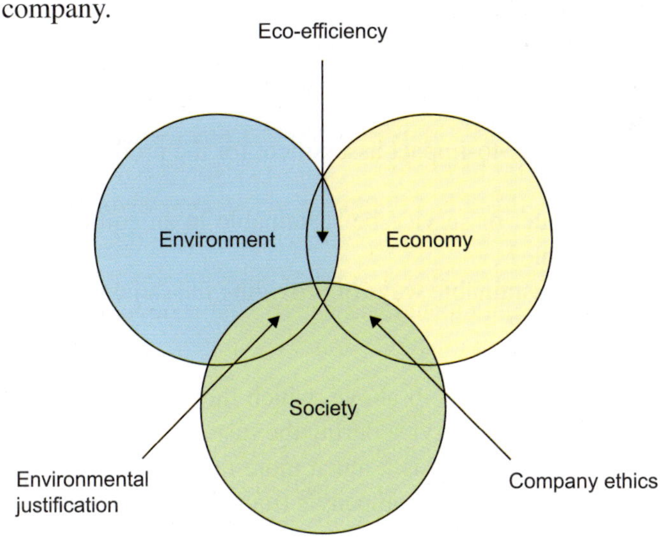

Fig. 11: The dimensions of sustainability

In the intersection between the environmental and the economic dimension lies the eco-efficiency (Figure 11) which has been discussed at length earlier in this chapter. Eco-efficiency is important but it only addresses the environmental impacts associated with the provision of the function that the product gives us. Solely looking at the eco-efficiency will not guarantee sustainability in the long run. Sustainability is also about justice and distribution of assets between individuals and between generations.

In the intersection between the environmental and the social dimension lies the environmental justification of the product. For an industry to be sustainable, it must also address the environmental justification of the product asking about the relevance of the product for the society. Are the impacts caused by this product justified by the service it provides to the user, and to society? Does society have a real need for this product or could the function it serves be fulfilled in a more sustainable way, or should the function be fulfilled at all, does it have a place in a sustainable society? Elkington [92] provides the following Need Test to establish whether product will be justified in a future sustainable society:

The Need Test

- What is the primary function of the product or service?
- What other benefits does the product or service offer?
- Is there likely to be a long-term need and/or demand for the product or service?
- How does the value-to-impact assessment for the product or service look today?
- Would this product or service be sustainable in an equitable world of 8 – 10 billion people?
- Are there more sustainable ways of providing the same function - or meeting the same need?

On the basis of points 1 – 6 above, which threats and opportunities will there be for the product or service during the sustainability transition?

Environmental justification is a much more radical requirement than eco-efficiency since it questions a company's (moral) right to produce the products it does. On the other hand, by addressing both the social impacts and the environmental impacts, it is crucial for the company's sustainability in the long run. If Elkington's Need Test or similar considerations come out with a negative answer, a company aiming for sustainability should take this quite seriously in the strategic planning of their product programme [82].

In the intersection between the economic and the social dimension of sustainability lie the *company ethics* concerning the way in which the company behaves towards the people it interacts with, e.g. its employees, neighbours, and suppliers. For a sustainable industrial company the traditional focus on the shareholders is broadened to include a wide range of stakeholders along its product chain, and issues like discrimination, child labour, corruption and fair working conditions all reflect the company ethics at the intersection between the economic and social dimensions of sustainability. Dreyer et al. develop a framework for assessing the social performance of companies in a life cycle perspective based on the Universal Declaration of Human Rights and the International Labour Organisation conventions translating the Human Rights into global workers rights for the company's employees around the world [41].

Welford et al. [93] summarize the performance of a sustainable industry in their "seven E's":

- *Environment,* covering the environmental impacts of the company's activities applying a life cycle perspective on all its products, assuming responsibility all along the product chain, and also looking into the environmental justification of the products,
- *Empowerment,* addressing the involvement of all employees in the work for sustainability and the conscience of the potential contribution from everyone,
- *Employment,* as a fundamental function of the company in society, creating the basis of an income and material welfare of the employees along the product chain and creating meaningful jobs contributing to the development and quality of life of the employees,
- *Economy,* as a prerequisite for the sustainability of the company. The company must be able to justify its economic actions also ethically, but if it does not generate sufficient profit to the investors, it will not be sustainable in the traditional meaning of the word.
- *Ethics,* covering the moral of the way the company operates. The ethics should be laid down in a clear code of conduct, and it is important that the company communicates in an open way about the economic, environmental and social impacts of its operations so stakeholders can judge it by its actions.
- *Equity* within the organisation through equal rights to employees and equal remuneration for equal work. Outside its own organisation, the company shall work for equity through its trade, paying a fair price for the amenities and services, it buys.
- *Education* of employees, customers and other stakeholders is a mean for the company to spread knowledge and instigate debate in sustainability.

3.1.7 Conclusion

The challenge of creating a sustainable society is considerable. The needed eco-efficiency improvements were estimated to be of a factor 4 – 10, and in parallel there are obvious requirements for an improved social performance. While society has the responsibility for infrastructure and legislation, and hereby for creating a level playing ground for a fair competition among companies, the improvements in environmental and social performance must be created by the industry who manages the production of material goods in our society today.

The globalisation of our economies poses great challenges to our societies but at the same time it presents an opportunity for addressing the sustainability challenge on the scale where it really belongs – the global scale. The environmental impacts do not respect national borders. In developing economies they are often poorly regulated by authorities, and the same holds true for the social impacts. Here is an important role for industries which establish new production sites in developing countries in respecting western standards for environmental protection and treatment of employees and other local stakeholders.

The globalisation thus emphasizes the need for a life cycle perspective which is already inherent in sustainability. The sustainable company must assume responsibility for all three dimensions of sustainability throughout its product chain from the cradle to the grave, and Life Cycle Assessment, Life Cycle Thinking and Design for Environment are central tools in this endeavour.

As described earlier in this chapter, apart from the social life cycle impacts, both the analysis tools (LCA) and the synthesis tools (DFE) for improvement of the life cycle performance of products are available today to support a widespread Life Cycle Engineering focus in the industry, but are they used?

Indeed, a number of large corporations have adopted life cycle-oriented management and aim at becoming sustainable. A few selected examples are:

- *Hartmann A/S* producing moulded paper products applying life cycle assessment and life cycle thinking as an integrated part of their environmental management system world wide and addressing both environmental and social impacts.
- *Philips* using life cycle assessment and life cycle thinking in the development of many of their consumer electronics products and aiming for "environmental champion products" within the different product groups.

- *Rank Xerox* closing their life cycle by providing photocopy services rather than selling photocopiers already back in the 1970s and today using life cycle assessment and life cycle thinking in the development of their new generations of products.
- *Novo Nordic* aiming for sustainability management including a continuous dialogue with representatives of stakeholders along their pharmaceutical product chains.
- *SmithKline Beecham* developing sustainable pharmaceutical business with sustainability metrics based on life cycle assessment, total cost assessment and green chemistry.
- *Volvo* supplying environmental product declarations based on life cycle assessment for their cars
- *Ericsson* developing design guidelines for their telecommunication hardware based on extensive life cycle assessments of all important elements of their systems [94].

These examples all involve large international firms which apply strategic planning with a long time perspective, but how is the situation for the large undergrowth of small and medium sized enterprises which performs a large part of the industrial production for many product types?

In Denmark, an integrated product policy has been in vigour for nearly a decade now [56, 57], and environmental authorities have invested a lot of effort, not to mention considerable funds, in pushing Danish industries to work with the environmental performance of their products in a life cycle perspective. The means have been the development of LCA tools and databases [24, 26, 95], tools for DFE [37, 39, 41, 79], and guidelines and manuals on life cycle management [96], and introduction of life cycle thinking in the decision making in small and medium sized enterprises [97, 98, 99]. The dissemination and implementation of the tools has been supported by the establishment of the Danish LCA Center.

The availability of the necessary tools and data is such that no industry should be prevented from embarking on Life Cycle Engineering, but in spite of all these efforts, the Danish experience shows that the dissemination of tools which have been developed and adapted, particularly for the SMEs, is slow for most trades and sometimes completely absent. It is the impression of the authors that, a similar trend can be observed in the rest of the industrialised world. Thus there seems to be an insufficient motivation, particularly for the small and medium-sized companies which operate with a rather short time-perspective in their planning, to undertake the endeavour of implementing a life cycle perspective on their operations.

Industry can be motivated by legal requirements forcing them to take interest in the life cycle of their products, but today, there are only few examples of hard legislation of relevance to the environmental performance of products. The most obvious example in the EU is a frame directive for Energy using Products (EuP) which creates a system where LCA-based criteria will be set for individual energy-using products requiring a minimum performance which the products must meet in order to obtain the CE label and hence be allowed to be marketed on the European Market [100]. This directive will (depending on how the frames are filled in) lead to the first (hard) legislative requirements on environmental performance of products within the EU. With its requirements to environmental assessment and documentation of the assessment, it is foreseeable that it will cause repercussions worldwide. Most of the products covered by the directive are produced in regions outside the EU, and the requirements are thus targeted on producers world wide, providing a fine example of the complexity of regulating products instead of processes or production plants.

As another example of hard regulation with relevance to the environmental performance of products, the end-of-life-vehicle directive [83] extends producer liability for vehicles to their disposal stage thus motivating product developers to design vehicles that will be easily dismantled, contain less problematic substances, weigh less and be easier to repair and upgrade – in short: to design and construct products with their disposal in mind.

Within the field of Electrical and Electronic Equipment (EEE), two directives on waste of EEE (the WEEE directive, [101]) and on restrictions on the use of some hazardous substances in EEE (the RoHS directive, [102]), have recently been adopted. The WEEE directive follows the line of the end-of-life-vehicle directive by extending producer's responsibility for the product beyond the gate.

Apart from these examples, there is very little hard regulation of the environmental behaviour of products today. If the product is safe, it can be manufactured and sold no matter how useless and environmentally inefficient, it may be. The environmental justification of a product is still far from entering into any legal requirements.

In the absence of sufficiently efficient sticks in the form of hard legislation, the development of soft regulation in the form of IPP is trying to present a number of carrots to motivate the industry to develop and market greener products. It is obvious that the industry is sensitive to the public perception of its brands and products, and in many countries authorities attempt to influence the demand of the market through introduction of ecolabelling systems, environmental product declarations and green public purchase guidelines instructing public (and other) purchasers in what to consider if they want to buy greener products.

The effect of such tools will vary among product types but the Danish experience shows that their effect is often limited, particularly for the small and medium-sized companies. It is the belief of the authors that for some product types, hard legislation is needed to help the propagation of life cycle thinking.

In a longer time perspective, a stronger future trend towards life cycle thinking in the industry must be founded now in the education of future generations of citizens, customers, employees and last but not least engineers who shall generate the solutions for a sustainable future. The development of environmental awareness must start in primary schools and be an integrated part of the learning all the way through the educational system in order to create citizens for whom the environmental (and social) impacts of products throughout their life cycle are as important as the more tangible physical quality parameters of the product.

Acknowledgements

Discussions with professor Jack Jeswiet, Queens University, Canada on design for environment and Life Cycle Engineering provided inspiration in the elaboration of this chapter.

References

1 Hauschild, M.; Alting, L.; Poll, C.: Life Cycle Engineering – From Methodology to Enterprise Culture. In: Hon, B. (ed.): Design and Manufacture for Sustainable Development. Engineering Publishing Ltd., London 2003, pp. 3-22.

2 Hunt, R.G.; Franklin, W.E.; Welch, R.O.; Cross, J.A.; Woodall, A.E.: Resource and Environmental Profile Analysis of Nine Beverage Container Alternatives. United States Environmental Protection Agency (US EPA), Office of Solid Waste Management Programs, EPA/530/SW-91c), Washington D.C. 1974.

3 Meadows, D.; Meadows, D. (eds.): The Limits to Growth. A Report for the Club of Rome's Project on the Predicament of Mankind, Universe Publications, 1972.

4 Bundesamt für Umweltschutz: Ecobalance for Packaging Materials (in German). Schriftenreihe Umweltschutz no. 24, Bern, Switzerland 1984.

5 Lundholm, M.P.; Sundström, G.: Resource and Environmental Impact of Tetra Brik Carton and Refillable and Non-refillable Glass Bottles. Tetra Brik Aseptic Environmental Profile, AB Tetra Pak, Malmö, Sweden 1985.

6 Franke, M.: Environmental Impacts from Beverage Containers – Methodology for Measuring the Environmental Impacts from Complex Processes with an Example on One-way and Returnable Beverage Containers (in German). Technische Universität Berlin, Institut für Technischen Umweltschutz, Berlin 1984.

7 Mekel, O.C.L.; Huppes, G.: Environmental Effects of Different Package Systems for Fresh Milk. CML-Mededelingen no. 70, Centrum voor Milieukunde, Rijksuniversiteit Leiden, the Netherlands 1990.

8 Pommer, K.; Hillersborg, A.S.; Hirsbak, S.; Hoffman, L.; Schmidt, A.: Environmental Assessment of Milk Packaging (in Danish). Environmental Project no. 168, Danish Environmental Protection Agency, Copenhagen, Denmark 1991.

9 Consoli, F.; Allen, D.; Boustead, I.; Fava, J.; Franklin, W.; Jensen, A.A.; de Oude, N.; Parrish, R.; Perriman, R.; Postlethewaite, D.; Quay, B.; Séguin, J.; Vigon, B.: Guidelines for Life Cycle Assessment: A Code of Practice. SETAC Press, Pensacola, Florida 1993.

10 Fava, J.; Jensen, A.A.; Lindfors, L.-G.; Pomper, S.; de Smet, B.; Warren, J.; Vigon, B.: Life Cycle Assessment Data Quality: A Conceptual Framework. SETAC Press, Pensacola, Florida 1994.

11 Udo de Haes, H.A. (ed.): Towards a Methodology for Life Cycle Impact Assessment. SETAC-Europe, Brussels 1996.

12 Udo de Haes, H.A.; Finnveden, G.; Goedkoop, M.; Hauschild, M.; Hertwich, E.; Hofstetter, P.; Klöpffer, W.; Krewitt, W.; Lindeijer, E.; Jolliet, O.; Müller-Wenk, R.; Olsen, S.; Pennington, D.; Potting, J.; Steen B. (eds.): Life Cycle Impact Assessment: Striving towards best practice. SETAC Press, Pensacola, Florida 2002.

13 Christiansen, K. (ed.): Simplifying LCA: Just a Cut? SETAC Press, Pensacola, Florida 1997.
14 ISO 14040: Environmental Management – Life Cycle Assessment – Principles and Framework. International Organisation for Standardisation, Geneva, Switzerland 1997.
15 ISO 14041: Environmental Management – Life Cycle Assessment – Goal and Scope Definition and Inventory Analysis. International Organisation for Standardisation, Geneva, Switzerland 1998.
16 ISO 14042: Environmental Management – Life Cycle Assessment – Life Cycle Impact Assessment. International Organisation for Standardisation, Geneva, Switzerland 2000.
17 ISO 14043: Environmental Management – Life Cycle Assessment – Life Cycle Interpretation. International Organisation for Standardisation, Geneva, Switzerland 2000.
18 Harrison, K.W.; Dumas, R.D.; Solano, E.; Barlaz, M.A.; Brill, E.D.; Ranjithan, S.R.: Decision Support Tool for Life-cycle-based Solid Waste Management. Journal of Computing Engineering (2001) January, pp. 44-58.
19 Björklund, A.; Bjuggren, C.; Dalemo, M.; Sonesson, U.: Planning Biodegradable Waste Management in Stockholm. Research and Analysis 3 (2000) 4, pp. 43-58.
20 Dalemo, M.; Sonesson, U.; Björklund, A.; Mingarini, K.; Frostell, B.; Jönsson, H.; Nybrant, T.; Sundqvist, J.-O.; Thyselius, L.: ORWARE – A Simulation Model for Organic Waste Handling System. Resources Conservation & Recycling (1997) 21, pp. 17-37.
21 Stewart, M.; Jolliet, O.: User Needs Analysis and Development of Priorities for Life Cycle Impact Assessment. Int. J. LCA 9 (2004) 3, pp. 153-160.
22 UNEP: Life Cycle Initiative homepage: http://www.uneptie.org/pc/sustain/lcinitiative/home.html (Update: 2002).
23 Hauschild, M.: Assessing Environmental Impacts in a Life Cycle Perspective. Environmental Science and Technology, 2005, pp. 81A-88A.
24 Wenzel, H.; Hauschild M.Z.; Alting, L.: Environmental Assessment of Products. Vol. 1 – Methodology, Tools, Techniques and Case Studies. Chapman & Hall, United Kingdom, Kluwer Academic Publishers, Hingham, MA. USA 1997.
25 Heijungs, R.; Guinée, J.; Huppes, G.; Lankreijer, R.M.; Udo de Haes, H.A.; Sleeswijk, A.W.; Ansems, A.M.M.; Eggels, P.G.; van Duin, R.; de Goede, H.P.: Environmental Life Cycle Assessment of Products. Guide, Report no. 9266, CML, Leiden 1992.
26 Hauschild, M.Z.; Wenzel, H.: Environmental Assessment of Products. Vol. 2 – Scientific background. Chapman & Hall, United Kingdom, Kluwer Academic Publishers, Hingham, MA. USA 1998.
27 Goedkoop, M.; Spriensma, R.: The Eco-indicator 99 – A Damage-oriented Method for Life Cycle Impact Assessment. Methodology Report, 2nd Edition, Version 17-4-2000. Pré Consultants, B.V. Amersfoort, the Netherlands 2000.

28 Guinée, J. (ed.): Handbook on Life Cycle Assessment. Operational Guide to the ISO standards. Kluwer Academic Publishers, Dordrecht, the Netherlands 2002.
29 Steen, B.: A Systematic Approach to Environmental Priority Strategies in Product Development (EPS). Version 2000 – General System Characteristics / Models and Data of the Default Method. CPM Report 1999:4 and CPM Report 1999:5. Chalmers University of Technology, Gothenburg, Sweden 1999.
30 Bare, J.C.; Norris, G.A.; Pennington, D.W.; McKone, T.E.: The Tool for the Reduction and Assessment of Chemical and Other Environmental Impacts. J. of Industrial Ecology 6 (2003) 3-4, pp. 49-78.
31 Itsubo, N.; Inaba, A.: A New LCIA Method: LIME has been Completed. Int.J. LCA 8 (2003) 5, p. 305.
32 Hauschild, M.; Wenzel, H.: The European Person Equivalent: Measuring the Personal Environmental Space. 2000 Annual Report NATO/CCMS Pilot Study, Clean Products and Processes (Phase I), Report no. 242. U.S. Environmental Protection Agency, EPA/625/R-01/002, 2001.
33 Bare, J.C.; Hofstetter, P.; Pennington, D.; Udo de Haes, H.A.: Midpoints versus Endpoints: The Sacrifices and Benefits. Int. J. LCA 5 (2000) 6, pp. 319-326.
34 Heijungs, R.; Goedkoop, M.; Struijs, J.; Effting, S.; Sevenster, M.; Huppes, G.: Towards a Life Cycle Assessment Method Which Comprises Category Indicators at the Midpoint and the Endpoint Level. Report of the First Project Phase: Design of the New Method, Netherlands Ministry of Housing, Spatial Planning and the Environment, Available from http://www.pre.nl/download/recipephase1final.pdf, 2003.
35 Norris, G.: Integrating life Cycle Cost Analysis in LCA. Int. J. LCA 6 (2001) 2, pp. 118-120.
36 Rebitzer, G.; Hunkeler, D.: Life Cycle Costing in LCA: Ambitions, Opportunities, and Limitations. Int. J. LCA 8 (2003) 5, pp. 253-256.
37 Schmidt, W.-P.: Life Cycle Costing as Part of Design for Environment. Int.J. LCA 8 (2003) 3, pp. 167-174.
38 Shapiro, K.G.: Incorporating Costs in LCA. Int. J. LCA 6 (2001) 2, pp. 121-123.
39 Krewitt, W.: External Costs of Energy – Do the Answers Match the Questions? Energy Policy (2002) 30, pp. 839-848.
40 O'Brian, M.; Doig, A.; Clift, R.: Social and Environmental Life Cycle Assessment (SELCA). Int. J. LCA, 1 (1996) 4, pp. 231-237.
41 Dreyer, L.C.; Hauschild, M.Z.; Schierbeck, J.: A Framework for Social Life Cycle Impact Assessment. Int. J. LCA ? (2005) ?, pp. ??-??.
42 SimaPro (http://www.pre.nl)
43 GaBi (http://www.gabi-software.de)
44 LCAiT (http://www.lcait.com)
45 Umberto (http://www.umberto.de)

46 TEAM (http://www.ecobilan.com).
47 http://www.nrel.gov/lci/about.html
48 http://www.ecoinvent.CH/en/index.htm
49 http://www.apme.org
50 http://www.worldsteel.org/lci
51 http://www.nickelinstitute.org/index.cfm/ci_id/114.htm
52 Graedel, T.: Streamlined Life Cycle Assessment. Prentice-Hall, Inc. 1998.
53 VROM: National Environmental Policy Plan, To Choose or To Lose. The Dutch Ministry of Housing, Physical Planning and Environment, The Hague 1989.
54 VROM: Note on Product and Environment (in Dutch), VROM 93681/h/1-94, The Dutch Ministry for Health and the Environment, The Hague 1994.
55 Scholl, G.: Sustainable Product Policy in Europe. European Environment (1996) 6, pp. 183-193.
56 DEPA: Intensified Product-oriented Environmental Initiative. Danish Environmental Protection Agency (in Danish, English version 1999), Ministry of Environment and Energy, Copenhagen 1996.
57 DEPA: Statement on a Product-orientated Environmental Initiative. Danish Environmental Protection Agency (in Danish, English version 1999), Ministry of Environment and Energy, Copenhagen 1998.
58 CEC: Green Paper on Integrated Product Policy, COM(2001) 68 final. Commission of the European Communities, Brussels, 2001.
59 CEC: Integrated Product Policy – Building on Environmental Life-Cycle Thinking, COM(2003)XX communication of 18th June 2003. Commission of the European Communities, Brussels, 2003.
60 ISO 14020: Environmental Management – Environmental Labels and Declarations – General Principles. International Organisation for Standardisation, Geneva, Switzerland 2000.
61 ISO 14024: Environmental Management – Environmental Labels and Leclarations – Type I Environmental Labelling – Principles and procedures. International Organisation for Standardisation, Geneva, Switzerland 1999.
62 http://www.gen.gr.JP
63 ISO/TR 14025: Environmental Management – Environmental Labels and Declarations – Type III Environmental Declarations. International Organisation for Standardisation, Geneva, Switzerland 2000.
64 http://www.environdec.com
65 ISO 14021: Environmental Management – Environmental Labels and Declarations – Self-declared Environmental Claims (Type II environmental labelling). International Organisation for Standardisation, Geneva, Switzerland 1999.
66 Knudsen, P.E.; Jensen, M.J.; Chabert, A.: Evaluating Product Panels. Working Report No. 14/2003, Danish EPA, 2003.
67 http://www.lcacenter.org
68 http://www.polymtl.ca/ciraig/reception.html

69 http://www.netzwerk-lebenszyklusdaten.de
70 http://unit.aist.go.JP/lca-center/English/top.htm
71 http://www.lcarc.re.kr/English/index.htm
72 http://doi.eng.cmu.AC.th/thailca
73 http://www.lca-center.dk
74 Hill B.: Industry's Integration of Environmental Product Design. IEEE International Symposium on Electronics and the Environment, 1993.
75 Hauschild, M.; Wenzel, H.; Alting, L.: Life Cycle Design – A Route to the Sustainable Industrial Culture? Annals of the CIRP 48/1/1999, pp. 393-396.
76 Alting, L.; Hauschild, M.; Wenzel, H.: Environmental Assessment in Product Development. Phil.Trans. R. Soc. Lond. A, 355, 1997, pp. 1373-1388.
77 Boothroyd, Dewhurst, Knight: Product Design for Manufacture and Assembly, Marcel Dekker, NY 1994.
78 Graedel, T.E.; Allenby, B.R.: Industrial Ecology. Prentice Hall, New Jersey 2003.
79 Olesen, J.; Wenzel, H.; Hein, L.; Andreasen, M.M.: Design for Environment (in Danish). The Danish Ministry of the Environment and Energy and the Confederation of Danish Industries, Copenhagen, 1996.
80 Luttropp, C.; Züst, R.: Ecoeffective Products from a Holistic Point of View. In Proceedings of the 5th International CIRP Seminar on Life Cycle Engineering, Stockholm, Sweden 1998.
81 Kaldjian, P.J.: Ecological Design: Source Reduction, Recycling, and the LCA. Innovation. Journal of the Industrial Designers Association of America, 11 (1992) 3. pp.11-13.
82 Hauschild, M.; Jeswiet, J.; Alting, L.: Design for environment – Wo e get the focus right? Annals of the CIRP, 53/1/2004, pp. 1-4.
83 P&CEC: Directive 2000/53 on End-of-Life Vehicles. The Parliament and the Council of the European Communities, Brussels, 2000.
84 Wenzel, H.; Alting, L.: Architecture of Life Cycle, keynote at Global Conference on Sustainable Product Development and Life Cycle Engineering, Sept. 29th – Oct. 1st, 2004, Berlin, Germany 2004.
85 World Commission on Environment and Development: Our Common Future. Oxford University Press, United Kingdom 1987.
86 Graedel, T.E.; Allenby, B.R.: Industrial Ecology. Prentice Hall, Englewood Cliffs, New Jersey 1995.
87 United Nations: Long-range World Population Projections. United Nation Publications, Geneva 1992.
88 von Weizsäcker, E.; Lovins, A.B.; Lovins, L.H.: Factor Four – Doubling Wealth, Halving Resource Use. Earthscan, UK 1997.
89 Schmidt-Bleek, F.: Factor 10 Club – Carnoules Declarations, Wuppertal Institute for Climate, Environment and Energy. Wuppertal, Germany 1995.
90 Elkington, J.:Cannibals with Forks – The Triple Bottom Line of 21st Century Business. New Society Publishers, Canada 1998.
91 Westkämper, E.; Alting, L.; Arndt, G.: Life Cycle Management and Assessment:

Approaches and Visions Towards Sustainable Manufacturing. Annals of the CIRP, 49/2/2002, pp. 501.
92 Elkington, J.: Who needs it? – Market Implications of Sustainable Lifestyles, SustainAbility Ltd., London 1995.
93 Welford, R.; Young, W.; Ytterhus, B.: Toward Sustainable Production and Consumption: A Conceptual Framework: In Roome, N.: Sustainability Strategies for Industry – The Future of Corporate Practice, The Greening of Industry Network Series, Island Press, Washington 1998, pp. 75-96.
94 Wenzel, H.; Molin, C.; Hauschild, M.; Knudsen, H.H. (eds.): Annual report NATO/CCMS Pilot Study, Clean Products and Processes (Phase I). Report no. 242. U.S. Environmental Protection Agency, EPA/625/R-01/002, 2001, 2000.
95 Pommer, K.; Bech, P.; Wenzel, H.; Caspersen, N.; Olsen, S.I.: Handbook on Environmental Assessment of Products. Environmental Report No. 813, Danish EPA, Copenhagen, 2003.
96 Remmen, A.; Münster, M.: An Introduction to Life-Cycle Thinking and Management. Environmental News No. 68, Danish EPA, Copenhagen, 2003.
97 Schmidt, K.; Christensen, F.M.; Juul, L.; Øllgaard, H.; Nielsen, C.B.: Manual on Product-oriented Environmental Work. Environmental News No 64, Danish EPA, Copenhagen, 2002.
98 Keiding, K.; Hanneslund, L.; Niemann, M.: Stimulation of SME's Focus on Cleaner Products (in Danish). Working Report No. 32/2002, Danish EPA, Copenhagen, 2002.
99 Niemann, H.; Møller-Jørgensen, A.; Nielsen, B.B.; Petersen, A.: Planning and Organising an Environmental Dialogue, Environmental News, No. 63. The Danish Agency for Trade and Industry and the Danish EPA, Copenhagen, 2001.
100 CEC: Draft Framework Directive for Eco-design of End Use Equipment (EuE). Commission of the European Communities, Brussels, 2002.
101 P&CEC: Directive 2002/96 on Waste Electrical and Electronic Equipment (WEEE). The Parliament and the Council of the European Communities, Brussels, 2002.
102 P&CEC: Directive 2002/95 on the Restriction of the Use of Certain Hazardous Substances in Electrical and Electronic Equipment (RoHS). The Parliament and the Council of the European Communities, Brussels, 2002.

3.2 Saving Resources by Reuse, Recycling, Recovery

Günter Fleischer, Julia Dose, Robert Ackermann, Berlin, Germany

In order to evaluate sustainability[1] several national and international indicators are defined. Saving resources is an important indicator for sustainability. Among others, environmental impacts are caused by resource consumption: The more resources are used for a certain product the more environmental burdens are produced. There is a general believe that Reuse (R), Recycling (R), and Recovery (R) themselves lead to saving resources and reduced environmental impacts. In some cases this to believe is wrong in other it is right. However, we don't know to what extend and how it can be influenced. Are consequently, e.g. abiotic non-renewable, resources[2] used, then their consumption should be optimal from an environmental point of view. In this context Disassembly Factories could provide successful solutions for saving resources by RRR. Saving resources leads to reduce environmental impacts and has significant impact on sustainability. In order to determine optimal solutions by RRR in relation to efficiency of resources and environmental impacts, the method of Life Cycle Assessment (LCA) according to [1, 2] shall be used. The determination whether RRR options contribute to saving resources shall also consider resources needed for the whole process chain, e.g. allocation, transportation, processing.

In addition to the determination of better RRR options for already existing products the development of new, environmental optimized products is another successful contribution to sustainability. Designing environmental friendly products is a crucial factor in the EU strategy on Integrated Product Policy (IPP) tightened by the EuP Directive [3].

3.2.1 Environmental Evaluation

Determination of Environmental Significant RRR Options

In the End-of-Life (EOL) stage, a distinction is made between the terms of reuse, recycling, and recovery and their links to production, consumption, and disposal [4] (Figure 1). Reuse is defined as the reuse of a product for

[1] Sustainable development is defined as development that meets the needs of the present without compromising the ability of future generations to meet their own needs [5].

[2] Abiotic resources are natural resources such as iron ore, which are regarded as non-living [6].

the same purpose without any structural changes. Recycling is specified as material recycling with reference to structural changes in products and recovery represents the processing of residues.

Different options for RRR can have completely diverse impacts on the environment. Therefore, a comparison of options is strictly necessary. Figure 2 illustrates one principal example. System 1 produces a secondary product out of scraped plastics, while system 2 produces useable energy. Until this stage, both systems deliver different products with different benefits. Therefore, the environmental impacts caused by these systems are not comparable. To assure that both systems are comparable with each other, they have to be expanded by means of complementary production in order to get the same benefit for both systems. This equality in system benefits is the basic requirement of an environmental evaluation.

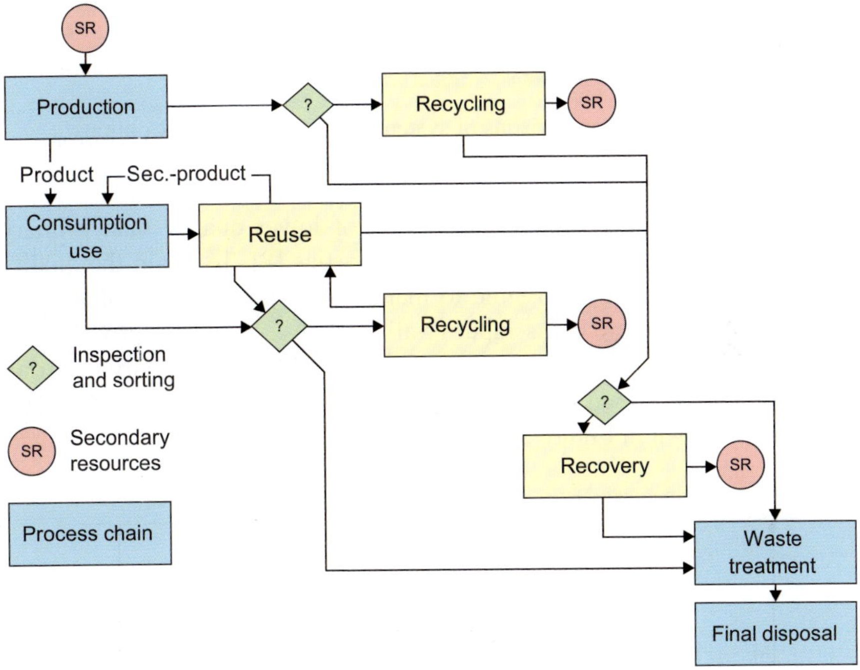

Fig. 1: Options of material flows in a life cycle system according to [7]

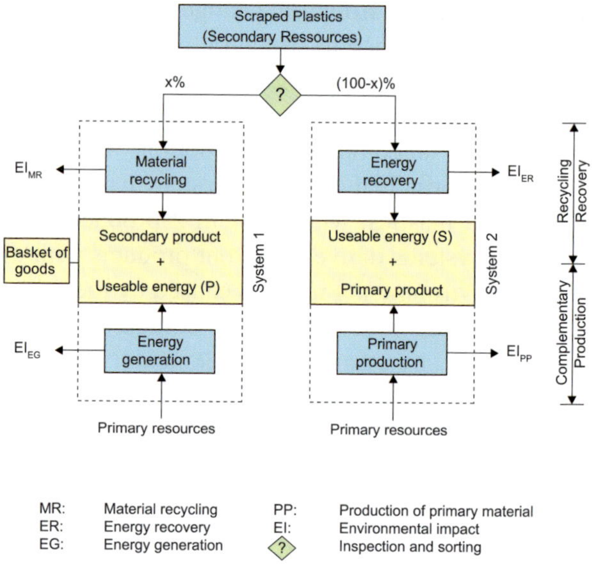

Fig. 2: Precondition: Comparisons of systems for evaluating RRR options according to [7]

Recycled secondary products are often of lower quality than primary products, e.g. recycled paper or recycled plastics [8]. To attain the quality of primary products it is possible to manufacture products out of secondary material with more mass input (wall thickness of plastic products for equivalent stability) or to use a mixture of secondary and primary material. The correlation between addition and substitution of primary and secondary materials is shown in Figure 3.

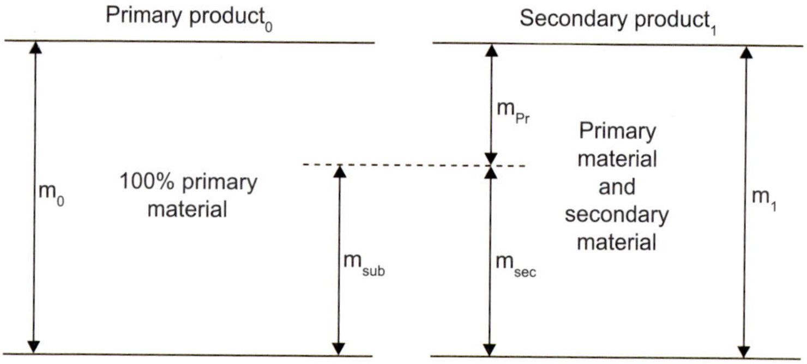

Fig. 3: Addition and substitution of primary and secondary material [7]

The recycling rate R^1 for one secondary product (m_1) in relation to one primary product (m_0) (Eq. 1) is presented in Figure 3. It is determined by the mass of the substituted material (m_{sub}) in relation to the mass of the primary product.

$$R^1 = \frac{m_{sub}}{m_0} \tag{1}$$

The recycling rate is often stated as addition factor B (Eq. 2), which stands for the fraction of secondary material mass in the secondary product (Figure 3) .

$$B = \frac{m_{sec}}{m_1} \tag{2}$$

The rate of substitution is given by substitution factor S (Eq. 3). It determines the ratio of the substituted material (Eq. 4) to the mass of secondary material.

$$S = \frac{m_{sub}}{m_{sec}} = \frac{m_0 - m_{Pr}}{m_{sec}} \tag{3}$$

$$m_{sub} = m_0 - m_{Pr} \tag{4}$$

$$R = \frac{n_1 * m_{sub}}{n_0 * m_0} = \frac{n_1 * m_{sec}}{n_0 * m_0} * S \tag{5}$$

The real recycling rate considers the relation of those secondary products made out of the corresponding primary product to the primary products themselves (Eq. 5). The recycling rate R for n_0 primary products and n_1 secondary products is given as function of factor S. The amount of substitutable raw material by secondary material depends on the quality of the secondary material and should be analyzed to clarify the environmental benefit of the recycling scenario. On the basis of the recycling rate R of a product it is possible to work out whether an admixture of secondary material to a product represents real recycling or whether it is just a material input into a product. In real recycling, primary material is saved by substitution.

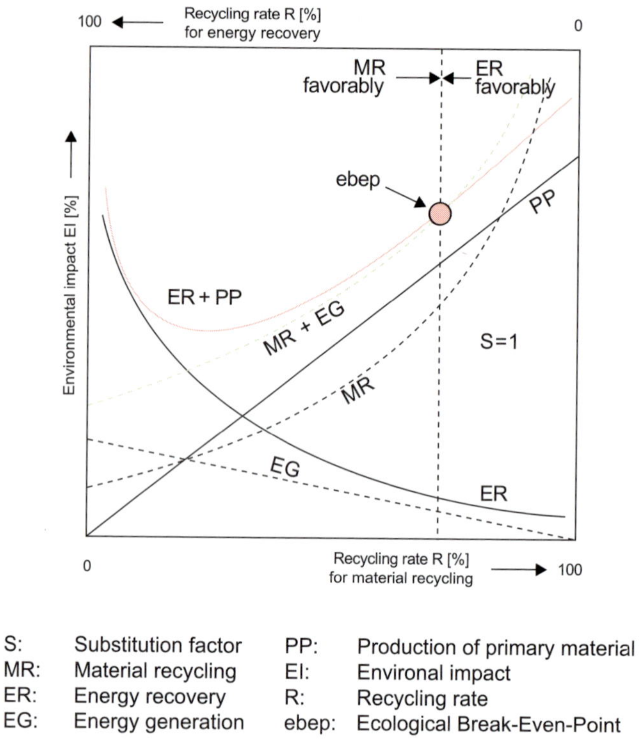

Fig. 4: Determination of environmental significant scenario (Figure 2) in relation to the recycling rate R [9]

The Ecological Break-Even-Point (EBEP) in Figure 4 determined by LCA is an adequate parameter to compare different RRR options in order to identify the optimal option – the one with the highest environmental benefit [9]. Generally speaking RRR options make sense when potential environmental impacts are lower than the potential environmental impacts caused by the EOL stage without RRR. Potential Environmental Impacts (EI) in Figure 4 resulting from consumption of primary and secondary resources are directly dependent of the mass of consumed primary resources and the mass of saved primary resources by recycled secondary material. In the range of low material recycling rates, the EOL stage with material recycling scenario (MR+EG curve) causes less environmental impacts than the EOL stage without material recycling scenario (ER+PP curve). In the range of material recycling rates above the EBEP, the material-recycling scenario offers less benefit than energy recovery which becomes the favourable EOL scenario. The green coloured curves in Figure 4 indicate environmentally significant

RRR options. The location of the EBEP is highly depended on the specific situations such as transport distances, available technology, quality of secondary material, consumer requirements, etc. It explains the importance of a detailed comparison between different options for identifying better EOL scenarios including any RRR option.

3.2.2 Application

In order to determine better EOL scenarios for already existing products, the entire life cycle of a product has to be analyzed. Disassembly sequences and technologies, e.g. depend – among others – on the materials and joints used for the product.

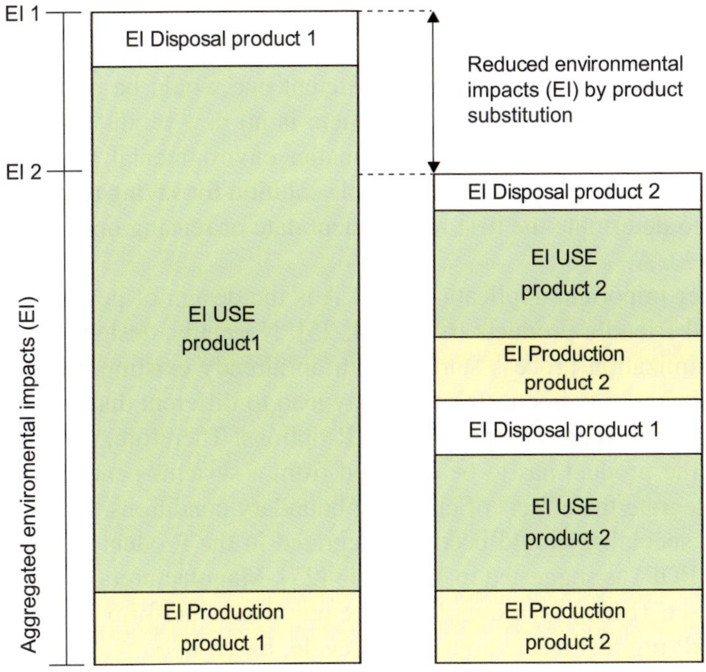

Fig. 5: Reduced environmental impacts by product substitution according to [10]

Conducted LCA studies for evaluating EOL scenarios including different RRR options based on [11, 12, 13, 14, 15] show that the disassembly of a washing machine clearly reduces the potential environmental impacts over the products' life cycle [16, 17]. Main reasons for environmental advantages are product characteristics and RRR options enabled by a disassembly

[18] on Disassembly Knowledge Platform. Disassembly factories provide a very good option for generating environmental friendly RRR options. Furthermore, they can be helpful to fulfil legal recycling quotas in the future, as they are stipulated in the WEEE Directive [19] on waste electrical and electronic equipment.

Another approach to calculate the optimised use-time of a product in terms of saving resources with regard to the optimal environmental replacement date is presented in [20]. The analysis of possible replacement of a product by a new, more innovative product is in general reasonable from an environmental point of view. This reasonability is justified, if the potential environmental impacts in the use stage are dominant compared to other life cycle stages ,and if the new product has a significantly lower environmental impact in the use stage (Figure 5) [21]. A ten years old refrigerator for example consumes about twice as much energy as a new refrigerator. Since energy consumption causes the main environmental impact of this product, substitution by a new, more energy-efficient one would be better from an environmental point of view [22]. If there is no new product or assembly available, which has a significantly smaller environmental impact during the use stage, reuse might be a reasonable solution for saving resources. The analysis to determine the best replacement date or reusing option has to be based on LCA.

Another important application of LCA is the design of new, more environmental friendly products [6, 23, 24, 25]. Normally, designing products is an optimization process starting with an already existing product [26]. Small changes, e.g. in construction, may lead to different disassembly processes and sequences and possible RRR options. Therefore, the whole life cycle of the product has to be analyzed from a systems perspective in order to integrate the effects of changing boundary conditions. Consequently the LCA specialist should become integrated in the Product Development Process (PDP) as suggested in VDI 2243 [27]. Manufacturers of Energy using Products (EuPs) for example will be required to perform an assessment of the EuP product throughout its life cycle, based upon realistic assumptions about normal conditions and purposes of use [3] from 2008 onwards. The assessment will be based on environmentally relevant product characteristics and inputs/outputs throughout the product life cycle expressed in physical quantities that can be measured. Manufacturers will make use of this assessment to evaluate alternative design options and the achieved environmental performance of the product. Assessing potential environmental impacts already during the design stage is recommended for improving environmental performance of a product and providing efficient strategies for saving resources. Potential environmental impacts caused during a product's life cycle as well as used resources are determined during the design phase and most of the costs involved are committed then depending on the kind of product up to 80 %. This underlines the importance of designing environmentally optimized products.

3.2.3 Conclusion

In order to analyze optimal solutions by RRR in relation to efficiency of resources and environmental impacts, the method of LCA has been used. Therefore, equality in system benefits is a basic requirement. EBEP determined by LCA is an adequate parameter to identify optimal RRR options with the highest environmental benefit. Disassembly Factories provide a very good option for generating environmental friendly RRR options. The amount of substitutable raw material by secondary material depends on the quality of the secondary material and should be analyzed in order to clarify the environmental benefit of recycling options. Determination of the best replacement date and reuse has to be based on LCA as well. For providing sustainable EOL scenarios for already existing products, optimal transportation distances, recycling rates, and RRR options are preconditioned. A life-cycle perspective has to be utilized for improving the environmental performance of new products and therefore providing efficient strategies for saving resources. Integrating LCA in the design process is an activity of increasing importance imposing specific requirements on the availability of life cycle information. Knowledge of the product's life cycle as well as LCA methodology is therefore one basic requirement.

References

1 EN ISO 14040: Environmental Management – Life Cycle Assessment – Principles and Framework, 2006.
2 EN ISO 14044: Environmental Management – Life Cycle Assessment – Requirements and Guidelines, 2006.
3 EuP Directive: Directive 2005/32/EC of the European Parliament and of the Council on Establishing a Framework for the Setting of Ecodesign Requirements for Energy-using Products and Amending Council Directive 92/42/EECm and Directives 96/57/EC and 2000/55/EC, 2005.
4 Fleischer, G.: Prevention of Waste. In: Ullmanns Encyclopedia of Industrial Chemistry, 6th Edition, Vol. B 8, VCH Verlagsgesellschaft, Weinheim 1995, pp 612-624.

5 World Commission on Environment and Development: Our Common Future. Oxford Paperbacks, 1987.

6 Guinée, J.B.; Gorrée, M.; Heijungs, R.; Huppes, G.; Kleijn, R:, de Koning, A.; van Oers, L.; Wegener Sleeswijk, A.; Suh, S.; Udo de Haes, H.; de Bruijn, H.; van Duin, R; Huijbregts, M.A.J.: Handbook on Life Cycle Assessment. Operational Guide to the ISO Standards. Kluwer Academic Publishers, Dordrecht, Bostan, London 2002.

7 Fleischer, G.: Recycling vs. Verwertung. Presentation, Abfalltage 2002, Stuttgart, 11. Juli 2002.

8 Fleischer, G.: Nachsorgende Abfallvermeidung – Recycling. Technische Universität Berlin, Vorlesungsskript, Berlin 2001.

9 Fleischer, G.: Der ökologische break-even-piont für das Recycling. In: Thomé-Kozmiensky, K.J. (ed.): Modelle für eine zukünftige Siedlungsabfallwirtscha ft. EF-Verlag für Energie- und Umwelttechnik GmbH, Berlin 1993.

10 Janusz, G.; Fleischer, G.: Ecological Product Lifetime Optimization. In: Proceedings of LCM 2005, Innovation by Life Cycle Management, Barcelona, Spain, pp. 125-129.

11 Fleischer, G.; Dose, J.; Ewers, H.J.; Schatz, M.: Disassembly Factories: Economic and Environmental Options. In: George Lee, C.S. (ed): Proceedings of the 2001 IEEE International Symposium on Assembly and Task Planning (ISATP 2001), Assembly and Disassembly in the Twenty-first Century. Purdue University, USA 2001, pp 447-452.

12 Perlewitz, H.: Planung und marktorientierter Betrieb von Demontagefabriken. Dissertation, TU Berlin, 2000.

13 Seliger, G.; Uhlmann, E.; Keil, T., Härtwig, J.P.: Realisierung eines Pilot-Demontagesystems. In: ZWF, Sonderbeilage Sfb 281 „Demontagefabriken“. Carl Hanser Verlag München 2000, pp 20-24.

14 Sfb 281 – Sonderforschungsbereich 281: Demontagefabriken zur Rückgewinnung von Ressourcen in Produkt- und Materialkreisläufen. Arbeits- und Ergebnisbericht. Berlin, 2000.

15 Sfb 281 – Sonderforschungsbereich 281: Demontagefabriken zur Rückgewinnung von Ressourcen in Produkt- und Materialkreisläufen. Arbeits- und Ergebnisbericht. Berlin, 2003.

16 Fleischer, G.; Dose, J.: Standard Entsorgung versus Demontage - ein ökologischer Vergleich. In: Thomé-Kozmiensky, K.J. (ed): Reformbedarf in der Abfallwirtschaft. TK Verlag Karl-Thomé-Kozmiensky, Neuruppin 2001, pp 1181-1190.

17 Fleischer, G.; Dose, J.: End-of-Life of Household Appliances: LCA for Evaluating End-of Life Scenarios. In: Feldmann, K. (ed): Proceedings of the 9th International Seminar on Life Cycle Engineering, Integrated Product Policy-Chance and Challenge. Universität Erlangen-Nürnberg, Erlangen 2002.

18 Dose, J.: Disassembly Knowledge Platform 2006, Economic and Environmental Options, http://demonty.ipk.fhg.de/end_of_life_of_household_appliances (Access: 21.04.2006).

19 WEEE Directive: Directive of the European Parliament and of the Council on Waste Electrical and Electronic Equipment, Directive 2002/96/EG, 2003.
20 Kunst, H.: Ökologische Optimierung von Substitutionsentscheidungen. Dissertation, TU Berlin, 2003.
21 Alonso, J.C.; Dose, J.; Greif, A.; Maas, H.; Eikelenberg, N.: Eco-Design Guidelines. Deliverable D8 of the project SEES of the EC, contract no. TST3-CT-2003-506075, 2006.
22 Kemna, R.; van Elburg, M.; Li, W.; van Holsteijn, R.: Methodology Study Eco-Design of Energy-using Products, Final report http://www.eupproject.org(product cases report.pdf, Delft 2005.
23 Dose, J.: Austauschbare Sachbilanzmodule auf Basis von Produktkomponenten. Dissertation, Institut für Technischen Umweltschutz, TU Berlin, 2005.
24 Fleischer, G.; Becker, J.; Braunmiller, U.; Klocke, F.; Klöppfer, W.; Michaeli, W: Eco-Design – Effiziente Entwicklung nachhaltiger Produkte mit euroMat. Springer-Verlag, Berlin, Heidelberg 2000.
25 Wenzel, H:, Hauschild, M.; Leo, A.: Environmental Assessment of Products, Vol 1, Methodology, Tools and Case Studies in Product Development. Chapman & Hall, London 1997.
26 Fleischer, G.; Dose, J.; Gries, B.; Blessing, L.: Entwicklung nachhaltiger Produkte unter ökobilanziellen Gesichtspunkten. In: Meerkamm, H. (ed): Design for X, Beiträge zum 14. Symposium. Lehrstuhl für Konstruktionstechnik. Friedrich-Alexander-Universität Erlangen-Nürnberg, Erlangen 2003.
27 VDI 2243: Richtlinie – Recyclingorientierte Produktentwicklung. VDI-Verlag, Düsseldorf 2002.

3.3 Minimizing the Overall Environmental Impact

Günter Fleischer, Julia Dose, Robert Ackermann, Berlin, Germany

Designing environmentally friendly products is a crucial factor in the EU strategy on Integrated Product Policy (IPP) [6]. Therefore, the integration of environmental aspects into the Product Developing Process (PDP) with the aim of improving the environmental performance of a product throughout its entire life cycle is getting more and more important. Customer expectations concerning better environmental performance as well as stronger legislation are leading to the necessity of integrating environmental information in PDP. Potential environmental impacts caused during a product's life cycle are mainly fixed in the PDP (depending on the kind of product up to 80 %) [9]. Life Cycle Approach (LCA) is the most important instrument for designing environmentally friendly products and one of the main tools for IPP [13]. Results of an LCA are based on the process model within the product system of a product's life cycle, which is developed in the Life Cycle Inventory (LCI) step [18]. Therefore, designing environmentally friendly products imposes specific requirements on the availability of life cycle information [12]. An LCA specialist should become an integrated part of PDP in order to identify optimal environmental solutions for e.g. material choice, processes or product structure. Currently LCA is mainly used retrospectively and LCA determines, e.g. the optimum Reuse, Recycling, Recovery (RRR) option for already existing products.

The importance of LCA as a decision tool for prospective studies is disadvantaged by its data-intensity and its comparatively difficult manageability. Gathering and processing necessary data as well as modeling the product life cycle are complex and time consuming activities.

Therefore, the Life Cycle Approach (LCAp) (Figure 1) presented in this chapter should accompany the PDP. LCAp for enhancing the environmental performance of products towards sustainability includes the structure of modeling in LCI for the (modular) LCA, determination of environmental "hot spots" by means of LCA results, and the modeling and evaluation of new design options. To accelerate the performance of an LCA as a precondition for integration this environmental measure in PDP, the modular LCA is developed. The modular LCA shall deliver reliable results with an adequate accuracy by means of exchangeable LCI modules based on product components.

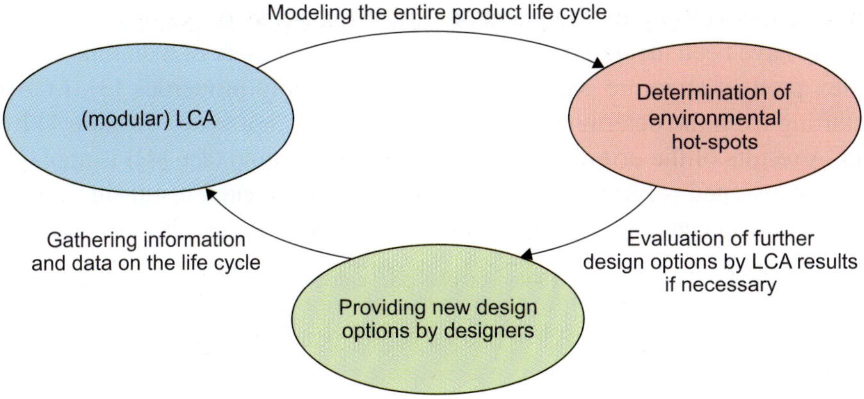

Fig. 1: Life Cycle Approach (LCAp) for enhancing sustainability

3.3.1 Life Cycle Approach

Considerations regarding product design for reuse and recycling will become increasingly important in the future for both, Electrical and Electronic Equipment (EEE) manufacturers and recyclers according to [6, 28]. Environmental aspects have to be considered and evaluated during the PDP [22] in order to develop environmentally friendly products and to support sustainability. Designing environmentally friendly products is an optimization process, in most cases based on an already existing product. Simultaneously, LCAs are conducted in decision situations at least for two alternatives (product or design options) and accordingly at least two product systems are defined and evaluated within impact assessment methods [30]. Following design guidelines, which aim e.g. at an easy disassembling and recycling, may enhance the environmental performance of products. The traditional evaluation of construction-methods can determine whether design guidelines have been considered during the PDP or not. But they cannot answer the question, if the new design (disassembly oriented) is better from an environmental point of view compared to the original product. Consequently, the method of LCA is proposed in [27] as a suitable evaluation tool and is applied in LCAp.

LCAp for Components of a Washing Machine

Based on an original valve pump of a washing machine, a redesigned valve pump with the objective of improving disassembly and recycling features has been developed by designers [4, 25]. Weak points of the original valve pump have been identified by designers, e.g. large variety of materials, complex product structure with ambiguous disassembly properties [3]. LCAp starting from the determination of environmental "hot spots" (Figure 1) by LCA results of the original and redesigned valve pump (see [4]) is applied. The redesigned valve pump obtained better environmental results in all analyzed impact categories, except for climate change (Figure 2).

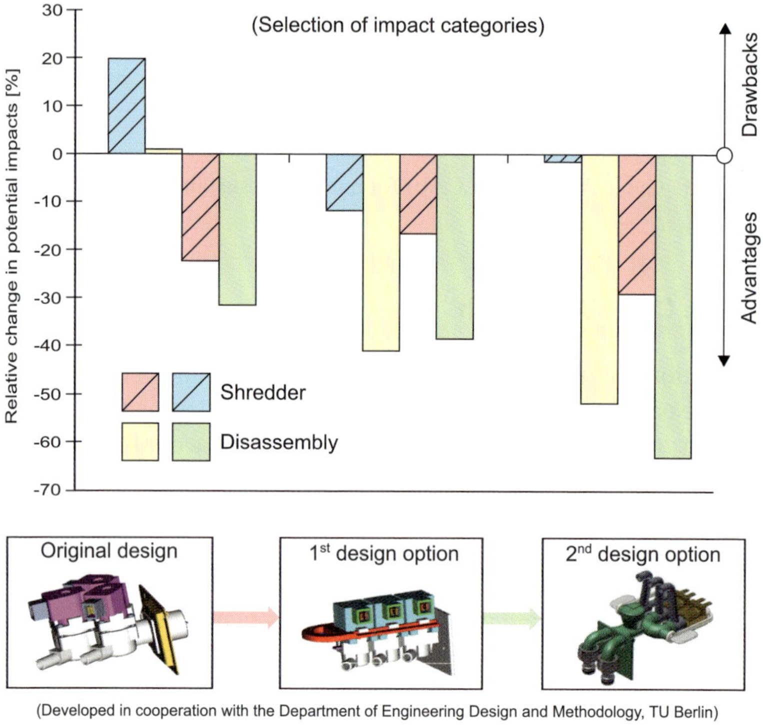

Fig. 2: Relative change in potential environmental impact for two redesigned options (reference component: original design = 100%)

The reason for this disadvantage was clearly identified by the LCA results as material based (content of Polyamide 6 – PA 6), which leads to higher potential environmental impacts in climate change. Consequently, PA 6 should be replaced by another material. The original valve pump contains a higher amount of metals which results in smaller advantages for the redesigned valve pump in current End-of-Life (EOL) processes (shredding with following material recycling [23] and burning of Shredder Residue (SR) in

waste incineration plants[1]). The results from the analyzed valve pumps cannot be interpreted as a general acknowledgement of present methodical approaches to design for disassembly and recycling. But it is possible to point out specific design weaknesses by LCA results in terms of environment and sustainability that could not be identified by evaluation processes usually applied in PDP.
Based on the constructive cognitions from the identified "hot spots", resulting from the LCA studies of the original and redesigned valve pump (avoidance of certain materials, e.g. PA 6, preferably light construction) a disassembly- and environmentally friendly lye pump as an assembly of a disassembly optimized washing machine was developed [4] using LCAp. The following generic product properties of the lye pump have been improved:

- mass reduction by 5,9 %,
- material diversity reduction by 12,5 %, and
- number of components reduction by 21,1 %.

In addition, new fasteners enable a disassembly of the component and a possible reuse of the motor [4].

Impact category indicator results of the original and redesigned lye pump (new design option) determined with the help of LCA are based on CML 2001 characterization factors [12]. In order to evaluate the potential environmental impacts caused by the components, the category indicator results of the life cycle models including the respective EOL scenarios are compared. The improvement in product design, based on constructive cognitions from LCA studies of a valve pump [4], offers an environmental advantage in all analyzed impact categories. The redesigned lye pump has a better environmental performance than the original component. The modeled EOL processes influence the values of category indicator results. The highest environmental advantage of the redesigned lye pump arises from an EOL scenario, in which the motor of the pump is reused.

LCAp leads to a disassembly- and environmentally friendly lye pump. Lightweight construction, modularity and the avoidance of certain materials were important design aspects in this case for achieving better environmental results. Beside these properties, a high functionality and easy disassembly of the redesigned product is aspired [4].

[1] From July 2005 onwards, only waste that has been subject to treatment is landfilled. SR goes to waste incineration plants according to [1].

3.3.2 Modular Life Cycle Assessment

Modeling the product system as a process model of the life cycle of a product and calculating the LCI results are key elements of the LCI step also [7, 12, 14, 29]. Development of these product systems and interpretation of their results require experience as well as expertise. LCI has high demands on product data in the early stages of PDP, but at the early stages of PDP data are only available in low volume and of low quality [9, 30]. Most designers hardly have enough adequate knowledge of the LCA methodology, the time and experience to perform detailed LCA case studies and to interpret a whole range of environmental impacts scores [5]. Therefore the LCA specialist should become an integrated part of the PDP.

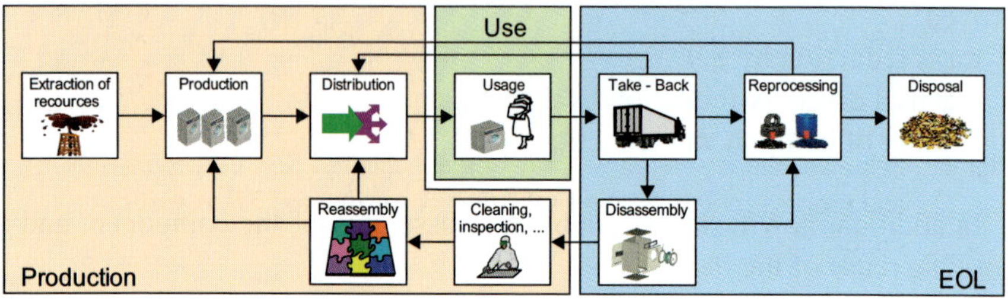

Fig. 3: Simplified product system covering the products' life cycle: Process model according to [26]

There are different approaches for modeling the products' life cycle in LCI and modeling the products' characteristics in the PDP. The process model in LCI includes all processes of the life cycle stages: production, use and EOL (Figure 3). The process model analyses the structure of relations between processes along the whole life cycle of a product [8, 24].

In PDP a product model is used [11, 19, 21]. The product model contains data about product hierarchies, material definition and structures of joining within the product. Different Computer Aided Design (CAD) tools are used for data processing. CAD systems are based on a product model. The designer defines both product- and implicit process-specific data (Figure 4). Data concerning the kind and mass of used materials as well as the composition structure and the products' use stage belong to product-specific data. Data concerning production or disassembly processes and other processes belong to the field of process-specific data.

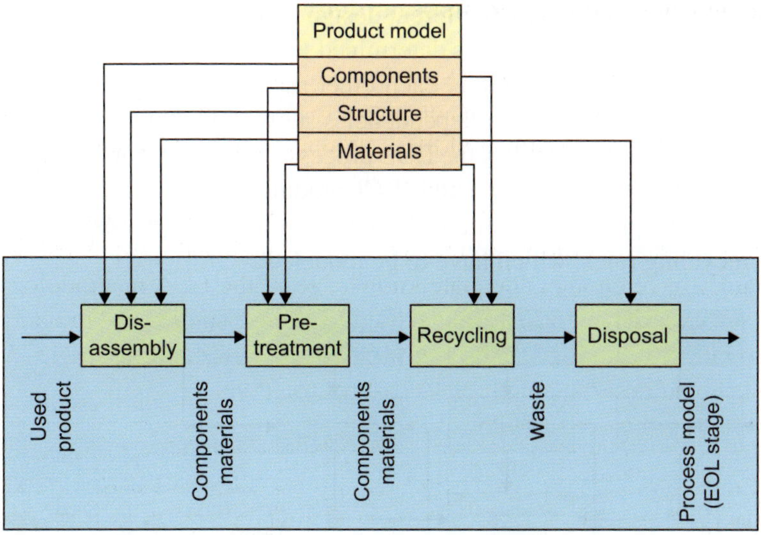

Fig. 4: Schematic representation of the interdependencies in the product- and process-model for the EOL stage according to [20] as part of the life cycle

The approach of LCI modules for product components integrates on the one hand modules for component parts and on the other hand modules for assemblies [2]. According to the product model of the Design Support System [11, 26] a component part consists of a single kind of material and is part of one or no assembly. An assembly contains at least one component part, but generally various. Therefore, the approach of product components' LCI modules is based on a logical detachment of the process model tree from the product system into different levels (Figure 5). The system levels in this figure relate to the hierarchy of the product (product model). System boundaries for the LCI module of a component part (system level III) contain process chains for material extraction, production processes and EOL processes. LCI modules for an assembly (system level II in Figure 5) contain process chains for component parts as well as assembly and disassembly processes for this assembly. System level I integrates all processes of system level II as well as assembly and disassembly processes of the product and processes of the use stage.

The interdependencies between the components of a product, which determine the specific disposal processes in the EOL stage, are analyzed by the determination of the coupling of the components and assemblies to each other, as well as the material types of these coupled components. This has to be described on the one hand via matrix representation, which provides an overview of the structure of different design options. On the other hand, the quantity of the material types is determined for the design options. Since the components of a product affect each other regarding their EOL processes, the determination of the interdependencies shows whether the insertion of a modified LCI module affects the EOL process chains (a group of interlinked processes) for an already existing LCI module. The mixture of materials combined in a product or assembly determines the EOL processes, e.g. possible recycling rates, which have to be modeled.

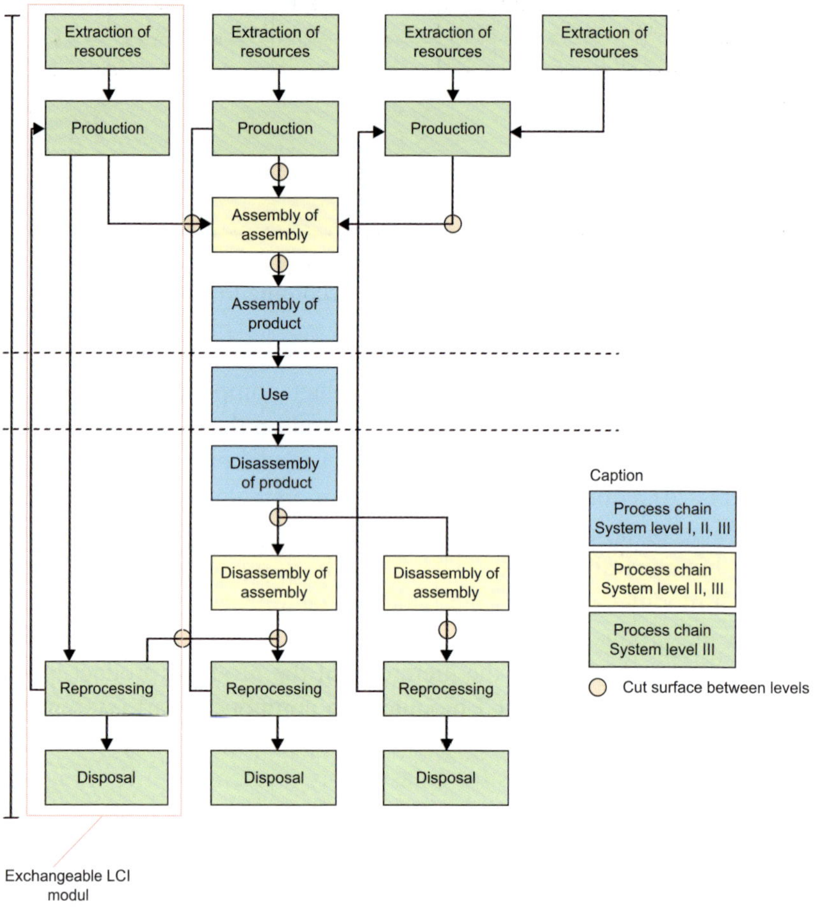

Fig. 5: Product system structure for exchangeable LCI modules based on product components on level III [2]

Consequently, the exchange of process chains from the product system that differ in material type and hierarchy is possible. The modular LCA supports decisions taken during product design. These decisions are reflected in a product model used in product design. The modular LCA (based on enhancement of the methodology in line with the ISO 1404x series [15, 16, 17, 18]) results in a common modeling structure for the product's life cycle.

The access to a joint product database simplifies the modeling process in the LCI step and requires a common structure for modeling the product system. The system for the support of the construction, which was developed at the Collaborative Research Center 281 [11], provides interfaces as well as a consistent product model, which can be used to represent product or process aspects.

Modular LCA for a Lye Pump of a Washing Machine

Exchangeable LCI modules based on product components for the modular LCA are modeled exemplarily by design options for a lye pump. Goal of the LCA studies was to analyze the environmental benefits that may be realized by a redesigned, disassembly orientated lye pump of a washing machine in comparison to the original one [4]. The reference flow according to EN ISO 14044 (the amount of a product, which is necessary to fulfill the functional unit) is one lye pump of a washing machine. The reference flow is used to calculate the inputs and outputs of the system. The amount of potential environmental impacts arising during the use stage of the redesigned lye pump is not affected by the arrangements for optimization of disassembly and recycling properties. Use time of the lye pumps is identical and longer than or identical to the use time of the product. Therefore, it is acceptable to neglect the use stage in the modeling process for comparison of the design options in relation to the goal of this study. Basically, the design options for lye pumps differ in material composition, structure and disassembly effort. All LCI data refer to life cycle processes of the reference flow. The mass fractions of all parts of the original component have been determined. For the disassembly optimized component, geometric data from the designers' CAD system have been used [26]. Due to the characteristics of the analyzed component and product, five different EOL scenarios were analyzed, modeled and integrated into the life cycle models.

Firstly, the lye pump (original design) is deconstructed in components and assemblies, mass and surface related variables were identified.

For an LCI module's multiple-shift usage, amongst others mass and surface related variables have to be identified to represent the interactions of the LCI modules of product components [10]. These variables are needed to include changes in the modeling process, because LCI modules based on product components contain surface related production processes, as well. The product structure of the original design was analyzed and the status of the product was determined: All components were joined. The interrelation of component parts (determined by the status of the product) defines the EOL processes, which are modeled for the specific LCI modules. All components are shredded resulting in the recyclable material streams of 95 % ferrous metals and 50 % Copper from the material input, which are separated for recycling. All plastics and residues are incinerated. In a second step, changed product characteristics for the redesigned product are integrated by the modular LCA: The system state of the redesigned product is determined. Exchangeable LCI modules enable the integration of changed product characteristics like component mass, materials used and types of fasteners used in a prospective way. The minimum of processes that must be remodeled for the respective components for the process model of the redesigned lye pump is determined. LCI modules for ten out of sixteen component parts have been reused, whereas five have been reused with modification in EOL processes. One new LCI module has to be modeled for applied new fasteners. The system state of the new designed pump shows, that for five LCI modules new EOL processes (pure recycling for the respective material) should be modeled. There are no changes in the assembly processes of the production stage and in the use stage. Therefore, these processes are omitted for the comparisons of product systems.

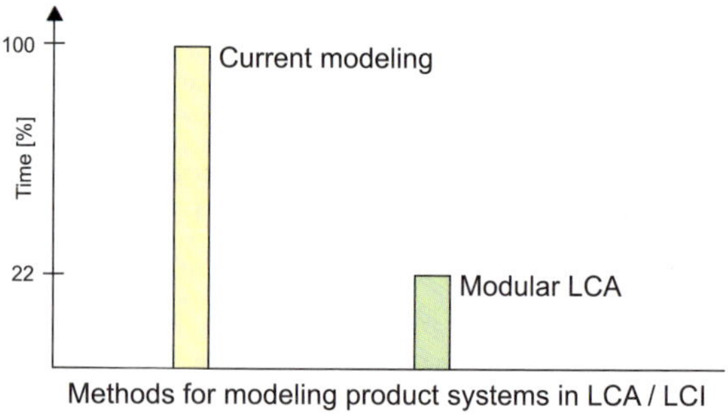

Fig. 6: Comparison of time consumption for modeling the lye pump by means of the original LCA method and the modular LCA

The developed methodology for the modular LCA enables the integration of changed product characteristics in a prospective way. Time consumption for a reused LCI module is 0 %, for modified LCI modules it is about 50 %

and 100 % for a new module. A comparison of the total time consumption shows: Where the traditional modeling will take 100 %, the modeling by the modular LCA will take about 22 % in this example. Product modeling in modular LCA is nearly five times faster than in traditional LCA (Figure 6).

3.3.3 Conclusion

LCAp delivers quantitative results for enhancing the environmental performance of products. LCAp is a very useful approach for designers, because a system perspective is used for analyzing the product and its optimization potential in relation to sustainable issues. The modular LCA as part of LCAp enables the quantitative evaluation of improvements, arising from e.g. arrangements of the recycling and disassembling properties. The conducted LCA studies support the environmental success of redesigned components. Furthermore, they disclosed further optimization potential, especially in material choice and overall weight by the determination of environmental "hot spots". Modular LCA by exchangeable LCI modules based on product components is applied for evaluating new design options already in the PDP. The redesigned lye pump of a washing machine (changes in terms of mass, materials and fasteners) shows significant environmental improvements in the stages of production and EOL for the analyzed example. The improvements in modeling time within LCI by means of the modular LCA lead to an efficient and objective oriented evaluation using the method of LCA. Therefore, the modular LCA based on a common structure of modeling provides a substantial time saving in the modeling process within the framework of LCI/ LCA without compromising reliability of results. Requirements laid down in [6], which have to be fulfilled by manufacturers of Energy using Products (EuP) from 2008 onwards, underline the importance of LCAp for integrating significant environmental aspects in product design.

References

1 AbfAblV: Verordnung über die umweltverträgliche Ablagerung von Siedlungsabfällen AbfAblV – Abfallablagerungsverordnung vom 20. Februar 2001 (BGBl. I S. 305; 24.7. 2002, S. 2807).

2 Dose, J.: Austauschbare Sachbilanzmodule auf Basis von Produktkomponenten. Dissertation, Institut für Technischen Umweltschutz, TU Berlin, 2005.

3 Dose, J.; Gries, B.; Fleischer, G.; Blessing, L.: Comparative Life Cycle Assessment of a Washing Machine Component Designed for Disassembly. In:

Seliger, G. (ed.): Proceedings Colloquium e-Ecological Manufacturing, 2003, pp 51-55.
4 Dose, J.; Gries, B.; Fleischer, G.; Blessing, L.: Entwicklung nachhaltiger Produkte unter ökobilanziellen Gesichtspunkten. In: Meerkamm, H. (ed.): Design for X, Beiträge zum 14. Symposium. Lehrstuhl für Konstruktionstechnik, Erlangen 2003, pp 61-70.
5 Eikelenberg, N.; Kok, I.C.; Tempelman, E.: Implementation ecodesign for EEE-products. Electronic Goes Green 2004+, pp 469-474.
6 EuP Directive: Directive 2005/32/EC of the European Parliament and of the Council on Establishing a Framework for the Setting of Ecodesign Requirements for Energy-using Products and Amending Council Directive 92/42/EECm and Directives 96/57/EC and 2000/55/EC.
7 Fleischer, G.; Hake, F.: Aufwands- und ergebnisrelevante Probleme der Sachbilanzierung. Schriften des Forschungszentrums Jülich, Reihe Umwelt, Vol. 30, 2002.
8 Fleischer, G.; Schmidt, W.P.: Ökologische Bilanzierung. In: Nickel, W. (ed.): Recycling-Handbuch, Strategien – Technologien – Produkte. VDI Verlag, Düsseldorf 1996.
9 Fleischer, G.; Becker, J.; Braunmiller, U.; Klocke, F.; Klöppfer, W.; Michaeli, W.: Eco-Design – Effiziente Entwicklung nachhaltiger Produkte mit euroMat. Springer-Verlag, Berlin, Heidelberg New York 2000.
10 Gerner, K.: Bestimmung potentieller Umweltbelastungen von Produkten durch die lineare Extrapolation der Ergebnisse ähnlicher Produkte. Dissertation, Institut für Technischen Umweltschutz, TU Berlin, 2005.
11 Gries, B.; Blessing, L.: A Software Based System To Support the Design For Disassembly. In: Marjanovic, D. (ed.): Proceedings of the 8th International Design Conference "Design 2004". Dubrovnik 2004, pp. 767-774.
12 Guinée, J.B.; Gorrée, M.; Heijungs, R.; Huppes, G.; Kleijn, R.; de Koning, A.; van Oers, L.; Wegener Sleeswijk, A.; Suh, S.; de Haes U. H.; de Bruijn, H.; van Duin, R.; Huijbregts, M.A.J.: Handbook on Life Cycle Assessment. Operational Guide to the ISO Standards. Kluwer Academic Publishers, Dordrecht, Boston London 2002.
13 Hauschild, M.; Fryddental, J.: LCA Center Denmark, status and perspective for dissemination of IPP and life cycle thinking in industry. In: Proceedings of the 13th CIRP international conference on Life Cycle Engineering, LCE 2006 Towards a closed loop economy, Leuven 2006, pp 39-42.
14 Heijungs, R.; Suh, S.: The Computational Structure of Life Cycle Assessment. Kluwer Academic Publishers, Dordrecht, Boston, London 2002.
15 EN ISO 14040: Environmental Management – Life Cycle Assessment – Principles and Framework, 2006.
16 EN ISO 14044: Environmental Management – Life Cycle Assessment – Requirements and Guidelines, 2006.
17 ISO/ TR 14047: Environmental Management – Life Cycle Impact Assessment - Examples of Application of ISO 14042, 2003.

18 ISO/ TS 14048: Environmental Management – Life Cycle Assessment – Data Documentation Format. ISO copyright office, Geneva, 2002.

19 Leibrecht, S.: Fundamental Principles for CAD-based Ecological Assessments. Centre for Design at RMIT University, Melbourne 2005.

20 Lichtenvort, K.; Barruetabena, L.; Kunst, H.; Stutz, M.; Nissen, N.; Boyce, J.; Pflieger, J.; Alber, S.; Johansson, G.; Alonso, J.C.: grEEEn D1 Report, Deliverable D1: Green Method, Report of the grEEEn project (G1RD-CT-2000-00355), January 2002.

21 Matthieux, F.; Roucoules, L.; Lescuyer, L.; Bouzidi, Y.: Opportunities and Challenges for Connecting Environmental Assessment Tools and CAD Software. In: Innovation by Life Cycle Management LCM 2005 International Conference. Barcelona, 2005.

22 Pahl, G.; Beitz, W.: Konstruktionslehre, Methoden und Anwendungen. Springer-Verlag, Berlin, Heidelberg 1997.

23 Reinhardt, T.; Richers, U.: Entsorgung von Shredderrückständen – ein aktueller Überblick. Forschungszentrum Karlsruhe in der Helmholtz – Gemeinschaft. Wissenschaftliche Beichte FZKA 6940, 2004.

24 Schmidt, W.P.: Ökologische Grenzkosten der Kreislaufwirtschaft. Dissertation Fachgebiet für Abfallvermeidung, TU Berlin, 1997.

25 Seliger, G. (ed.); Baumgarten, H.; Bayerer, P.; Blessing, L.; Fleischer, G.; von Hirschhausen, C.; Krause, F.-L.; Reichel, H.; Uhlmann, E.: Sonderforschungsbereich 281: Demontagefabriken zur Rückgewinnung von Ressourcen in Produkt- und Materialkreisläufen, Arbeits- und Ergebnisbericht. Berlin 2000.

26 Seliger, G, (ed.); Baumgarten, H.; Bayerer, P.; Blessing, L.; Fleischer, G.; von Hirschhausen, C.; Krause, F.-L.; Reichel, H.; Uhlmann, E.: Sonderforschungsbereich 281: Demontagefabriken zur Rückgewinnung von Ressourcen in Produkt- und Materialkreisläufen, Arbeits- und Ergebnisbericht. Berlin 2003.

27 VDI 2243: Richtlinie – Recyclingorientierte Produktentwicklung. VDI-Verlag, Düsseldorf 2002.

28 WEEE Directive: Directive of the European Parliament and of the Council on Waste Electrical and Electronic Equipment, Directive 2002/96/EG, 2003.

29 Wenzel, H.; Hauschild, M.; Leo, A.: Environmental Assessment of Products, Vol 1, Methodology, Tools and Case Studies in Product Development. Chapman & Hall, London 1997.

30 Werner, F.; Scholz, R.W.: Ambiguities in Decision-Oriented Life Cycle Inventories, The Role of Mental Models. International J Life Cycle Assessment 7 (2002) 6, pp. 330-338.

3.4 Assessment of Recyclability

Jürgen Hesselbach, Christoph Herrmann, Tobias Luger, Braunschweig, Germany

3.4.1 Process Steps for Closed-loop Material Flows

Recycling of complex products with the objective of closing material loops takes place in multiple successive process steps (Figure 1). The use phase is followed by redistribution, i.e. the logistic tasks of collction, transport, handling, storage and supply of the products for recycling [1, 2].

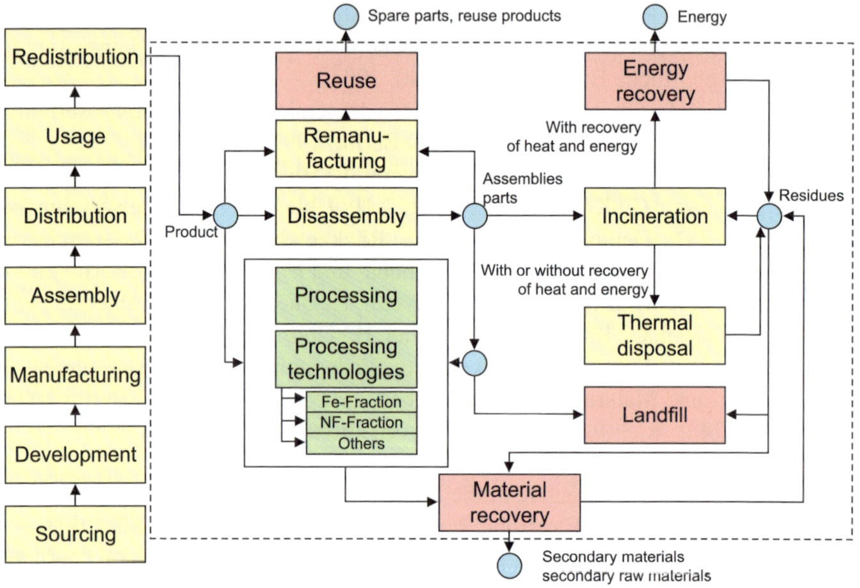

Fig. 1: Process steps in the end-of-life phase

Remanufacturing processes focus on products and parts of products. Unlike repair of products solely, remanufacturing not only extends a product life but also has the goal to bring the product into an as-new state [3]. Sundin [4] defines a generic remanufacturing process comprising the steps of disassembly, cleaning, inspection, reprocessing, storage, testing and reassembly in any applicable order. After conducting remanufacturing, products are reused in another use cycle.

Disassembly as the next step inheres often a particular importance for the results of recycling. Its goals are the separation of hazardous material or components containing hazardous substances, the non-destructive extraction of components for reuse, the separation of material which cannot or can hardly be separated procedurally and which may disturb any mechanical preparation processes or recycling processes, and the extraction of any valuable material for subsequent preparation of recovery processes [5, 6, 7]. Although in practice disassembly is carried out mainly by manual operations, the development of purpose-built universal and mechanized tools lead to a significant decrease of disassembly times and increase in flexibility [8].

Subsequent to disassembly mechanical processing steps aim at the extraction of material fractions and comprise processes of comminution, separation and sorting of various output fractions which have already been relieved of pollutants and interfering substances. Advantages of mechanical processing include a high material throughput while requiring little manual operations [9].

The output fractions of disassembly and mechanical processing are supplied to material recovery processes, in which secondary raw materials are produced. Materials recovery processes can be classified in:

- recovery processes for metals:
  smelting, pyro- and hydro-metallurgical processes,
- recovery processes for plastics:
  feedstock recovery and material recovery,
- recovery processes for other materials, e.g. paper and glass.

Besides the recovery of energy, incineration aims at the destruction of toxic, organic compounds, accumulation of inorganic pollutants and the reduction of the volume for landfill. The caloric value of the input material as well as the energy cycle efficiency of the incineration process is authorative for the distinction between energy recovery and thermal disposal [10].

As a matter of principle, material loops can never be closed entirely. In any application, a share of a raw material is mixed so strongly with others that it has to be taken out of the loop [11].

3.4.2 Recyclability of a Product

Stakeholders from government, non-governmental organizations and environmentally sensitive society make demands on companies to contribute to a closed-loop economy. The concept of Extended Producer Responsibility (EPR) imposes at least part of the responsibility and the costs for the environmental impact and performance of the product during its later life cycle stages to the manufacturer.

Consideration of requirements from a closed-loop economy has to take place within the scope of the product development phase. Changes in the product design affect the product performance in later life cycle stages and, thus, offer the potential of cost reduction and increase in efficiency also for the end-of-life stage [12]. This is especially due to the fact, that only a minor share of costs over the life cycle actually occur in the product development stage but in later stages [13]. Thus, fairly low additional costs in product development may result in strong cost reduction throughout the life cycle.

To develop these potentials a feedback loop from the end-of-life stage to the product development is much needed, as designers do not generally have the experience in disassembly and recycling to determine the impact of various design aspects on the product disassembly and recycling as depicted in Figure 2 [14].

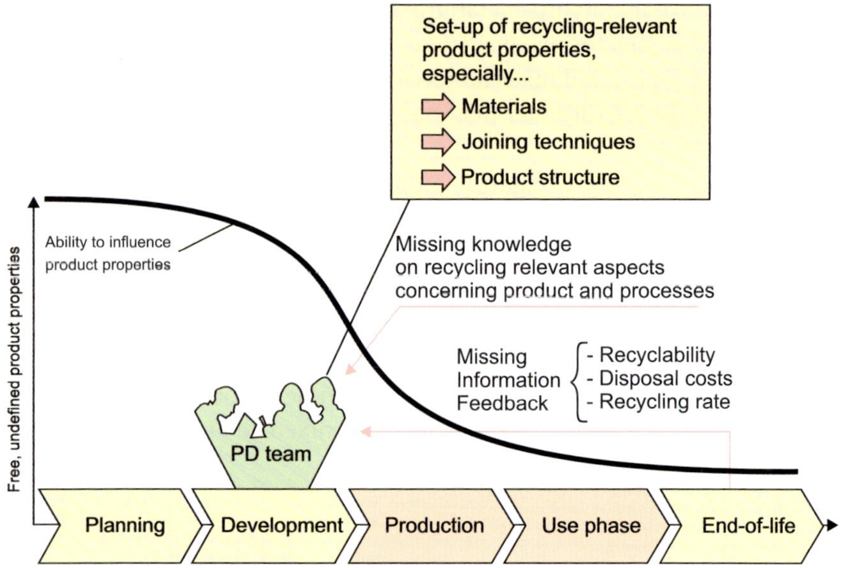

Fig. 2: Importance of product development

Goals for the end-of-life stage for each product have to be set and the product design has to be evaluated against these requirements. Therefore it is necessary to assess the end-of-life relevant product properties by determining process characteristics:

- disassembly time, level and cost,
- recycling cost and profit, respectively,
- recycling and recovery quotas and
- environmental impact related to different treatment scenarios.

Due to the complexity of the problem and the high level of interdependencies between the impact factors a methodological support is needed to assess these characteristics. Thereby, calculation of the assessment characteristics should be conducted as far as possible independent of the experience based knowledge of the user.

Concepts for the evaluation of a product's end-of-life behavior can be differentiated in product-oriented, process-oriented and integrated approaches. They vary from a series of basic questions and checklists based on a set of criteria used by design teams to methods to calculate quantitative measures for the end-of-life performance of the product.

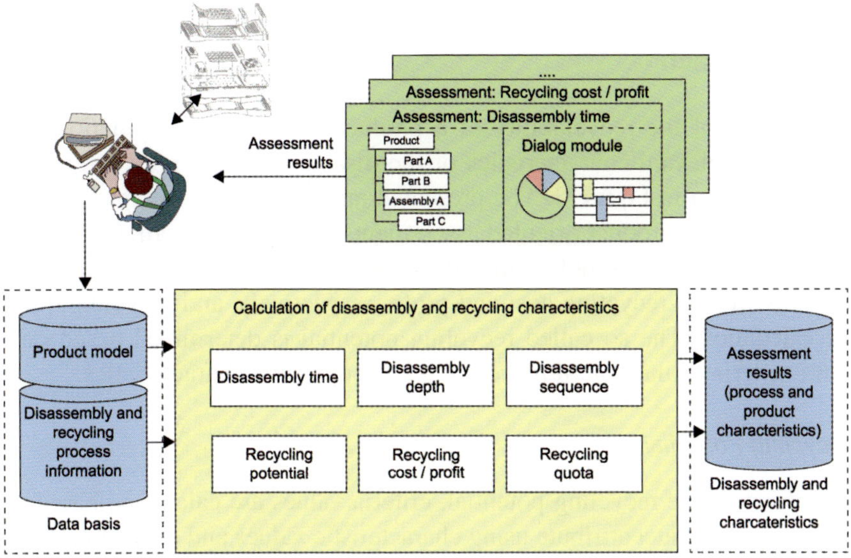

Fig. 3: Elements of the assessment of recyclability [28]

Product oriented approaches evaluate the effect of a product's design properties on its recyclability. This evaluation takes more consideration of the product structure, materials composition and used joining techniques [15], but does not incorporate a simulation of the actual behavior of the product in the end-of-life stage and, thus, does not deliver quantitative process related characteristics. Representatives of product-oriented approaches have been developed, e.g. by Wende [16], Eversheim et al. [17] and Kroll et al. [18]. On the other side, process oriented methods evaluate the effects of the product design on the disassembly and recycling processes efficiency by taking into account the effects of product properties on the treatment and recycling processes. Therefore, methods, e.g. ReStar (Recycling-Star) [19], DEMROP (Design and Evaluation Method for the Recyclability of Electromechanical Products) [20] and „Design for Environment" by Boothroyd and Dewhurst [21], are based on product models as well as models of the applied processes. Integrated approaches, such as REKON by Spath, Hartel et al. [22, 23] and ReGrEd/DisPlay [24, 25, 26] combine product- and process-oriented approaches to a holistic concept.

A detailed overview and analysis of critical points of existing approaches can be found in [27, 28]. To facilitate a comprehensive holistic assessment of the disassembly and recycling characteristics of a product integrated approaches are to prioritize.

3.4.3 Approach

Assessment of the recyclability of a product requires the calculation of a number of relevant characteristics of the disassembly and recycling of the product shown in Figure 3. The presented approach comprises process oriented as well as product oriented assessments [27, 28]. In detail, the process characteristics disassembly time, disassembly sequence and disassembly depth as well as recycling cost and profit, respectively, and recycling quota are calculated. The so called recycling potential is determined based on a multi-criteria evaluation of recycling relevant product characteristics.

Recycling Potential

To determine the recycling potential, criteria values are calculated based on the relevant product attribute using characteristic values and evaluation functions [29]. The criteria are divided in groups: material and marking, product structure and joining technologies. Subsequently, group values are established by averaging the criteria values. The aggregation of the group values to one product value represents the so called recycling potential, which describes quantitatively the product's general ability to be recycled and is the concentrated result of the product assessment (Figure 4). A detailed presentation of the conceptual and methodical basis of the disassembly and recycling assessment as well as a description of the necessary data pool can be found in [27].

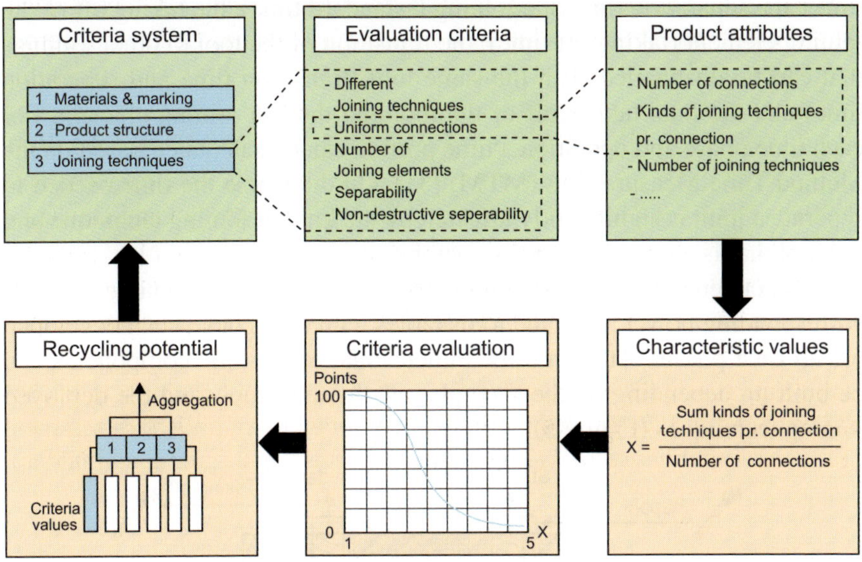

Fig. 4: Calculation of the product potential [27]

Disassembly Time

The disassembly times are the required times to perform the disassembly operations. Calculating necessary disassembly times analytically based on rough product information is a key factor in evaluating the product's ability to be disassembled and recycled. Therefore, the entire disassembly process is decomposed into singular disassembly steps and into different phases. This decomposition into steps of low complexity facilitates the determination of time components [30]. The following time phases are distinguished [14]:

- Supply Time: All motions required to grasp, move, orient and arrange the tool in the correct position for detaching the joining elements and components.
- Standard Time: Motions required to separate the joints.
- Transition Time: Motions to take off the previous joining element, to move and to place the tool for the removing of the next joining element if more than one joining element has to be detached with the same tool.

- Taking off Time: All motions to move and place the joining elements and components at a defined place.

A disassembly operation is composed of the supply of the tool (supply time), the separation time of a joining (sstandard time), the taking off of the joining element (taking off time), the transition of the tool to other joinings that can be dismantled with the same tool (transition time and separation time) and the final taking off of the tool and the last joining element. For each time phase and operation a time portion can be calculated based on the Method Time Measurement (MTM) [31]. On the whole the time needed to separate a joint is significantly affected by the kind of joining technique and the tool that is chosen for the disassembly.

In dependence of the most commonly applied joining techniques and the corresponding power and hand tools tables with times have been developed for all time phases. Thus, a three dimensional set of time components can be built up depending on the time phases, the used tools and the deployed joining techniques (Figure 5).

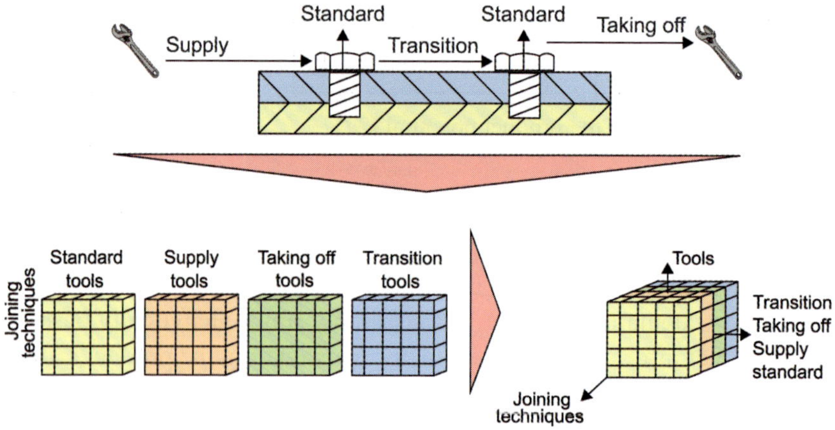

Fig. 5: Disassembly time phases

The whole MTM analysis is verified by Time Exposure. The validation of the model confirms that the time calculated using this approach does not significantly deviate from real disassembly data. The model accuracy is checked against the actual disassembly time. The disassembly cost can be calculated by multiplying the overall time determined for the disassembly process with a labor cost factor for the analysis scenario.

Recycling Quota

The calculation of recycling and recovery quotas for a product as well as recycling cost and profits, respectively, entails the modelling of processes of the end-of-life phase. Hoberg et al. [32] define a technical recycling quota $R_{q,t}$, as the ratio of the amount of secondary material produced in a recovery process to the amount of material made available the process. The system boundaries for this calculation of recycling quotas have a strong impact on the calculated quota [28]. In the process model recycling and mechanical treatment processes have to be differentiated. Mechanical processes that split up one material input stream into two or more output streams are modelled as a distribution matrix representing the allocation of the different materials of the input stream to the output material streams. Recycling processes in contrast, represent the sink of materials in this regard.

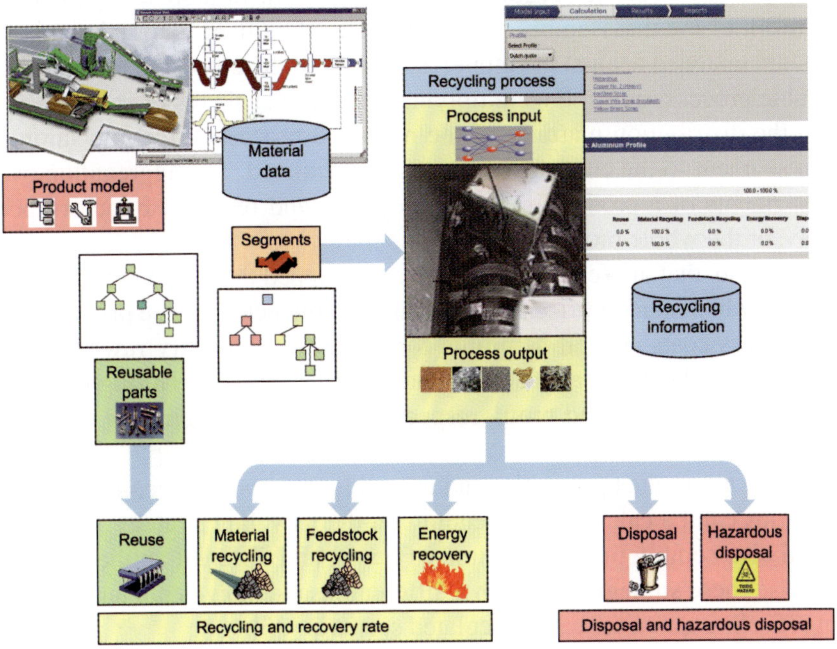

Fig. 6: Process model

The process itself is viewed as a black box. On the input side, each process is defined by input restrictions, i.e. shares of materials suitable for the process and prohibited in a process, respectively. On the output side, each process is defined by distribution matrix of the input materials on the different end-of-life options, as depicted in Figure 6:

- material recycling,
- feedstock recycling,
- energy recovery,
- disposal and
- hazardous disposal.

Here, it is not relevant which product is generated on the output side (secondary raw material, slag, dust), as in e.g. [33], but the distribution on the different recovery and disposal options [34]. Using different matrices of process models allows the differentiation of different system boundaries and of alternative analysis scenarios, such as process availability in different countries.

For the purpose of calculation of process characteristic values, such as quotas, costs and profits, segments consisting of one or more components as the result of the disassembly, which are defined by their material composition, are matched against the input restrictions of the processes to determine suitable process routes. By multiplying the material shares of each fraction with the distribution matrices of the modelled processes, the allocation of each segment to the end-of-life options is calculated. The aggregation over all segments of a product gives the recycling and recovery quota, respectively, of the whole product.

Segments that are fed into processes are mapped to tradable market fractions, to which market prices of materials are connected. These prices reflect the process costs as well as profits for produced secondary raw materials and costs for the disposal of waste materials, respectively. Through this, the actual recycling cost and profit, respectively, for the recycling of a product is calculated.

Taking all product properties and recycling options into account different scenarios can be calculated depending on the recycling strategy, e.g. minimized disassembly time or maximized quota, and different applied process models. These strategies result in a different disassembly depth and a different recycling quota for each calculated scenario [35].

The complexity of the assessment of disassembly and recycling relevant product characteristics presented in this approach requires the computer-based implementation of the methods. Therein, a representation of the product data in a product model and the provision of the process data in a process model (see Figure 3) are necessary.

The presented approach for the assessment of a product's recyclability has been implemented in a commercially available software tool named ProdTect [35]. The tool is distributed by KERP (Center of Excellence

Electronics & Environment) in Vienna, Austria, and developed in cooperation with the IWF (Institute of Machine Tools and Production Technology) at the Technical University Braunschweig, Germany.

The software is based on a five step workflow (Figure 7) comprising the above described calculation and presentation of process oriented characteristics and determination of the product's recycling potential. The workflow is completed by the input of the product model as the first step and the final reporting of the assessment results.

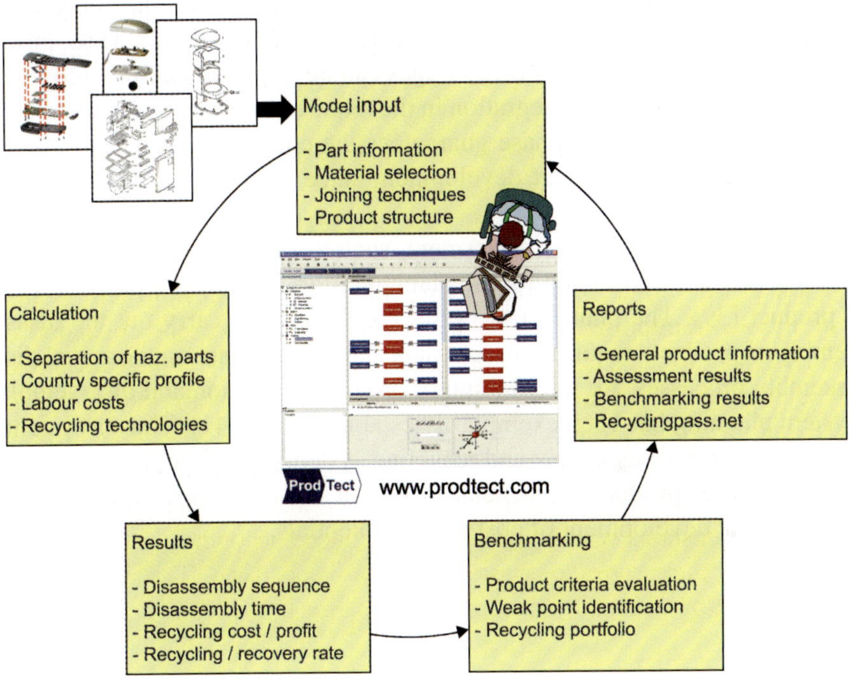

Fig. 7: Workflow

In the product model input module, the product structural information is composed of:

- parts information, such as material composition, disassembly movement, dimension, shape and accessibility,
- connection information composed of the different joining elements in the product as well as
- priorities information, which gives an order of the parts inside the product. A part is prior to another if it needs to be dismantled to get access to the other.

3.4.4 Integration of Product Assessment

A key factor in the successful application of the presented product assessment methods is the integration in the product planning process. During the product development a variety of different requirements from several sources have to be taken into account and balanced against each other [36]. To support this process in the conceptual design phase the software tool FOD (Functional Oriented Design) was developed by Krause et al. [37]. An interface between the assessment software ProdTect and the conceptual design support tool FOD has been successfully established [38]. Thus, product data already from early design stages can be imported and utilized for a product assessment. The integration in the requirement management already in the conceptual design phase guarantees the consideration of end-of-life requirements in the product development stage with only little additional complexity.

The resulting data of the assessment, such as disassembly times and sequence, can also be utilized for the planning of the end-of-life processes for a product [39]. The planning of optimized systems to carry out the disassembly of products requires product data and process data on the actual disassembly processes. An interlinking of product evaluation and disassembly system planning has been exemplarily established [40, 41]. Thus, a systematic planning of disassembly and recycling systems based on results of end-of-life oriented product assessments enables a successful implementation of EPR and the development of cost saving potentials.

References

1 Perlewitz, H.: Planung und marktorientierter Betrieb von Demontagefabriken. Ph.D. thesis, Technical University Berlin 2000.

2 v. Westernhagen, K.: Planung und Steuerung der Retro-Produktion. Vulkan-Verlag, Essen 2001.

3 Steinhilper, R.: Remanufacturing – The Ultimate Form of Recycling. Fraunhofer IRB Verlag, Stuttgart 1998.

4 Sundin, E.: Product and Process Design for Successful Remanufacturing. Ph.D. thesis, Linköpings Universitet 2004.

5 Nickel, W.I.: Recyclinghandbuch – Strategien, Technologien, Produkte. VDI Verlag, Düsseldorf 1996.

6 Bullinger, H.-J.; Menrad, W.; Dietrich, J.: Schwerpunkte zukünftiger Forschungsaktivitäten in der Kreislaufwirtschaft – Ergebnisse einer Unternehmensbefragung. Industrie Management 16 (2000) 1, pp. 43-48.

7 Meißner, S.; Schöps, D.; Herrmann, C.: Optimale Ausbeute – In Elektronikschrott-Zerlegebetrieben können Informationssysteme den Werkern Angaben zur Steigerung der Prozesswirtschaftlichkeit machen. Müllmagazin (1999) 3, pp. 21-23.

8 Seliger, G.; Basdere, B.; Keil, T.; Rebafka, U.: Innovative Processes and Tools for Disassembly. In: Annals of CIRP, vol. 51/1/2002, pp. 37-40.

9 Tiltmann, K.O.; Schüren, A.: Recyclingpraxis Elektronik. Springer-Verlag, Berlin, Heidelberg, New York 1994.

10 Giesberts, L.; Posser, H.: Grundfragen des Abfallrechts. Verlag C.H. Beck, München 2001.

11 Rechberger, H.: Ein Beitrag zur Bewertung des Stoffhaushaltes von Metallen. Technologiefolgenabschätzung – Theorie und Praxis 11 (2002) 1, pp. 25-31.

12 Fleischer, G.: EDV – Elektronik – Abfallwirtschaft. EF-Verlag für Energie- und Umwelttechnik, Berlin 1993.

13 Kahmeyer, M.; Rupprecht, R.: Recyclinggerechte Produktgestaltung. Vogel Verlag, Würzburg 1996.

14 Hesselbach, J.; Kühn, M.: Disassembly Evaluation of Electronic & Electrical Products. In: Proceedings of Int. Symposium on Electronics & the Environment, IEEE, May 4-6 1998, Chicago, IL USA 1998, pp. 79-81.

15 N.N.: VDI 2243: Recycling-oriented Product Development. VDI-Verlag, Düsseldorf 2002.

16 Wende, A.: Integration der recyclingorientierten Produktgestaltung in den methodischen Konstruktionsprozess. VDI-Verlag, Düsseldorf 1994.

17 Eversheim, W.; Hartmann, M.; Linnhoff, M.: Zukunftsperspektive Demontage. VDI-Z 134 (1992) 6, pp. 83-86.

18 Kroll, E.; Carver, B.S.: Disassembly analysis through time extimation and other metrics. Robotics and Computer Integrated Manufacturing, 15 (1999) 3, pp. 191-200.

19 Navin-Chandra, D.: ReStar – A Design tool for Environmental Recovery

Analysis. In: Proceedings of the 9th International Conference on Engineering Design IECD 93. Schriftenreihe WDK 22, Edition Heurista, Zürich 1993, pp. 780-787.

20 Kaase, W.; Baier, C.: DV-gestützte Recyclingoptimierung komplexer Produkte. VDI Berichte, Nr. 1400. VDI Verlag, Düsseldorf 1998, pp. 125-140.

21 Okada, H.; Ono, H.; Yamano, A.: Disassembility Evaluation and Designing for Environmentally Conscious Products. In: Proceedings of the 1st International Symposium on Environmentally Conscious Design and Inverse Manufacturing, Tokyo, Japan 1999, pp. 281-284.

22 Hartel, M.: Kennzahlbasiertes Bewertungssystem zur Beurteilung der Demontage- und Recyclingeignung von Produkten. In: Weule, H. (ed.): Forschungsberichte aus dem Institut für Werkzeugmaschinen und Betriebstechnik, Vol. 72, Karlsruhe 1997.

23 Spath, D.; Scharer, M.; Trender, L.: Bewertungsmethoden zur Unterstützung einer lebenszyklusgerechten Konstruktion. Industrie Management 16 (2000) 1, pp. 9-13.

24 Feldmann, K.; Trautner, S.: Design for Environment – Methodology, Implementation and Industrial Experience, Part 1: Computer Based Disassembly Planning and Design for Recycling. In: Proceedings of the Joint International Congress on Electronics Goes Green 2000+, VDE Verlag, Berlin 2000, pp. 231-236.

25 Meerkamm, H.; Rosemann. B.: Ökonomisch-ökologische Produktoptimierung durch recyclinggerechte Konstruktion. Industrie Management 16 (2000) 1, pp. 14-18.

26 Trautner, S.: Technische Umsetzung produktbezogener Instrumente der Umweltpolitik bei Elektro- und Elektronikgeräten. Meisenbach Verlag, Bamberg 2002.

27 Kühn, M.: Demontage- und recyclingorientierte Bewertung. Vulkan-Verlag, Essen 2001.

28 Herrmann, C.: Unterstützung der Entwicklung recyclinggerechter Produkte. Vulkan-Verlag, Essen 2003.

29 Lenk, E.: Zur Problematik der technischen Bewertung. Konstruktionstechnik (1994) 13, pp. 523-528.

30 Olbrich, R.: Aufbau einer Zeitwirtschaft: Beschreibung der Vorgehensweise. Wirtschaftsverlag Bachem, Köln 1993.

31 REFA - Verband für Arbeitsgestaltung, Betriebsorganisation und Unternehmensentwicklung: Methodenlehere der Betriebsorganisation: Datenermittlung. Carl Hanser Verlag, München, Wien 1997.

32 Hoberg, H.; Wolf, S.; Meier-Kortwig, J.: Modelling the Material Flow of Recycling Processes for Aluminium Alloys by Means of Technical Recycling Quotas. In: Proceedings of the REWAS '99 Global Symposium on Recycling, Waste Treatment and Clean Technology, San Sebastian, Spain 1999, pp. 1023-1033.

33 Abele, E.; Anderl, R.; Birkhofer, H.: Environmentally-Friendly Product

Development. Springer, Berlin Heidelberg London 2005.

34 Herrmann, C.; Antrekowitsch, H.; Luger, T.; Seebacher, H.; Stachura, M.: Product Development and Metallurgy – Combining two views on the Calculation of Recovery Rates in WEEE Recycling.In: Proceedings of R'05 - 7th World Congress on Recovery, Recycling and Reintegration, Beijing, China 2005.

35 Herrmann, C.; Frad, A.; Revnic, I.; Luger, T.: Product Architect – A New Approach for Transparency and Controlling of the End-of-Life Performance. In: Proceedings of the EcoDesign2005, Tokyo, Japan 2005.

36 Luttropp, C.: Eco-Design in early product development. In: Barrage, A.; Edelmann, X. (eds): Proceedings of the R'99 Recovery Recycling Re-Integration Volume III, February 2 – 5, 1999, Geneva, Switzerland 1999, pp. 270-275.

37 Krause, F.-L.; Baumann, R.; Kaufmann, U.; Kühn, T.; Leemhuis, H.; Ragan, Z.; Swoboda, F.: Computer Aided Conceptual Design. In: Proceedings of the 36 CIRP-International Seminar on Manufacturing Systems, June 3 – 5, 2003. Saabrücken, Germany 2003.

38 Krause, F.-L.; Herrmann, C.; Frad, A.; Ragan, Z.; Luger, T.: Integrating End-of-Life Evaluation in Conceptual Design. In: Proceedings of the 2006 IEEE International Symposium on Electronics & the Environment, May 8 – 11, 2006, San Francisco, CA USA 2006, pp. 245-250.

39 Hesselbach, J.; Herrmann, C.; v. Westernhagen, K.: Datenbanken als Informationsträger von der Produktentstehung bis zur Demontageplanung. In: Proceedings of Umweltinformatik zwischen Theorie und Industrieanwendung, 13th International Symposium, Magdeburg, Germany 1999.

40 Herrmann, C.; Luger, T.; Ohlendorf, M.; Ciupek, M.; Kernbaum, S.: Planning Approaches for Manual and Automated Disassembly Systems. In: Seliger, G.; Nasr, N.; Bras, B.; Alting, L. (eds): Proceedings Global Conference on Sustainable Product Development and Life Cycle Engineering. September 29 – October 1, 2004, Berlin, Germany, uni-edition, Berlin 2004, pp. 127-134.

41 Herrmann, C.; Spengler, T.; Luger, T.; Walther, G.; Schmid, E.: Design and Control of Material Flow Networks for the Recycling of WEEE. In: Proceedings of the 2006 IEEE International Symposium on Electronics & the Environment May 8 – 11, 2006, San Francisco, CA USA 2006, pp. 340-345.

3.5 Efficient Economic Organization

Carl-Friedrich Elmer, Berlin, Germany

The improvement of techniques and planning instruments employed in the end-of-life phase as well as in the product design process is not sufficient to foster a sustainable closed loop economy. What is also required is a legal and institutional framework that provides reasonable environmental targets and appropriate incentives to reach these targets in an efficient way.

Therefore, some essential economic issues to achieve a closed loop economy will be highlighted in the following pages. The economic analysis conducted in this chapter is split into two parts. In the first part we analyze an elementary question of environmental economics: the definition of the environmental target and the choice of the means to realize it. The three basic approaches of environmental regulation to be distinguished are direct administrative controls, and the higher efficiency promising market based instruments taxes and tradable quantities. Nonetheless, when it comes to recycling obligations the most commonly used instrument are neither taxes nor absolute quantities, but relative quotas. Relative quotas set the mandatory amount of waste to be recycled in relation to a certain reference parameter, in general a certain output. Due to an incomplete producer responsibility a lack of efficiency is inherent to these relative quotas. For that reason we present an approach which combines a relative recycling quota with a tax on tradable recycling certificates, which is thereby capable of offsetting the aforementioned efficiency problem.

The second part of the chapter deals with issues crucial for the actual implementation of a take-back obligation in practice. Five central issues and their impact are exposed and subsequently applied. Assets and drawbacks of the European Directive on Packaging and Packaging Waste (94/62/EC) and the European Directive on Waste Electrical and Electronic Equipment (WEEE)[1] as well as their transposition into national legislation in Germany (and the UK) are subject to a critical consideration in the light of the afore formulated criterions. The chapter closes with concluding remarks and an outlook on further research requirements.

[1] 2002/96/EC; minor amendments are carried out in directive 2003/108/EC.

3.5.1 Basic Issues for an Efficient Closed-Loop-Economy

First of all we identify the allocative optimum, using a straightforward model comprising production costs, recycling costs, and environmental damages. The latter two are dependent on the aggregate amount of pollutants sent to recycling and to the environment respectively, while the production costs are determined by the output level and the pollutant intensity. The relevant variables to determine are the aggregate production level of the polluting commodity, the amount of this commodity to be recycled, and the pollutant intensity, e.g. the amount of pollutants per commodity. After introducing potentially efficient instruments of environmental policy we show the immanent inefficiency of relative quotas, employing the introduced model. Then the capabilities of a relative recycling obligation supplemented by a tax on tradable recycling certificates are set out.

Optimal Extent of Recycling Activities

The fundamental justification for environmental policy is the existence of externalities.[2] External costs arise when an activity by one agent, here environmental pollution, causes an uncompensated loss of welfare to another agent. The crucial point is the compensation; there is no need for regulation if the polluter pays for the damage he inflicts upon others. A lack of internalization of the entire costs associated with the polluting activity yields in exceeding of the optimal extent of this activity since it is "too cheap". For a social optimum the polluters have to consider all costs in their individual maximization calculus. External costs related to waste disposal and recycling may occur both directly through emissions within the disposal chain (leaching of toxins, emissions due to combustion, etc.) and indirectly through environmental damages during the extraction and refinement of virgin materials (energy consumption implying CO_2 emissions, surface degradation, etc.).[3]

For simplicity of the argumentation we assume in the following that the recycling costs also comprise the additional costs for separate collection, waste treatment and the revenues for secondary materials. Therefore, the term Recycling Costs (RC) represents the net costs for an environmentally sound disposal way deviating from the disposal way of normal household waste (i.e. mainly landfilling and incineration). Hence, the recycling cost function can be interpreted as the cost function for "downstream" pollution

2 See Pearce, D.W.; Turner, R.K., chapter 4 [1]; Tietenberg, T., pp 60 et seqq [2].

3 See Linscheidt, B., pp 45 et seqq [3].

reduction. There also exists a direct way, here named abatement, of lowering environmental damages caused through waste disposal. The Abatement Costs (AC) represent the costs for reducing emissions already in the production process, either by a reduction of the output or by lessening the pollutant intensity.[4]

An optimum level of recycling activities requires maximizing the difference of benefits[5] from the use of the products subject to recycling obligations less the sum of Production Costs (PC), recycling costs, and remaining (external) Environmental Damage (ED). We define the two latter depending only on the amount of pollutants to be recycled or to be disposed by other means, respectively. α represents the share of pollutants to be recycled, β the pollutant intensity, and X the aggregate number of commodities produced. Then we get for the Social Welfare (SW) that should be optimized:

$$SW = \int_{0}^{x*} p(X)dX - PC(X,\beta) - RC(\alpha * X * \beta) - ED((1-\alpha) * X * \beta)$$

The first order conditions are:

$$p(X) = MPC(X,\beta) + \alpha * \beta * RC'(\alpha * X * \beta) + (1-\alpha) * \beta * ED'((1-\alpha) * X * \beta) \quad (1)$$

$$PC_{\beta} = -\alpha * X * RC'(\alpha * X * \beta) - (1-\alpha) * X * ED'((1-\alpha) * X * \beta) \quad (2)$$

$$RC'(\alpha * X * \beta) = ED'(1-\alpha) * X * \beta) \quad (3)$$

$$\text{with } MPC(X,\beta) > 0,\ RC'(\alpha * X * \beta) > 0, PC_{\beta} < 0, ED'((1-\alpha) * X * \beta) > 0$$

A social planner would set the overall production X according to Equating (1) so that the market price, i.e. the marginal benefit, equals the sum of changes in the production costs, recycling costs, and (monetized) environmental damages due to a marginal variation of X. Equating (2) states that in the optimum the production cost savings of a marginal slackening of β have to be exactly outweighed by the rise of recycling costs and environmental damages resulting from the increased emissions. Finally, according to Equating (3) α has to be chosen so that it assures equality of the Marginal Recycling Costs (MRC) and the Marginal Environmental Damages (MED) with respect to a marginal variation of the amount of pollutants to be re-

4 The abatement costs comprise actual expenditures, e.g. R&D expenses and substitution costs, as well as opportunity costs in the form of reduced revenues due to decreased output.

5 The benefits are expressed through the willingness to pay, i.e. the inverse demand function (or the price function) p(X).

cycled or to be absorbed, respectively.[6] Otherwise social cost savings are possible through a variation of α which implies a shift of pollutants from the more expensive to the relatively cheap alternative.

Direct Control vs. Prices vs. Quantities

In the previous section we examined the conditions for allocative efficiency and looked at the aggregate cost functions. To reach the allocative optimum it is indispensable to fulfill another criterion decisive for evaluating instruments of environmental policy: cost efficiency means that a given environmental target, which is not necessarily the allocative optimum, is achieved at least cost. This requires that the individual Marginal Abatement Costs (MAC) of all regulated pollution sources are identical.[7] Otherwise shifts of individual abatement efforts from sources with relatively high MAC to sources with lower MAC can yield the same aggregate reduction at lower aggregate costs or additional environmental protection at the same costs. We will now look at the three fundamental types of environmental policy instruments with respect to the criterion of cost efficiency.[8]

Direct Controls

Direct controls or command-and-control measures assign individual pollution reduction targets to each source. That may take the form of permissions for a fixed amount of emissions, permissions relative to output, individual recycling quotas, obligations to implement certain technologies (best available technology), etc. Theoretically command-and-control measures may provide an efficient allocation of a given amount of pollution, yet this requires the environmental authorities to have almost unattainable information. This includes knowledge of the individual MAC functions of all polluting facilities in order to assign the individual obligations in a way that equalizes the MAC at all sources. Such extensive information is virtually impossible to obtain, not at least because the polluters have an inter-

6 It is important to note that the abbreviations MRC, MED, and MAC represent the reaction on a marginal variation of emissions/pollution, while the Marginal Production Costs (MPC) represent the reaction of the production costs on a marginal output variation. See also annex 1 for mathematical definitions.

7 The MAC represent the marginal costs for reducing emissions/pollution in the production process.

8 The internalization of external effects is generally also possible by liability law or negotiations, which require clearly defined property rights. Nevertheless, we do not consider them here, because in the case of environmental damages through inappropriate waste disposal they have prohibitive transaction costs.

est to exaggerate their MAC to get less ambitious reduction obligations. Furthermore, adjustments of the allocation plan are regularly necessary due to persistent economic and/or technical developments affecting the MAC. Direct controls withhold flexibility to the market participants as shortfalls of the individual environmental target are forbidden and its exceeding is not rewarded. Thus, command-and-control measures are generally expected and empirically proven to be less cost efficient than more flexible market based instruments.[9]

Price Instruments

There are two approaches for market based environmental regulation: price and quantity instruments. Both prices and quantity solutions are cost efficient instruments to implement pollution control. Price instruments oblige polluters to pay a certain charge for every unit of the taxed polluting activity. In the case of pollution related to waste disposal the assessment basis may be the use of toxins or of primary resources whose extraction causes uncompensated environmental damages. To promote downstream measures of environmental protection, i.e. treatment and recycling, the tax burden has to be lessened by credits for recycling and treatment activities that prevent the entry of toxins into the environment. Thereby such a price instrument would rather be a deposit-refund scheme than a straight tax scheme, but this distinction has only minor impact on its general properties. In both schemes the polluters reduce their pollution as long as the tax rate exceeds their individual MAC. At the end of the adaptation process all polluters have equal MAC, which again equal the tax rate. Thus, cost efficiency can be achieved by the rational behavior of the obligated firms. A shortcoming of tax solutions, compared to quantity instruments and command-and-control measures, is their insufficient precision. If the authorities have incorrect information about the slope of the aggregate MAC curve, then the corresponding tax will miss the environmental target. If the aggregate MAC are underestimated, then the chosen tax rate will be too low and the actual pollution will exceed the target, and vice versa.

[9] However, a precise achievement of environmental targets, on an aggregate level as well as on the level of ambient targets, is in principle attainable with command-and-control measures. For that purpose absolute target values have to be set for each polluting facility, relative standards or the obligation to employ the best available (abatement) technology cannot provide for such a precision.

Quantity Instruments

In contrast to taxes, which only define the price of pollution, quantity instruments set a fixed pollution limit. If done within a framework based on market incentives and not by direct controls, tradable permits are usually employed. These permits certify the right to conduct a certain activity harmful to the environment. They are issued and distributed by the environmental authorities, which limit their aggregate amount with respect to the aimed environmental standard. After their initial allocation permits are freely tradable.[10] Each firm will buy permits as long as their current MAC exceed the current permit price, and vice versa.

This behavior is analogue to the firm's behavior in a tax regime. In the equilibrium identity of all individual MAC with the permit price and consequently cost efficiency is reached. While the amount of pollution is determined ex ante, the ultimately resulting permit price is subject to market mechanisms and will equal the value of the aggregate MAC function at the pre-defined pollution level. Though the achievement of the environmental target is ensured by limiting the number of permits, uncertainty regarding the MED function or the MAC function may result in welfare losses compared to the allocative optimum, because the environmental target set by the authorities is not optimal any longer if these functions deviate from the estimated slopes.

Effects of Uncertainty

For an incorrectly estimated MED function, the welfare losses of a system of tradable permits are identical to those of a tax solution since both realize the same false level of pollution. Consequently, a lack of information regarding the MED function does not matter for the assessment whether taxes or permits have relative advantages. Another situation occurs if there is uncertainty about the MAC function. Then price and quantity instruments will result in different amounts of pollution, and typically in different welfare losses. In cases where the slope of the MED function is steeper than the slope of the MAC function quantity instruments should be employed, and vice versa: In a tax scheme the problem of uncertainty gains importance with the "steepness" of the MED function and the elasticity or the "flat-

[10] There are two fundamental mechanisms for the initial allocation of pollution rights. They can be distributed free of charge according to each firm's previous pollution or they can be auctioned by the authorities. While auctioning may imply politically not enforceable severities to existing polluters, grandfathering benefits the previously most polluting firms and may establish market entry barriers.

ness" of the aggregate MAC function.[11] A flat aggregate MAC implicates a strong quantitative reaction on a suboptimal or "false" tax rate, which causes according to a steep MED function a substantial loss in social welfare. Therefore, tax solutions should not be implemented in cases where an environmental or health related threshold exists, whose transgression will cause rapidly increasing damages. On the other hand a steep MAC function can put tremendous burdens on the producers if the amount of issued permits is too small.[12] Thus, permits should be preferred in cases, where many reduction opportunities are available so that the MAC do not rise rapidly.[13]

Inefficiency of Relative Quotas

So far, in public waste management direct controls are predominant. Even if market based instruments are employed, they are mostly neither pure price instruments nor pure quantity instruments; there are only few examples for permit or tax systems.[14] Especially in the field of recycling legislation another instrument is widely spread. Relative quotas prescribe the amount of waste to be recycled in relation to a certain reference parameter, usually an output parameter. Regarding their allocative effects such relative recycling quotas are located between price and quantity regulations. In contrast to emission allowances they do not set limits to the emission of pollutants, but demand a certain amount of protective activity. However, in the end both definitions are almost identical if there are only two options for disposal. The prevalence of relative standards may be owed to insufficient information of the authorities about the MAC, MED and MRC. Relative quotas have no absolute caps, but still limit the emissions even if the MAC are considerably underestimated by the regulation authorities. Therefore, the decision in favor of relative quotas can be interpreted as an attempt to limit potential losses of social welfare. Beside the intention of risk reduction, political considerations regarding the feasibility of measures of environmental protection may have played a decisive role.[15] Nevertheless, even in theory a

[11] A "steep" MED function means that the environmental damages rise disproportionately with increasing pollution loads. An elastic or "flat" MAC function implies a strong quantitative reaction on varying price signals, here taxes.

[12] If the amount of permits is too high, there may also occur detrimental consequences on social welfare since environmental damages are accepted, which could have been avoided at relatively low costs.

[13] For a more detailed derivation of the results of this paragraph see Weitzman, M.L. [4], and also Fishelson, G. [5], Adar, Z. and Griffin, J.M. [6].

[14] See the Dutch landfill charge or the landfill allowance trading scheme in the UK. For an economic analysis of the latter see Barrow, M. [7].

[15] See Quirion, P. [8].

relative quota can never be cost efficient; this is shown in the following.[16]

In the first instance we deduce the optimization calculus of a single producer of the polluting commodity. We assume perfect competition on the market for the commodity and on the market for recycling services. The producers maximize their profits π being subject to a relative recycling obligation. This obligation is defined as a percentage α of the amount of pollutants contained in the products. The compliance with this obligation is proven by the purchase of a corresponding amount of recycling certificates; thus, there is no physical take-back and recycling obligation. The market price of the recycling certificates c equals the marginal recycling costs. Since we assume an atomistic market, a single firm's individual production x and its individual pollutant intensity χ, which determine its Individual Production Costs (IPC), have no significant impact on the aggregate values X and β. Then we get for the maximization calculus of a single producer:

$$\begin{aligned} Max\ \pi &= p(X)*x - IPC(x,\chi) - x*\chi*\alpha*\chi \\ &= p(X)*x - IPC(x,\chi) - x*\chi*\alpha*RC'(\alpha*X*\beta) \end{aligned}$$

$x*\chi*\alpha$ determines the firm's amount of pollutants to be recycled. We get the first order conditions:

$$p(X) = MIPC(x,\chi) + \chi*\alpha*RC'(\alpha*X*\beta) \tag{4}$$

$$IPC_{\chi} = -x*\alpha*RC'(\alpha*X*\beta) \tag{5}$$

Now we compare by means of partial analysis the individual optimum with the social optimum. At first we look at the production magnitude. We assume that the value assigned to α is socially optimal. Furthermore, we assume that the manufacturers produce with the optimal pollutant intensity, i.e. $\chi = \beta_{opt}$.

Under optimal α it holds for the social optimum:

$$p(X) = MPC(X,\beta) + \beta*\alpha*RC'(\alpha*X*\beta) \tag{6}$$

Equating (4) and (6) we get:

$$MIPC(x,\beta) = MPC(X,\beta) + (1-\alpha)*\beta*RC'(\alpha*X*\beta) \tag{7}$$

---

[16] For other derivations of the inefficiency of relative quotas and other kinds of output-based rebating environmental policy see Ebert, U. [9]; Fischer, C. [10].

Since (1-α)*β*RC'(α*X*β) > 0, it holds MIPC(x,β) > MPC(X,β). The Marginal Individual Production Costs (MIPC) of a producer exceed the socially optimal marginal production costs. Supposing a convex production cost function and transferring this result to the entire economy it becomes obvious that the aggregate production under a relative recycling obligation exceeds the optimal value. Consequently, the aggregate pollution also exceeds the optimum. This result is intuitive given that the producers do not bear the entire financial liability for their emissions. They only have to pay for the share of their emissions that has to be recycled, the remaining pollution stays uncompensated. At the actually realized production level optimum condition (1) is violated; the MAC fall below the MRC and the MED, respectively.[17]

Not only is the output too high under a relative standard, but also the pollutant intensity is suboptimal. For simplicity we now assume that the optimal output level is realized, and again that α is set optimal.[18] We only focus on χ and β. Under these assumptions we obtain from Equating (2):

$$PC_{\beta} = -X * RC'(\alpha * X * \beta) \tag{8}$$

Equating (5) and (8) we get:

$$\begin{aligned} & PC_{\beta} + X * RC'(\alpha * X * \beta) = IPC_{\chi} + \chi * \alpha * RC'(\alpha * X * \beta) \\ & \Leftrightarrow IPC_{\beta} + \chi * RC'(\alpha * X * \beta) = IPC_{\chi} + \chi * \alpha * RC'(\alpha * X * \beta) \\ & \Leftrightarrow IPC_{\chi} = IPC_{\beta} + (1-\alpha) * \chi * RC'(\alpha * X * \beta) \end{aligned} \tag{9}$$

$$with\ \ IPC_{\beta} = PC_{\beta} * (\chi/X), \frac{\partial^2 IPC}{\partial c^2} > 0, \sum_{i} \chi_i = X_{opt}$$

We can see from Equating (9) that the optimal pollutant intensity β will be missed, even if the optimal output level is realized. The actually chosen individual intensity χ exceeds the optimal value β, i.e. the pollution prevention in the production process falls short of the optimal level.

17 We make the conventional assumption that the RC function and the ED function are convex, and that the AC function is concave. For a mathematical definition of the MAC see Annex I.

18 We further assume similar cost structures of the producers.

As already mentioned above the incomplete financial producer responsibility is the crucial reason for this failure; the producers are released from their responsibility in the extent of (1-α), or in financial terms $(1-\alpha)*\chi*x*c$.

Normally there will not be an overshoot of either the optimal output or the optimal pollutant intensity, but rather a surpassing of both; the price for the commodity will be below the optimum price. It becomes clear that allocative efficiency cannot be reached with relative quotas unless α is set to 100 percent.[19] Due to an insufficient producer responsibility the manufacturers' objective of individual profit maximization leads to excessive emissions.

Supplementary Taxation of Recycling Certificates

Let us now modify the relative recycling obligation with a supplementing price instrument. The purchase of recycling certificates will be taxed with the rate γ, applied to the price of the recycling certificates c. Thereby the optimization calculus of the producers changes to

$$\begin{aligned} Max\ \pi &= p(X)*x - IPC(x,\chi) - x*\chi*\alpha*\chi*(1+\gamma) \\ &= p(X)*x - IPC(x,\chi) - x*\chi*\alpha*RC'(\alpha*X*\beta)*(1+\gamma) \end{aligned}$$

Then we have the following first order conditions:

$$p(X) = MIPC(x,\chi) + \chi*\alpha*RC'(\alpha*X*\beta)*(1+\gamma) \tag{10}$$

$$IPC_\chi = -x*\alpha*RC'(\alpha*X*\beta)*(1+\gamma) \tag{11}$$

We still act on the assumption of an optimally chosen α and that the producers choose their individual pollutant intensity χ in accordance with the optimal value β_{opt}. Equating (6) and (10) we get:

$$\begin{aligned} &MIPC(\chi,\beta) + \beta*\alpha*RC'(\alpha*X*\beta)*(1+\gamma) = \\ &MPC(X,\beta) + \beta*\alpha*RC'(\alpha*X*\beta) \\ &\Leftrightarrow MIPC(\chi,\beta) = MPC(X,\beta) + (1-\alpha*(1+\gamma))*\beta*RC'(\alpha*X*\beta) \end{aligned} \tag{12}$$

It becomes apparent that the aggregate production individually chosen by the producers and the social optimum are equal, i.e MIPC(x,β) = MPC(X,β), if it holds for γ:

[19] A recycling quota of 100 percent is typically not optimal since the MRC of such a quota are expected to exceed the MED of the first pollution unit.

$$\gamma = \frac{1}{\alpha} - 1 \tag{13}$$

In a similar way to the former analysis we are now looking at the pollutant intensity χ, still assuming an optimally chosen α, and an optimal aggregate output X, i.e.

$$\sum_i x_i = X_{opt}$$

Equating (8) and (12) leads to the same result:

$$PC_\beta + X * RC'(\alpha * X * \beta) = IPC_\chi + \chi * \alpha * RC'(\alpha * X * \beta) * (1+\gamma)$$
$$\Leftrightarrow IPC_b + \chi * RC'(\alpha * X * \beta) = IPC_\chi + \chi * \alpha * RC'(\alpha * X * \beta) * (1+\gamma)$$
$$\Leftrightarrow IPC_\chi = IPC_\beta + (1 - \alpha * (1+\chi)) * \chi * RC'(\alpha * X * \beta) \qquad 14)$$

Again, a tax rate $\gamma = (1/\alpha\text{-}1)$ yields an adoption of the socially optimal values, here the optimal pollutant intensity β.

It is shown that the allocative optimum is achievable even with a relative standard if the relative recycling obligation is complemented by an appropriate taxation. The tax has to be applied to the purchase of recycling certificates and must be set reciprocally to the recycling quota. Thereby the producers of the polluting commodity are compelled to consider the environmental damages caused by the quotient of their emissions that is not subject to the recycling obligation. If α is set optimally, i.e. MRC and MED are equalized, then the marginal tax burden due to an additional unit of emissions is exactly equal to the marginal environmental damage caused by the not recycled portion of this emission unit. Consequently, this combined instrument burdens the producers with the entire external costs of their production and thus induces their socially optimal behavior.

Optimal Adjustments in a Dynamic Market

Environmental policy regularly suffers from various kinds of uncertainty; this may comprise insufficient information for the authorities about the developing of the environmental damages, the recycling costs, or the abatement cost. These uncertainties make it difficult to determine the optimal tax rate, the optimal amount of permits, or the optimal recycling quota. While the marginal recycling costs may be read off from the market prices of recycling certificates, one can expect that the environmental authorities have no well-founded information about the marginal abatement costs.[20] Even if the producers have adequate information, they are not likely to have an interest in revealing them.

Besides uncertainty about the current state of the cost functions there also exists uncertainty about the future trends, which could require adjustments of the parameters price, quantity, or quota. We now focus on unforeseen changes of the MAC and their impact on the efficiency properties of pure taxes, pure quantity instruments, and the approach presented above.[21] Changes of the MAC can result either from changes in the costs of the substitution technology for the pollutant, in the production costs, or from shifts of the demand for the polluting commodity.[22] The latter is characteristic for dynamic markets, as it applies for instance to the market for electronic equipment with its constant growth over the recent years.

We start with the reaction of the producers to, ceteris paribus, increasing MAC (MAC_1 to MAC_2 in Figure 1), e.g. owing to an increasing demand, if they are subject to a tax scheme. As their MAC do now exceed the tax t (and the equal price of the recycling certificates), they expand their production and thereby their emissions, unless the MAC are again equal to the tax. There will be no increase in recycling activities, because that would cause the price of the recycling certificates to exceed the costs of paying the tax; optimality condition (3), i.e. the equality of MED and MRC, is violated. As a result, the emissions $[X_t*\beta_t]$ and in particular the actual pollution $[(1-\alpha_t)*(X_t*\beta_t)]$ exceed the new social optimum $[X_2*\beta_2$ and $(1-\alpha)*(X_2*\beta_2)]$.[23] The actually realized recycling quota α_t falls short of the optimal recycling quota α.[24] The net loss of social welfare is given by the vertically brindled area less the horizontally brindled area. The first area represents the disproportionately high increase in environmental costs due to a rise in emissions at a constant level of recycling activities. The latter represents the additional welfare generated by these emissions, comprising of augmented consumer utility and possibly lessened production costs due to an increased pollutant intensity.

20 It will be almost impossible to find consensus about the extent of environmental costs since there exists a number of different approaches to determine and monetarize environmental damages.

21 The impacts of such shifts, which do not allow for a reaction due to political reasons or administrative inertia, are similar to uncertainty regarding the MAC.

22 Shifts of the demand curve/production costs imply changes of the opportunity costs for reducing pollution by curbing the production of the polluting commodity.

23 The growth of emission $[(X_t*\beta_t) - (X_1*\beta_1)]$ equals the increase of actual pollution $[(1-\alpha_t)*(X_t*\beta_t) - (1-\alpha)*(X_1*\beta_1)]$.

24 α_t and α_q are endogenous parameters, while α is exogenous.

The effects of rising MAC are converse in a scheme of tradable permits with an absolute cap, but there also occur welfare losses compared to the optimum. Facing increased opportunity costs of reducing emissions the producers have an interest to expand their emissions. Since the amount of pollution permits is strictly limited, every additional unit of emissions has to be fully neutralized. In consequence, for the producers the marginal costs of additional emissions are identical to the MRC, which will exceed the constant MED after an emission expansion.[25] Thus, the actual amount of emissions and the actual pollution $[(1-\alpha_q)*X_q*\beta_q = (1-\alpha)*X_1*\beta_1]$ fall short of the optimum. This will normally go along with a production level of the polluting commodity falling short of the optimal level. On the other side the actual extent of recycling activities $[\alpha_q*X_q*\beta_q]$ exceeds the optimal value $[\alpha*X_2*\beta_2)]$. The thereby induced welfare loss is represented by the dotted area. It consists of losses owing to fewer emissions than optimal, implying lesser consumer utility and possibly higher MPC due to concurrently reduced pollutant intensities, and a suboptimal allocation of these emissions to the alternatives recycling or disposal to the environment.

In contrast, the policy of a relative quota combined with taxation of the recycling certificates is capable of maintaining an efficient result in spite of uncertain MAC. While quantity instruments with absolute caps and pure taxes hold either the pollution or the extent of recycling activities constant, a relative quota will adjust both to variable MAC. Thereby it is possible to keep the MED and the MRC equalized. At best the reactions of the producers to variations of the MAC are identical to the reaction of an imaginary central planner optimizing social welfare:

$$MRC(\beta * X) * (\frac{1}{\alpha} - 1) = MAC(\beta * X) \overset{!}{=}$$

$$\min_{\alpha'} MRC((\alpha') * \beta * X) + MED((1-\alpha') * \beta * X)$$

$$\text{with } 0 \leq \alpha' \leq 1$$

[25] This means again a violation of optimality condition (3).

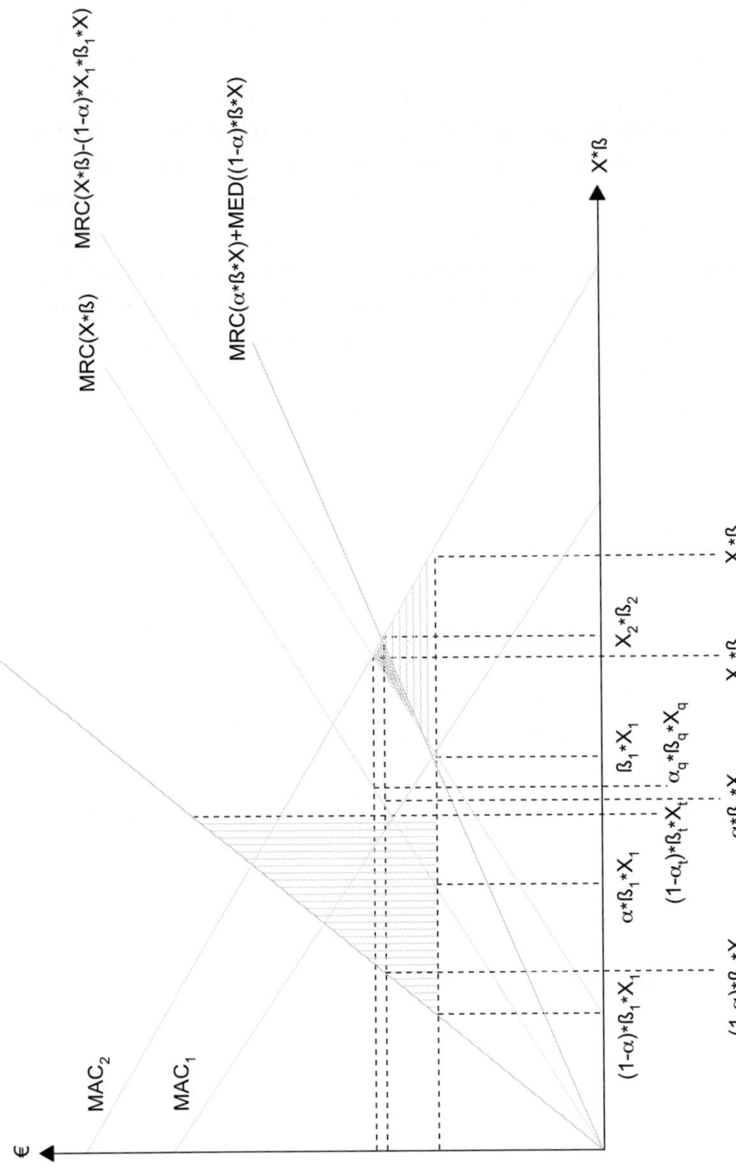

Fig. 1: Welfare losses due to incorrect expectations on the MAC

Prerequisite for such an optimal adjustment to a dynamic market or to uncertainty is constancy of the optimal recycling quota α',[26] and that it corresponds with the quota α chosen by the environmental authorities. If the slopes of MED and MRC do not allow specifying a constant recycling quota that steadily equalizes MED and MRC, the relative recycling obligation will also yield welfare losses. However, as long as the deviation of α' from α due to changes of the MAC is kept within certain limits, a tax supplemented relative recycling quota may still achieve better results than pure price or quantity instruments. As a rule of thumb one can generalize that the more (less) the slope of the MRC is increasing relative to the slope of the MED, the better price (quantity) instruments perform compared to tax supplemented relative quotas.

3.5.2 Practical Implementation of Take-Back Schemes

We now act on the assumption that the underlying instrument choice has been decided in favour of a take-back obligation relying on relative quotas. Therefore, in this subchapter we focus on institutional issues crucial for the successful practical implementation of an efficient take-back and recycling scheme. In view of deduced normative recommendations we assess the essential provisions of the WEEE and its transposition into German law via the Elektro- Elektronikgerätegesetz (ElektroG). When designing and assessing a new take-back scheme one should also hark back to the experiences made with such schemes for other waste fractions. Thus, we will also include in the following examination the most important results of an analysis that compared the take-back systems for packaging waste in the UK and in Germany with respect to their economic and ecological properties.[27]

Addressee of the Take-Back Obligation

The first decision that arises when initiating a new take-back system is the assignment of the addressee of the take-back and recycling obligation. This has to be done with respect to the question which addressee bears the best prospects of achieving the environmental targets, i.e. a decrease in the pro-

---

26 α' is the socially optimal recycling quota, minimizing the sum of recycling and environmental costs for a given amount of emissions $\beta * X$. In Figure 1 we assume an optimally chosen recycling quota so that it holds $\alpha = \alpha'$.

27 These results are primarily taken from Schatz, M. [11] and Ewers, H.J. et al. [12]; the analysis of the WEEE and the ElektroG essentially relies on Elmer, C.F. et al. [13].

duction of the polluting goods and/or a more eco-friendly[28] product design; the crucial point is the internalization of external effects. The potential addressees are producers, retailers and consumers.

Consumers that have to bear the subsequent disposal costs of their purchases have an incentive to buy environmentally sound products, thereby transmitting an impulse to the producers to consider the disposal costs in the product design process. However, consumers should not be the addressee for two reasons. Firstly, consumers may not have adequate information on the relevant product attributes or they may not be able to use them appropriately. Secondly, consumers have in general the possibility of illegal disposal in order to evade payments. Such illegal disposal patterns have counterproductive effects and will even increase the external costs.

Retailers are a more suitable addressee since they normally do not have such evasion possibilities; they do also have better information on the relevant product attributes. Their market power is a means to induce the producers to design their products more eco-friendly. Nevertheless, under normal circumstances the producer of the commodity is the best addressee. They have immediate control on the relevant product attributes, such as polluting/toxic contents, durability, weight, packaging, etc.; and, if legally responsible, they have an immediate incentive to allow for the consequences of these attributes on the disposal costs and to (re-)configure their products accordingly. The disposal costs will reduce their margin or, if they can shift the additional costs to the consumers, the demand for their products.

The German packaging ordinance (Verpackungsverordnung) does not clearly assign the take-back obligation to a recipient; it can be seized either by the producer or by the retailer. The British transposition of the packaging waste directive via the PROPW[29] splits the financial responsibility to even four parties: manufacturers of raw materials, producers of packaging material (converters), the producers of the packed product, and the retailers. Such a divided responsibility is economically justifiable in the case of packaging waste, where the environmental targets (reduction of environmental damages through landfilling, and resource protection) should be achieved by a reduction of the arising amount of packaging waste.

In the case of waste electronic equipment the design of the products and the contained substances matter. Therefore, the take-back obligation should be clearly assigned to the producers. The German ElektroG shows a lack of such clarity: The municipalities are operationally and financially responsible

[28] This may also imply a more disassembly friendly product design.

[29] Producer Responsibility Obligations (Packaging Waste) Regulations.

for the collection of the appliances, the producers for the following steps. As a result two kinds of inefficiencies occur. The price impulse, intended to induce eco-friendly innovations, and to lessen the demand for hazardous or non-durable products, performs only incompletely. In addition, the municipalities have only weak incentives to collect and separate the returned products accurately, because they do not have to bear the financial consequences of an inappropriate storage.

Physical or Financial Take-Back Obligation

Take-back obligations can have either a physical or a financial form. In the first case the producer (or another addressee) has the duty to actually take-back used products and to arrange for their treatment/recovery; i.e. the waste products have to be passed into its sphere of influence. A financial obligation dispenses with such a physical assignment; it only demands that the producer has to bear the financial burdens for the collection and recovery of its products. Financial take-back obligations normally require less logistic efforts and therefore have lower operating costs. Physical take-back obligations are generally appropriate only for products, which can be re-used completely or in parts by the producer they are returned to.

Taking into account that packaging waste is been recycled or energetically recovered and not reused, Germany as well as the UK have chosen financial obligations to implement the European packaging waste directive. In the UK obliged firms prove their compliance with the regulations by the purchase of Packaging waste Recovery Notes (PRN). These are bought from recyclers, which physically operate the take-back and recovery process. In Germany obliged firms can transfer their obligation to dual systems, which manage the collection and recovery of packaging waste. The dual systems, financed through licence fees, do normally not collect and recycle by themselves, but put these services to tender.

In contrast, waste electrical and electronic equipment may contain reusable components or may be entirely reusable; consequently, a physical take-back obligation could be appropriate. The German ElektroG establishes such a physical obligation, but does it in a manner that in practice does not allow for substantial reuse. The municipalities, who are in charge for the collection, pass on a mixture of different products from different manufacturers to the producers. As a result, the producers will not search for reusable components, but commission external firms to recycle this mixture according to the minimum standards. Reuse will primarily carried out as far as municipalities use the possibility to recover the collected goods on their own. The use of this option, often combined with the involvement of social organizations, will be mainly focused on white goods.

Individual or Non-Individual Take-Back Obligation

An issue related to the former matter is the question whether the producers should take back their own products or whether it is sufficient to prove the recovery of an equivalent amount. This question is crucial with respect to the producers' incentives to design their products in an environmentally sound and disassembly facilitating manner. The producers will invest in an eco-friendly product design only if they benefit from the resulting end-of-life cost savings. If their returned products are distributed to all producers of that kind of good, there are no incentives to accept higher costs for environment-friendliness since the savings resulting from the lowered treatment and recovery costs disperse over all producers. On the other hand, the producer-individual take-back process is usually associated with higher collection, sorting and logistic costs. Therefore, the products subject to the take-back obligation have to possess a certain innovative potential to justify these additional costs of an individual take-back scheme.

In the case of packaging waste such a potential is lacking, the primary problem is the amount of waste accruing. Consequently, the corresponding European and national legislation does not provide for an individual take-back system. In contrast, electrical and electronic equipment offers a significantly larger scope for differentiation regarding their environmental properties; appropriate innovations can bring down the end-of-life costs. The WEEE does not make a clear-cut decision on the question of individual or non-individual take-back systems; the European member states have a wide range of options for the implementation. The German ElektroG defines non-individual take-back as the norm. However, producers can take their own products back or sort them out at the municipal collection points. Notwithstanding, producers that want to use the opportunity to recycle their own products are discriminated, because they have to bear the costs of setting up their own take-back system or of sorting out their own products, while the mixed collection is paid by the municipalities. In order to have a fair competition and to foster incentives for eco-friendly product design both alternatives should be treated equally.

An approach to strengthen these incentives without bearing the costs of individual take-back is the use of weighting factors. In a take-back scheme including weighting factors each product is given a value indicating its potential environmental harmfulness, i.e. if disposed of inappropriately. The aggregate recycling obligation of a producer is calculated with respect to these factors, i.e. by multiplying the number of sold products with their particular value.[30] If feasible weighting factors can be employed, even a non-individual take-back obligation may provide incentives for the production of environmentally sound products since the producers are interested to reduce their recycling obligation. Yet, the definition of feasible and broadly accepted weighting factors is the major problem of this approach. Moreover, the administrative costs of such a scheme are likely to be significantly higher compared to a scheme without weighting factors. In the end, one has to balance the potentially better achievement of environmental targets through appropriate incentives and the costs they incur.

Definition of Homogeneous Quotas

The definition of the recycling quotas should be closely linked to the targeted environmental impact. Otherwise it is possible that the quotas are formally satisfied, but the intended environmental improvement is not achieved. This issue is more critical in the case of recycling obligations than in other fields of environmental policy. While there is a direct link in climate policy between the combustion of fossil fuels and CO_2 emissions, the relation between recycling targets and environmental protection and/or the diminishment of resource depletion is less straightforward.[31] Therefore, special care has to be taken to face the lack of a clear relationship.

In view of this issue the WEEE and its national transpositions can be even seen as a step backwards compared to the packaging waste directive and its national transpositions. The latter define different recovery and recycling quotas for the various packaging materials. Thereby, one main objective of the directive, the protection of natural resources, is controllable relatively precisely. However, the superordinate objective, i.e. the protection of the natural environment, is generally difficult to address by means of recycling quotas: Different recycling techniques may have different ecological properties, although all comply with the legal requirements.

Regarding the WEEE it is doubtful for two reasons whether it will bring about any environmental improvements. Firstly, the WEEE defines relative recycling targets for groups of appliances, i.e. the segregative criterion is

30 Additionally, in such an approach recyclers should get variable "environmental damage neutralization credits" that depend on the kind of products they have recycled: the higher the weighting factor of the product, the higher the credit.

31 For instance energetic recovery is more efficient for some materials than recycling both from an economic and from an environmental perspective.

their function and not the materials they consist of. Secondly, the recycling quotas are quantitatively defined in relation to the amount collected and not to the amount of equipment sold of each group. The only absolute quantitative specification is the obligation of each member state to collect at least 4 kilograms per year per inhabitant over all groups of electrical and electronic equipment. Thus, it is possible to fulfill the requirements of the directive solely by the treatment and recovery of relatively environmentally innocuous white goods; the disposal of the generally significantly higher polluted small appliances together with municipal household waste could be continued without violating the directive.[32] This will mean no noticeable change of the pre-WEEE situation since white goods had already been collected and recycled before the WEEE came into force.

On the other hand it is important not to define the quotas/target sectors too narrow in order to seize all efficiency potentials, i.e. equalizing the marginal reduction costs over all sources of similar environmental pollution. Hence, a major failure of the German transposition of the European packaging waste directive is the distinction between transport and retail packaging, which is not prescribed in the directive. While the quotas of the PROPW are applied to all kinds of packaging, the Verpackungsverordnung sets concrete quotas only for retail packaging. As a consequence, the British packaging waste recyclers focus on transport packaging, which is inexpensive to collect and cheaper to recycle due to its more homogeneous material. In contrast, the German dual system(s) have to establish a relatively costly household collection for retail packaging.

The Need for Competition

When designing a take-back scheme one should not disregard the efficiency fostering effects of a competitive institutional framework. Monopolistic structures generally bring about higher prices, neglect realising cost saving potentials, and often tend to show innovative inertia. Furthermore, they are normally more vulnerable to favouritism.

The ignorance of these concerns could be observed in the initial German transposition of the packaging waste directive. In its first version the Verpackungsverordnung demanded from each dual system to meet the relative recycling quotas relating to the entire amount of generated packaging waste and not only relating to the amount of waste licensed by the respective

[32] This disposal pattern for small appliances is likely to continue despite the possibility of returning them at no charge as the indirect costs (transport, time costs) may prevent consumers from returning them to municipal collection points.

dual system.[33] Thereby a monopolistic market for dual systems was created as it was not possible for more than one system to comply with the quotas. The collection, sorting, and recovery of packaging waste from households were assigned to a sole provider, the Duales System Deutschland (DSD). The DSD coordinates these procedures and commissions them to external firms. Retailers, bottlers, packaging producers, and waste management enterprises were involved in the management of DSD and held shares. Especially the participation of the latter proved to be problematical and an essential cost driver. When the activities collecting, sorting, and recovery were tendered, the remitter and the consignee were to some extent one and the same. As a consequence, the contracts turned out to be very beneficial for the successful tenderers.[34] Finally, the excessive prices had to be born by the consumers via license fees. The antitrust authorities eventually intervened; they enforced that companies involved in the disposal of packaging waste had to withdraw from the management of DSD.

The antitrust authorities also rejected some other practices of the DSD, which was abusing its quasi monopoly position in different markets. Competitors in the market for collection and recovery services were hardly put under pressure,[35] even by means of a boycott. Due to the bundled collection of substantial amounts of secondary resources, DSD and its subsidiary companies also became an important player with significant market shares in the market for secondary raw materials.[36]

Similar problems could be prevented in the UK on both markets. The British packaging waste recovery system does not include dual systems; it is a two-step system with competition in each market. The first market is the market for recovery services or rather for PRNs: Recyclers acquire packaging waste, recover it, and issue PRNs, which are sold in a transparent market. The second market is the market for compliance schemes: Compliance schemes can prove the required recovery of packaging waste (by the purchase of PRNs) instead of the initially obliged firms. They are commissioned with this task by obliged firms that want to lessen their administrative

[33] This design fault was corrected 1998 by an amendment of the Verpackungsverordnung. Since that amendment each dual system and each individual disposer has to comply with the quotas only relating to its (licensed) amount of packaging.

[34] Particularly the contracts for the recycling of plastics were rejected by the Federal Cartel Office. See Holm-Müller, pp 191 et seqq [14].

[35] Competition between different dual systems was possible after the novel of the Verpackungsverordnung in the year 1998. In addition, cooperative communities of firms that collect and recover their packaging waste on their own can compete with dual systems.

[36] See Schatz, M. [11]; Monopolkommission [15].

burdens and because the compliance schemes' better knowledge of the recycling market enables them to prove compliance at lower costs. The obliged firms can choose between a variety of compliance schemes or choose to prove compliance by themselves. Therefore, no unduly market power accrues.

The German environmental authorities learned their lessons and avoided in the ElektroG the structural failures of the Verpackungsverordnung. The producers of electrical and electronic equipment cannot jointly transfer their obligations to another firm or organisation comparable to the DSD. Although they can subcontract the operational accomplishment to kinds of compliance schemes[37], they remain legally responsible. These compliance schemes are subject to competition law, a central assignment to a single compliance scheme is not allowed. A certain restriction of the competition between different compliance schemes results from the allocation algorithm that requires compliance schemes to collect the waste equipment throughout the entire country. Thereby small recyclers may be squeezed out of the market or be forced to cooperate with other recyclers.[38]

3.5.3 Conclusion

In the first part of this chapter we assessed the general assets and drawbacks of the basic instrument types of environmental policy. It was shown within a straightforward model that take-back schemes relying on relative quotas will not bring about the allocative optimum unless additional measures are taken. A possible efficiency enhancing supplementation is the levy of a tax on recycling certificates, which is set reciprocally to the recycling quota. Since neither the European packaging waste directive nor its transpositions in the UK and Germany nor the WEEE and the ElektroG provide for such a supplementation, the take-back schemes for both waste fractions cannot achieve the results of a first-best solution.

However, in reality details of the actual legal regulations are often more important than theoretical considerations concerning the choice of instrument. We have outlined five issues crucial for the efficiency and the practical success of take-back schemes, and we have applied them to the provi-

37 In contrast to the packaging waste scheme in the UK, the here mentioned compliance schemes manage only the recovery and the documentation of compliance with the obligations. The collection of waste electrical and electronic equipment is assigned to the municipalities; at the municipal collection points the appliances are picked up by the carriers of the compliance schemes.

38 See Elmer, C.F.; Schatz, M. [16].

sions for packaging waste and waste electrical and electronic equipment.

Although no concrete figures for the costs of implementing the WEEE and the ElektroG respectively are available yet, two major criticisms become obvious: Firstly, we expect negative effects as a consequence of setting recycling quotas for different groups of appliances according to their function instead of material orientated recycling quotas. Secondly, there is no link between the amount of electrical and electronic equipment sold and the quantitative recycling requirements; the only quantitative specification is a minimum collection limit of 4 kilograms per inhabitant per year. This means that member states and producers can fulfill their requirements solely through the treatment and recovery of environmentally innocuous white goods, while the more hazardous small appliances remain in their traditional disposal pattern, i.e. the disposal together with municipal household waste.

In the field of packaging waste more experiences and data exist. A comparison between the take-back schemes in Germany and the UK shows that the British system incurs only a small fraction of the costs that the German system imposes on the consumers: In the year 2002 the cost per inhabitant just for the DSD amounted to 23 Euro, while the cost in the UK amounted to approximately 3 Euro. At the same time the achieved recycling quota was roughly 80 percent in Germany and 50 percent in the UK. It becomes apparent that these cost differences cannot be explained by the different recycling results. Two fundamental reasons for this discrepancy have been revealed: On the one hand the lack of competition in every market related to the recovery of packaging waste has proven to be an essential cost driver. On the other hand the legal distinction of retail and transport packaging turned out to be very costly. The obligations for dual systems to provide countrywide curbside collection and to accomplish specific recycling quotas for retail packaging waste constitute the dominant part of the costs.

Setting incentives for environmentally sound product design and to simultaneously limit the administrative and logistic costs became obvious to be the major challenge of implementing efficient take-back regimes. In order to mitigate this trade-off, future research work should focus on the development of feasible weighting factors. These should be employed to assess the potential hazardousness of products subject to take-back obligations as well as to evaluate the environmental quality of different disposal patterns, i.e. treatment and recovery procedures. At the same time, the economic concepts for the implementation of such weighting factors have to be meliorated and customized for their potential applications.

References

1 Pearce, D.W.; Turner, R.K.: Economics of Natural Resources and the Environment. Harvester Wheatsheaf, Baltimore 1990.
2 Tietenberg, T.: Environmental Economics and Policy. 4th edn, Pearson Addison Wesley, Boston et al. 2004.
3 Linscheidt, B.: Ökonomische Anreizinstrumente in der Abfallpolitik. Analytica, Berlin 1998.
4 Weitzman, M.L.: Prices vs. Quantities. Review of Economic Studies (1997) 41, pp. 477-491.
5 Fishelson, G.: Emission Control Policies under Uncertainty. Journal of Environmental Economics and Management (1976) 3, pp. 189-197.
6 Adar, Z.; Griffin, J.M.: Uncertainty and the Choice of Pollution Control Instruments. Journal of Environmental Economics and Management (1976) 3, pp. 178-188.
7 Barrow, M.: An Economic Analysis of the UK Landfill Permits Scheme. Fiscal Studies (2003) 24, pp. 361-381.
8 Quirion, P.: Relative Quotas: Correct Answer to Uncertainty or Case of Regulatory Capture. Fondazione Eni Enrico Mattei, Nota Di Lavoro 33, 2003.
9 Ebert, U.: Relative Standards: A Positive and Normative Analysis. Journal of Economics (1998) 67, pp. 17-38.
10 Fischer, C.: Rebating Environmental Policy Revenues: Output-Based Allocations and Tradable Performance Standards. Resources for the Future, Discussion Paper 01-22-2001.
11 Schatz, M.: Wettbewerbliche Ausgestaltung von Rücknahmepflichten – Theoretische Analyse und Systemvergleich der Verwertung von Verpackungen in Deutschland und Großbritannien. LIT, Münster 2005.
12 Ewers, H.J.; Tegner, H.; Schatz, M.: Ausländische Modelle der Verpackungsverwertung: Das Beispiel Großbritannien. http://wip.ww.tu-berlin.de/publik/uk-bericht.pdf (Stand: 2002).
13 Elmer, C.F.; Schatz, M.; von Hirschhausen, C.: Effizienzanalyse der europäischen Elektroaltgeräte-Richtlinie (WEEE) sowie ihrer nationalen Umsetzungsmöglichkeiten. Zeitschrift für Umweltpolitik und Umweltrecht (ZfU) (2005) 28, S. 513-542.
14 Holm-Müller, K.: Ökonomische Anreize in der deutschen Abfallwirtschaftsp olitik. Physica, Heidelberg 1997.
15 Monopolkommission: Wettbewerbsfragen der Kreislauf- und Abfallwirtschaft. Sondergutachten der Monopolkommission gemäß § 44 Abs. 1 Satz 4 GWB (Survey of the German Monopoly Commission), 2003.
16 Elmer, C.F.; Schatz, M.: Das ElektroG wird die Entsorgungslandschaft nachhaltig verändern. Recycling Magazin (2005) 1, S. 14-15.

Annex: Mathematical Definitions

AC Abatement Costs

MAC Marginal Abatement Costs with Respect to a Variation of (X*β)

$$MAC = \frac{\partial Z}{\partial M}$$

for the optimization problem

$$Z = max \int_0^{x^*} p(X)dX - PC(X, \beta)$$

s.t. $X * \beta \leq M$

ED Environmental Damage

MED Marginal Environmental Damage with Respect to a Variation of X*β

$$MED = ED'((1-\alpha) * X * \beta) = \frac{\partial ED((1-\alpha) * X * \beta)}{\partial((1-\alpha) * X * \beta)}$$

PC Aggregate Production Costs

MPC Marginal Aggregate Production Costs with Respect to a Variation of X

$$MPC = \frac{\partial PC(X, \beta)}{\partial X}$$

PC_β Marginal Aggregate Production Costs with Respect to a Variation of β

$$PC_\beta = \frac{\partial PC(X, \beta)}{\partial b}$$

IPC Individual Production Costs

MIPC Marginal Individual Production Costs with Respect to a Variation of x

$$MIPC = \frac{\partial IPC(x, \chi)}{\partial x}$$

IPC_χ Marginal Individual Production Costs with Respect to a Variation of χ

$$IPC_\chi = \frac{\partial IPC(x, \chi)}{\partial \chi}$$

IPC_β Marginal Individual Production Costs of a Representative Producer with Respect to a Variation of β

RC	Recycling Costs
MRC	Marginal Recycling Costs with Respect to a Variation of $X*\beta$ $MRC = RC'(\alpha * X * \beta) = \frac{\partial RC(\alpha * X * \beta)}{\partial(\alpha * X * \beta)}$
X	Aggregate Production of the Polluting Commodity
x	Individual Production of the Polluting Commodity
p	Price of the Polluting Commodity
c	Price of Recycling Certificates
α	Recycling Quota
β	Aggregate Pollutant Intensity
χ	Individual Pollutant Intensity
γ	Tax Rate on the Purchase of Recycling Certificates
π	Profit of a Single/Individual Producer of the Polluting Commodity

3.6 Reverse Logistics

Helmut Baumgarten, Christian Butz, Nils Pietschmann, Berlin, Germany

Throughout the last years german recycling industry has undergone drastic change. Especially the enactment of a new law (Kreislaufwirtschafts- und Abfallgesetz) in 1996 has created a new setting. The focus no longer lies on the simple disposal of waste, that was organized and carried out by a great number of small regionally operating firms that are usually driven or financed by the community. They have been replaced by large autonomous companies that contribute to an entirely new sector of german economy [5].

The continuous development of this new sector has produced various topics of interest. In the past focus of attention used to lie primarily in finding ways to dispose waste the cheapest and ecologically most neutral manner. Lately logistical processes have reached great importance. Reasons for this development are new methods of processing waste. Modern facilities require large quantities of waste in order to operate with low variable costs. Additionally legislation has passed laws which express that a differenciation of waste is necessary in order to establish a cycle economy. This in return implies that logistics have to handle a great variety of differing and smaller streams of waste. The different characteristics of waste in return require different logistic technologies. The challenge for logistics lies on the one hand in organizing a structure to handle the huge amount of electronical waste which will be transported to a variety of processing plants and on the other hand in integrating the reverse logistics processes in existing logistic structures.

A large amount of the products that are generated in these facilities can be reused. This as well creates a logistical task. The distribution has to guarantee supply of production plants with the correct amount, in the demanded quality at the stipulated time. Nowadays logistical processes cause about 40 % – 60 % of total disposal cost [6]. For businesses operating in logistic services industry this opens a wide range of options. By restructuring and developing intelligent logistical process chains in the field of waste disposal and redistribution of used consumer products for resource recovery the companies have the opportunity to reduce costs and gain strategic competitive advantages.

During the last ten years german legal situation concerning ecology-supporting laws has experienced a great deal of innovation. The above mentioned "Kreislaufwirtschafts- und Abfallgesetz" contains three important novelties. One major aspect of the new law is to provide a revised, simplified definition of waste that is coherent with the European understanding. Waste is now defined as "all movables which the owner gets rid of, wants to get rid of, or has to get rid of" [10].

According to the new definition the law states an order of priority for how to treat waste. The first option is to prevent creation of waste completely. The second option to handle waste is to recycle it or recover it in order to make it less harmful for environment or even prepare the reuse of certain materials. Only if material recycling cannnot be technically realized an energetic recyling is permitted. The last option stated by law is only legal in case none of the above can be realized. It is the mere disposal of waste.

The third major novelty with strong impact on producers, retailers and consumers is a product responsibility that has been introduced by law. Product responsibility implies that everyone who develops, produces, processes or sells a product has complete responsibility for minimizing waste during utilization and securing an ecologically friendly treatment after use. The nature of the discussed law has one problem though, it only states these rules generally and does not specifically name any product or group of products for which product responsibility is valid. Therefore legislative measures (as an enactment of decrees) have to be taken in the near future in order to make law effective.

Through decrees ecologically harmless disposal of used technical consumer products will eventually become mandatory for the producers. Disposal policy of all companies and retailers shall therefore focus on product and material recycling. A further reason for this is decreasing space on public dumps. Additionally, dumps are only permitted for specifically identified waste streams respectively pretreated types of waste. Both tendencies lead to increasing disassembly activities. Dissassembly as a recyling technique is superior to the wide spread shredding and sorting technology. On the one hand, different materials can be extracted and processed so high quality secondary materials are recovered. On the other hand, disassembly enables to recover complete products and components which can be reused as replacement parts or even as components for production of new products. Eventually disassembly represents the best method to achieve a cycle-economy which in return is the primary goal of the "Kreislaufwirtschafts- und Abfallgesetz". The actual problem consists in the low degree of automation of the disassembly processes; for consequence disassembly is highly labour intensive and therefore expensive.

In order to meet above mentioned requirements, specific logistical structures are necessary. Challenge lies in obtaining as many complex consumer products as possible and redistribute them to the disassembly facilities. This has to occur in a way that transportation is optimized regarding two aspects. First, the planning of transportation routes has to be made in respect to existing collection points as well as population density. Secondly, means of transportation have to be selected according to economic and ecologic efficiency, i.e. break-even points for when to use heavy goods vehicles and when to employ alternative carriers as railway or ship have to be calculated. A third important aspect that logistics have to consider is that legal restrictions exist for dangerous and contaminated goods. Special logistical technologies have to be introduced and integrated in existing concepts of redistribution.

3.6.1 Redistribution Processes

So far it has been made clear that the establishment of a well structured redistribution system is the first element in the chain to achieve cycle economy. Generally speaking redistribution can be divided into strategic and operational processes. Strategic tasks include the long-term planning of a holistic system. This entails choice of locations for collection and disassembly activities. It also includes a decision on the way of collection as well as the decision on the means of transportation.

Furthermore the establishment of a general information system is included in strategic planning. These activities are primarily nonrecurring decisions. The redistribution process, however, includes dynamic activities as well. These are collection and registration, transportation, interim storage, and handling processes.

Collection is defined as the physical picking up of used consumer products. Prior to the pick up there has to be a flow of information from the consumer to the service provider. This informational process is defined as registration. Only if registration of the products has occurred, a reasonable collection process can be planned [11]. This means optimal planning of collection routes and optimal choice of means of transportation. Today most collection activity is done with regular trucks or delivery vans. Trucks usually have a container where all appliances are loaded on top of each other. This procedure already reveals potential weaknesses. The chance for demolition during transport is very high and the process of loading the container can not be automized. Another problem with the container is that they usually are not used to capacity which in return causes empty trips and higher costs.

Transportation generally follows the collection processes. If disassembly facilities are identical with collection locations, the two processes occur parallel. As already explained, the distances are at times very large, so that transportation costs increase. The logical consequence is that collection and transportation processes are separated. By doing this bundeling effects can be used. Cost-effective means of transportation that have been genuinely designed for long-distance (and not for collection) will be used to capacity. Depending on the available quantity of used products and on the geographic situation at the final destination, there is also the option to employ train or ship for long-distance transportation. Because used household appliances are not critical concerning transportation time, the more ecological transportation by ship seems to be appropriate.

The shipping time, generally non-productive time, can be used for value adding processes like sorting or disassembly. Another advantage of bundeling up the products at interim storage areas is that used consumer products can be sorted, identified, and suitability for disassembly can be registered so that disassembly factories have more time and certainty in advance to plan and control their internal processes. The described potentials through separation of collection and transportation will gain importance as soon as the capacity of disassembly facilities increases or when the catchment area increases so that cooperations between logistic service providers become necessary.

Handling means a change of the logistical aid. Within redistribution processes handling of the products occurs two times: the first time when the collection trucks unload at interim storage areas, the second time when products are unloaded at their final destination. Today, handling is not automated at all. Most of the handling activities are carried out manually which makes especially this process extremely time and labour intensive. Mechanized logistical aids as carts do not improve productivity a lot. Another problem caused by insufficient handling is the demolition of the products that can be found frequently. It is of extreme importance to establish handling techniques within the scope of multi-modal transportation, that prevents workers from touching every single machine. This automization is closely linked with the development of a container especially desigend for the transportation of household appliances. Missing strategic orientation and nonexistent advanced scheduling demands very high flexibility of present-day redistribution systems. The development of a general container system, a coherent transportation organization as well the choice of alternative means of transportation will result in efficient processes.

3.6.2 Cycle Economy and Disassembly

Through decrees ecologically harmless disposal of used technical consumer products will eventually become mandatory for the producers. The tendency has been proven by a study that reveals that increasing attention is given to the recycling of used consumer goods by industrial companies whereas the disposal of production waste and packaging materials has a decreasing priority (Figure 1) [3].

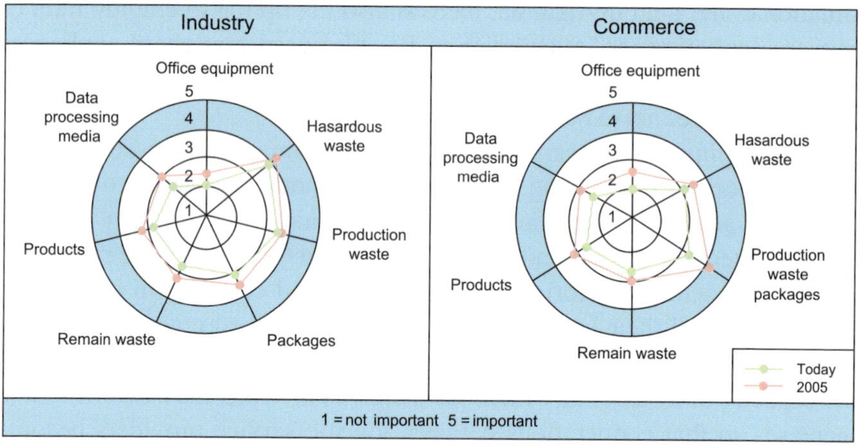

Fig. 1: Areas of waste disposal with high priority [3]

Disposal policy of all companies and retailers shall therefore focus on product and material recycling. Additionally, dumps are only permitted for specifically identified waste streams respectively pre-treated types of waste. Both tendencies lead to increasing disassembly activity. Disassembly as a recycling technique is superior to the wide spread shredding and sorting technology. On one hand different materials can be extracted and processed. So high quality secondary materials are recovered. On the other hand disassembly enables to recover complete products and components which can be reused as replacement parts or even as components for production of new products [8]. Eventually disassembly represents the best method to achieve a cycle-economy.

However, the problem presently is the low degree of automation. Disassembly is still labour intensive and therefore expensive. The whole process of recycling has to be managed so that costs are minimized and the process becomes efficient. This includes redistribution with its logistical requirements as well as disassembly with its technological requirements. Further a market for disassembled products or components has to be established [7]. A stable market will only exist if disassembled products constantly have reliable quality. Two aspects result from that:

First, a large amount of used products has to be collected constantly. The expected amount of old household appliances in Germany adds up to 12 million pieces (630,200 tons) per annum. This large number implies that there are potentially enough used products in circulation. The challenge is to find logistical methods to recover as many of the objects as possible [2].

Secondly, the used products have to be treated carefully, so that no major damage is done. Today most of the damage of old household appliances is being caused during the redistribution. For an efficient and automated disassembly the condition of used products should be as good as possible.

As one can see, the logistics plays a very important role in the cycle economy. Redistribution, treatment and reintegration are three main processes of disposal logistics (Figure 2).

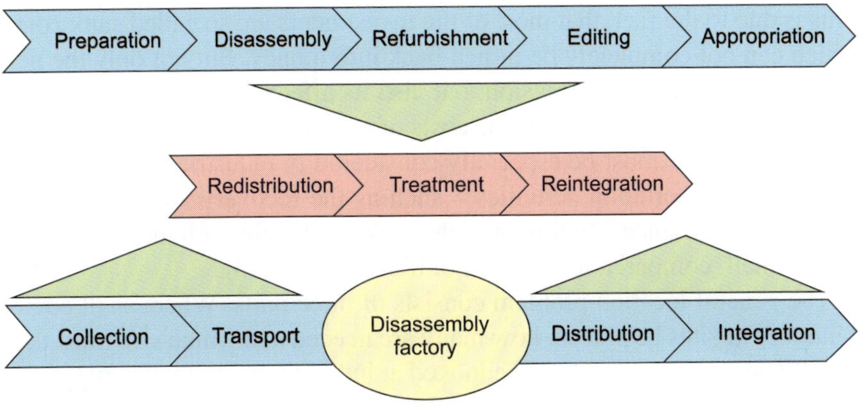

Fig. 2: Reverse logistics process chain model [2]

Collection and transport of used household appliances belong to the process of redistribution. This process is concerning the amount of collected appliances responsible for efficiency of the following processes and activities. After redistribution the collected used products have to be treated in a recycling plant resp. disassembly factory. The cycle economy is realized through the process of reintegration of disassembled products in the economy [1, 12]. Development of suitable logistical concepts and marketing strategies for the optimal distribution of disassembly products is of increasing importance. The market portrays a significant planning component for disassembly companies.

Before decisions about the process of organization of disassembly factories or distribution politics of recycling plants are made, costs of recovery, dismantling, assessment of component value and disposal of non-recycling materials have gaining profit in contrast to the net profit. On the one hand, these profits consist of acquisition profits, that come from acceptance fees, on the other hand, of sales profits, which come from selling the disassembled products. The object of the described project is the second profit share. According to different types of disassembly products, there can be differentiated between markets for used products, recycled elements and components as well as those of raw materials [1, 4].

3.6.3 Network of Disassembly Factories in Europe

The planning of locations is a major strategic problem of industrial businesses. Once a decision is made and executed, it can not be revoked easily. Restructuring a location network is always combined with great efforts. This is due to the fact, that most of the investments are so called sunk costs, which can not completely be turned back into money. But not only the perpetualness of a location decision, but also its affects on the operating costs are very serious. It determines the costs of almost all interplant and external processes. Latter must be especially considered in industries which require intensive transportation activities – such as the recovery and recycling of used home appliances. In this case, the added value through the treatment is rather small, compared to the costs of transportation [8].

The general location problem consists of three parts: Where shall one or a number of sites be placed, how many are needed and which size must they have? These aspects must not be looked at individually. In fact, they have to be solved together at the same time. When the overall capacity is given, the number of plants in a production network and the size of each site behave antipodal. The site and the planned capacity of a factory also correspond to each other, but often in very different ways. Factories for instance which require a great amount of local commodities can only grow together with the availability of such goods. On the other hand it is often not possible to set up big factories in very populous areas. Finally there is also a strong dependency between the geographical location and the number of plants in a network structure. Often a single location can not cover the whole area because of transport or other restrictions.

Due to the complexity of the location problem, it should always be solved in several steps. These can be distinguished into so called macro and micro planning steps. Beginning with very global but significant input parameters the first decisions should narrow the wide range of alternatives. As the planning process advances more and more the precision should also increase.

There is a wide variety of proceedings of location planning. They can be distinguished by all sorts of classifications, such as intuitively or systematically, qualitatively or quantitatively, and optimizing or heuristic. These methods can be applied to again a lot of different models of location problems. Often it is necessary or wise to evaluate a problem through diverse techniques in order to validate a solution. It is also recommendable to balance between the efforts and results of the different proceedings.

Since the creation of a network of disassembly factories in Europe must be at this stage seen as a rather uncertain planning process, the solving of the location problem should be restricted only to more strategic aspects. For this purpose one has to look at the important specifics of a location problem.

The recovery of used home appliances is a multilevel logistical problem. There are two major processes which involve the movement of goods: redistribution and reintegration. The former consists again of two activities: collection and transport. Between these steps the used devices have to be stored in collecting points, which form the starting point of transportation movements to the disassembly plants. On the other side there has to be considered the distribution of disassembly products from the factory to the customers. In this study the level of disassembly factories shall be discussed first, since the number of collection points will be very large and almost unpredictable. After having determined the covering area of the factories it will be possible to specify the locations of the collection points as well. This approach can be labeled as top-down planning because the different levels of locations are not planned hierarchically from the bottom level to the top [8].

Since the disassembly factories are both destination and origin of transports, the location problem is still extensive. This holds particularly in the case of an Europe-wide planning. In order to obtain feasible solutions a great amount of information and reasonable assumptions are necessary. But at this point of time no such things can be truly assured for the outlet market of disassembly factories. Therefore the problem has to be again simplified by focusing just on the side of procurement. This reduced model only considers the sources of old home appliances to plan the locations for the disassembly factories. All processes which only regard to the distribution of disassembly products will be neglected. The model can now be described as a single-stage location problem. But since the planning shall be based on the real geographical situation and population distribution in Europe it still is very intricate. Therefore it is inevitable to use a sophisticated software tool, in order to solve such a problem [7].

Most of these tools make use of Geographical Information Systems (GIS) and were primary developed for distribution and marketing problems in the retail industry. The digital map includes road networks, cities and administrative areas. Imported marketing data which can be linked to specific points (addresses) or territories (administrative areas) can be used as a basis for a wide variety of planning features.

The integrated location planning module is one of those. It rests upon the potential value of so called basic elements. The potential of an element (an area or a point) can be described as its weight. Elements with a high potential carry during the planning process more weight than others. As a basic principle the program will try to minimize the distance between such a heavy element and a planned location. Since it has to consider not only one but many elements, it has the agenda to minimize the weighted average of all routes between the elements and the planned location.

Since each of these routes matches the distance which a truck has to cover, it makes sense to use the number of trips, which are needed to redistribute all collected large home appliances of an area, as a value for the potential of that area (basis element). This approach will minimize the total number of transport activities. In order to determine the number of annual transports from one area, we need to look at the accumulation potential of the area and the transport capacity of a truck. Former can as shown above be derived from the number of households, the ratio of possession and the average lifetime of the home appliances. Latter is determined by the type of trucks which are used. Because of the size and the weight of large home appliances it is recommended to use large trucks with a gross weight of max. 40t. They have a loading capacity of max. 120 large home appliances.

By looking at the whole area of continental Europe one can estimate an total accumulation potential of more than 53 Mio. large home appliances per year, this leads to almost 2000 truck trips per day. map&market is able to link this trips to 1238 different European provinces, which can be seen as the sourcing areas of supply. Each of these provinces can be weighted by the expected annual accumulation potential resp. number of necessary transports [8].

After having calculated the potential of each area one has now to set the following parameters: criteria of distance, maximum distance allowed by this criteria, minimum and maximum aggregated potential of one planned site, and number of sites to be created.

As the criteria of distance an optimisation by transport costs qualifies best. Other criteria such as linear distance, stretch of way, and journey time are less accurate but can reduce the computing time. Especially the use of linear distances makes all calculations very simple since no information about the real geographic conditions is needed at all. An optimisation under the observation of costs suits the economic approach best, but also requires a lot of estimations and assumptions, which are often the source of inaccurateness. Such assumptions are mostly related to the question: How much does a trip cost? There are three aspects to consider: the fixed costs of a trip, the costs which are depending from the pure traveling distance, and the costs of time. map&market can only take linear cost rates in account. In this study they were estimated as follows: 25 € per trip, 0.75 € per km, and 25 € per hour.

In the next step the allowable peak value needs to be determined. Only round trips were tolerated, which would cost less than 500 € (approx. 5 € per unit). The value has been derived from a simple restriction: The maximum traveling time per day must be less than 7 hours. Under this requirement the distance between a collecting area and the disassembly factory is limited to approx. 350 km.

Determining the threshold values in respect to the capacity for each site and the number of planned sites emerge to be the most difficult problem. map&market is not able to take the operating costs of each site into consideration. Therefore it was inevitable to determine the optimal number of sites by comparing the unit costs of treatment with those of transport, which show a negative resp. positive correlation to the size of a plant. For this purpose it was necessary to graph both functions in a range from 1 to 100 factories [9].

The costs of treatment can be calculated by an excel spreadsheet which pays attention to overhead and direct unit costs of the treatment processes of large home appliances. Identifying the different elements of product costing emerged to be quite difficult. Especially the consideration of jumping overhead costs was complicated. For this reasons some simplifications were necessary, e.g. differing only between overhead costs, declining variable costs, and linear direct costs. Through this the solver can give rough calculations for the total annual operating costs of disassembly factories in respect to their capacities.

The costs of transportation were identified by applying the map&market location planning tool to 12 models which only differed from the number of sites it be planned (n = 1, 5, 10, 20, 30…100). Each of these models were optimized by taking the transport costs and the potential of each area in to account.

Figure 3 shows the final results – 50 disassembly factories with capacities from low as 160.000 to high as 2.7 Mio units per year were planned.

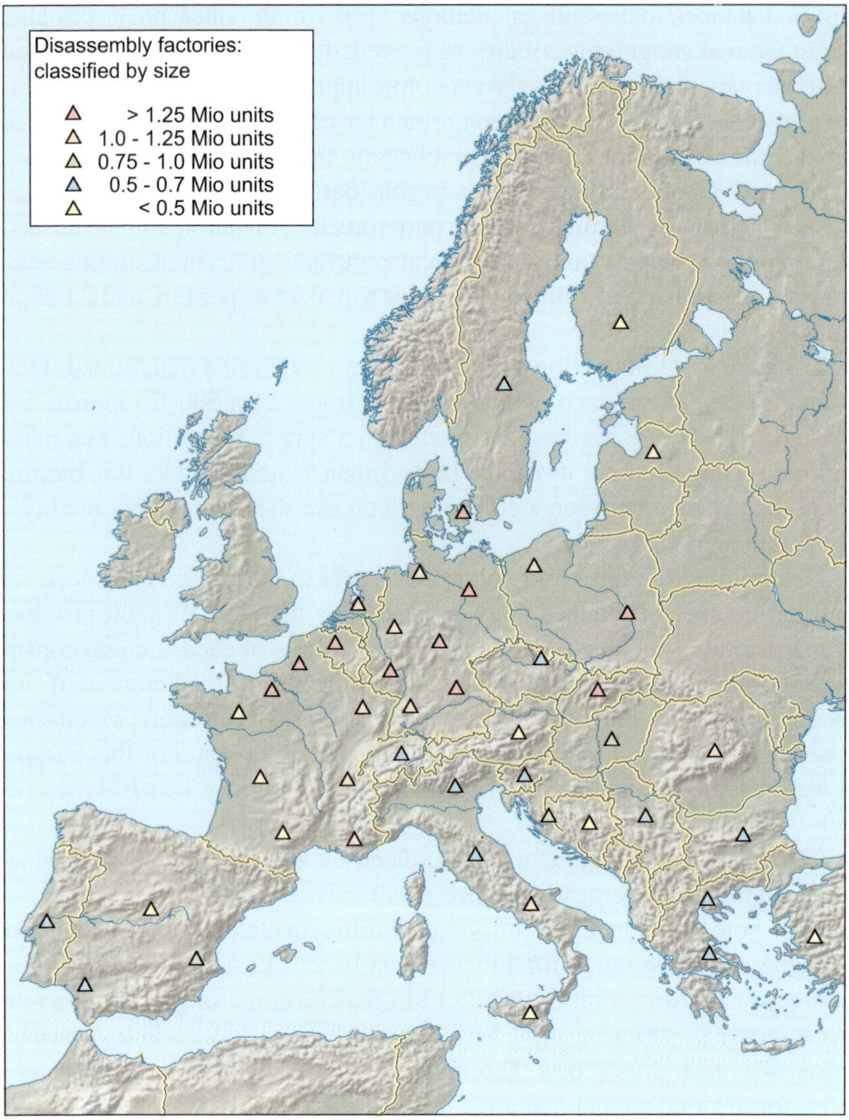

Fig. 3: European network of disassembly factories for used large home appliances [9]

They will cause transportation costs of around 224 Mio. €, which result from a transport activity of around 64 Mio. driven km. An average trip will have a distance of 133 km (one way) and cost 400€. This matches units transportation costs of approx. 5 €. The total costs of treatment will be approx. 1 Billion €.

The total recovery and recycling costs resulting from this are 27 € per unit of large home appliances [9]. The basic idea for the investigation is to determine the amount of planning needed to establish an efficient reintegration system for electric appliances on the accumulation potential of home appliances. The availability of information on the accumulation potential of old home appliances enables in estimating the back transportation costs (transportation costs to the disassembly factories), the planning of the transport structure and also the layout planning for the disassembly factory.

3.6.4 Conclusion

An efficient organisation of collection and redistribution processes of used products open possibilities for logistic service providers as well as producers to gain market share in an extremely competitive market. Since the market is characterized by many small disassembly businesses all fighting for larger quantities of used products, a well structured redistribution system should guarantee high efficiency, high functionality and a high customer orientation concerning the disposal of used household appliances. It should entail environmentally friendly solutions concerning energy consumption, emission of pollutants and noise as well as secure transportation.

Lastly it should provide a superior service for the clients. Superior service comprises characteristics as flexibility, reliability, quick response, high availibility, general provision of information, and additional performance. Producers as well as logistic service providers should implement the described strategies in order to achieve a strong market position and realize a functioning cycle economy.

References

1 Baumgarten, H.; Ivišić, R.-A.: Struktur und Potentiale des Absatzmarktes für Demontageprodukte der Weißen Ware – Ansätze zur Entwicklung kundenorientierter Marketing- und Logistikstrategien. UmweltWirtschaftForum (2000) 10, S. 63-67.

2 Baumgarten, H.; Sommer-Dittrich, T.: Prozeßkettenbezogene Auslegung logistischer Systeme. In: Technische Universität Berlin (Hrsg.): Sonderforschungsbereich 281 „Demontagefabriken zur Rückgewinnung von Ressourcen in Produkt- und Materialkreisläufen", Finanzierungsantrag 2001-2003, Berlin 2000, S. 303-336.

3 Baumgarten, H.; Walter, S.: Trends und Strategien in der Logistik 2000+: Eine Untersuchung der Logistik in Industrie, Handel, Logistik-Dienstleistung und anderen Dienstleistungsunternehmen. TU Berlin, Bereich Logistik, Berlin 2000.

4 Dural, J.; l'Ortye, V.; Deckers, E.; Huijben, A.; Melissen, F; de Ron, A. J.: A Decision Support Model for the Recovery and Re-Use of Products and Components. In: Flapper, S.D.P.; de Ron, Ad J. (Edit.): Proceedings of the Second International Working Seminar on Re-Use, March 1-3, 1999, Eindhoven, the Netherlands, 1999, pp. 31-43.

5 Emmermann, M.: Beitrag zur Entwicklung der Prozeßkette Entsorgung auf der Basis einer management-orientierten ganzheitlichen Entsorgungslogistik. Dissertation, TU Berlin, Berlin 1996.

6 Haberland, S.; Mager, A.; Waltemath, A.: Logistics as a Central Task for Establishing a Cycle Economy. SOLE, 1997, pp. 906-908.

7 Ivišić, R.-A.: Market Research for Disassembly Products for the Development of Logistical Concepts for the Distribution of Disassembly Products. In: R'99 Recovery, Recycling, Re-integration, Collected papers of the R'99 World Congress, Geneva Switzerland, February 2-5, 1999, Volume III, Sankt Gallen 1999, pp. 30-35.

8 Ivišić, R.-A.: Management kreislauforientierter Entsorgungskonzepte – Erfolgsfaktoren und Gestaltungsinstrumente. Dissertation, TU Berlin, Berlin 2001.

9 Ivišić, R.-A.: Expected Accumulation of Used Home Appliances as the Basis for the Conception an European Network of Dis-assembly Factories. Journal of Environmental Assessment Policy and Management (JEAPM) (2001) 12, pp. 368-372.

10 Kreislaufwirtschafts- und Abfallgesetz: Gesetz zur Vermeidung, Verwertung und Beseitigung von Abfällen vom 27.9.1994. In: Bundesgesetzblatt Teil I, Z 5702 A, Bonn 1994.

11 Rennert, S. : Rahmenkonzepte für ein Informations- und Kommunikationssystem zur kreislauforientierten Altproduktentsorgung. Dissertation, TU Berlin, Berlin 2000.

12 Waltemath, A.: Altproduktrückführung als logistische Dienstleistung – Entwicklung eines kundenorientierten Rück-führkonzeptes. Dissertation, TU Berlin, Berlin 2001.

4 Product Development

Product development for ease of disassembly and recycling requires methodical toolboxes and respective infrastructures of information and communication technology. Under conditions of globalization the international perspective has to be taken into account. Blessing, Gries present design guidelines applied on the example of a solenoid valve for a washing machine with respective IT-support. Blessing, Klett analyze disconnecting-supporting connections for disassembly. Regarding the whole solution space the designer is led in a step by step process of selecting. Krause, Rohman introduce the concept of functional mock-up beyond geometrically based digital mock-up. Wear and resulting product conditions are modelled for applications in maintenance, repair, dismantling and training. Krause, Jungk describe an information technological infrastructure for product life cycle planning and tracking. It helps for continuous adaptation by maintenance and repair within a usage phase and by changing product features between usage phases. Multiple usage phases enable for increasing the use productivity of resources. A product condition model created during the product development phase to be updated in product life time helps for guiding processes of adaptation. Rozenfeld and Oliveira give Brazilian perspectives on product development and cleaner production.

4.1 Design Guidelines

Lucienne Blessing, Bruno Gries, Berlin, Germany

The efficiency of disassembly and recycling processes is strongly determined by the design of the product, depending on decisions that were made a long time before the first product is disassembled. Therefore, today's designers have a very high level of product responsibility as disassembly is a major cost factor of product recycling. However, successful design for disassembly can have additional positive effects on a product, e.g. increasing its ease of maintenance as Table 1 shows.

Tab. 1: Relative costs of disassembly operations in maintenance and product recycling [1]

Cost distribution			
Maintenance		Product recycling	
Error finding	10 %	**Disassembly**	**40 %**
Disassembly	**35 %**	Cleaning	5 %
Reassembly	25 %	Inspection	4 %
Part costs	25 %	Reconditioning	10 %
Overhead costs	5 %	Reassembly	40 %

As Figure 1 illustrates, disassembly- and recycling-related issues need to be considered throughout the entire design process. Within this process, a great challenge lies in reconciling the concept of an environmentally-friendly post-use phase with "traditional" design aims such as low costs, high quality, good functionality and simple manufacturing.

The various interactions and possible design conflicts need to be considered thoroughly and early. As with all design measures, possibilities for considering disassembly- and recycling-related aspects decrease as the product development process progresses. The product aspects with the highest impact on disassembly and recycling are materials, product structure and connections. The recommendations for their optimization can briefly be summarized as follows:

Materials:

- avoid recycling-critical substances, pollutants and hazardous materials,
- use recyclable/recycled materials,
- reduce material variety,
- combine into assemblies/modules of same or similar materials and/or materials with same or similar recycling properties,
- identify materials.

Product structures:

- reduce the number of parts,
- create modular, horizontal product structures,
- allow for the reuse of standardized components and assemblies.

Connections:

- reduce the number of connections,
- reduce the variety of connection types,
- allow for a minimum of different disassembly directions,
- prefer fasteners that can be opened with standard tools,

- avoid connections where destructive separation is necessary, e.g. glued, riveted, welded or soldered connections.

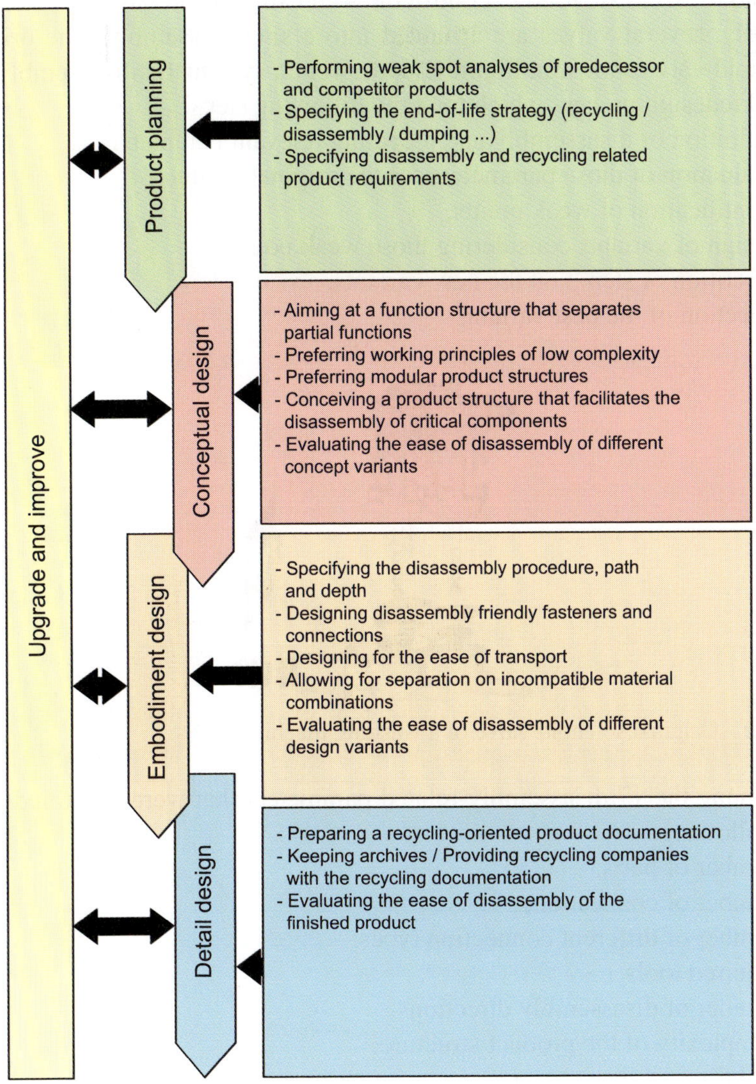

Fig. 1: Disassembly-related tasks and strategies within the design process [1]

Designers are supported in implementing those design recommendations, e.g. by tables that contain positive and negative examples like in [2]. Also, the recently adopted VDI guideline 2243 [3] summarizes current recycling-related design knowledge, such as recommendations, decision aids and product examples.

4.1.1 Redesign of a Washing Machine Solenoid Valve

Solenoid valves are used to control the water streams of washing machines. Usually, several valves are arranged into a single assembly. The original solenoid valve shown in Figure 2 underwent a systematic disassembly-oriented redesign process which took the following steps:

- definition of disassembly and recycling relevant parameters,
- evaluation of those parameters for the original product,
- identification of weak points,
- design of variants considering those weak points,
- repetition of step 2 on the new variants,
- selection of the best variant.

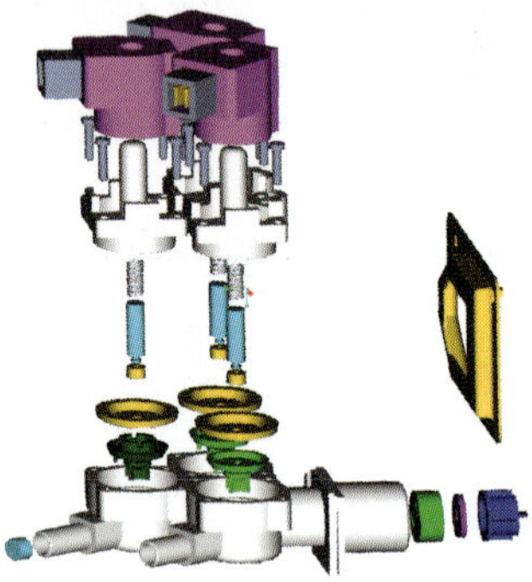

Fig. 2: Original solenoid valve of a washing machine

Among the 17 disassembly-related parameters that were identified were the following:

- number of parts,
- number of connections,
- number of different connection types,
- required tools,
- number of disassembly directions,
- complexity of the product structure,
- tool accessibility.

In order to evaluate the ease of disassembly of the existing assembly, the parameter magnitudes were correlated to a value scale between 0 (unsatisfactory) and 4 (ideal) [4]. Table 2 contains the correlations for some of the above parameters.

Tab. 2: Correlations between value scales and magnitudes of selected parameters

Value scale		**Parameter magnitudes**		
Pts.	Meaning	# of parts	# of connec-tions	# of different connection types
0	unsatisfactory	> 40	> 9	> 4
1	just tolerable	30 - 40	7 - 9	4
2	adequate	20 - 39	5 - 6	3
3	good	10 - 19	2 - 4	2
4	ideal	< 10	1	1

The evaluation of all parameters of the original assembly resulted in a score sum equivalent to 51 % of the maximum score (Table 3). The major disassembly-related weak points were identified as follows:

- great variety of materials and use of insoluble composite materials,
- large number of small parts made of different materials due to extreme separation of functions,
- complex product structure with ambiguous disassembly possibilities,
- small, irregularly shaped parts that easily get caught in one another.

Tab. 3: Comparison of the original solenoid valve with its redesign optimized fordisassembly for selected parameters

	Original design		**Optimized design**	
Parameter	Value	Score	Value	Score
# of parts	38	2	16	3
# of connections	112	0	42	3
# of different connection types	4[a]	2	1[b]	3
Required tools	screwdriver, pliers	0	none	4
…	…	…	…	…
Total score (relative to maximum)		51 %		80 %

[a] screwed: 12, snap-in: 3, clamped: 1, interference fit: 3
[b] clamped

The first step in the development of an assembly without these weaknesses was extending the manufacturer's requirements list with recycling- and disassembly-related issues. Based on these requirements and a functional analysis, several concepts were created that evolved into five variants of a disassembly-optimized solenoid valve. Repeating the initial evaluation process resulted in scores between 69 % and 80 %.

The variant with the highest score shown in Figure 3, consists of less parts, involves less connections and less different connection types than the original assembly (Table 3). Probably the biggest advantage, however, is that the optimized valve can be disassembled without any tool thanks to the central bracket (red) that attaches the solenoids (blue) to the main housing (white).

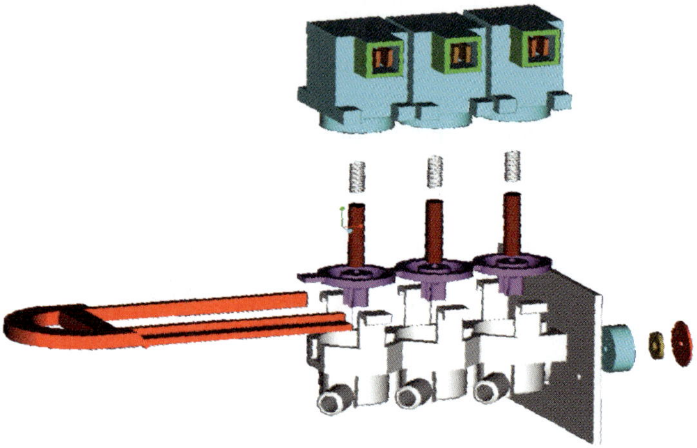

Fig. 3: Optimized solenoid valve – No tools are required for disassembly

4.1.2 IT-based Design Support

When creating disassembly- and recycling-friendly products, the designer defines materials, structure and connections, aspects critical to disassembly and recycling. Although software tools that deal with each of these aspects exist, each usually represents an individual solution with limited possibilities to exchange data with other applications. Looking up a material from a materials database, defining a material with the same properties within the CAD application and assigning the self-defined material to a part is therefore a common scenario.

The efficiency of such practice, however, becomes questionable e.g. when a product is undergoing a Life Cycle Assessment (LCA) in order to evaluate its environmental impacts [5]. The retrieval of the material quantities within the product is a process that requires considerable manual effort because available LCA software cannot access the product data of existing CAD applications. This example points at a typical issue in CAD: the problem of exchanging product data between applications that are based on different product models.

Gries and Blessing [6] have developed a Design Support System which is based on a disassembly- and recycling-oriented partial product model. This model represents all information related to a product which is relevant to disassembly and recycling. In order to provide a platform-independent and network-compatible access to the product data – as described by the partial product model – the system architecture of the Design Support System is based on a CORBA client/server environment (Figure 4).

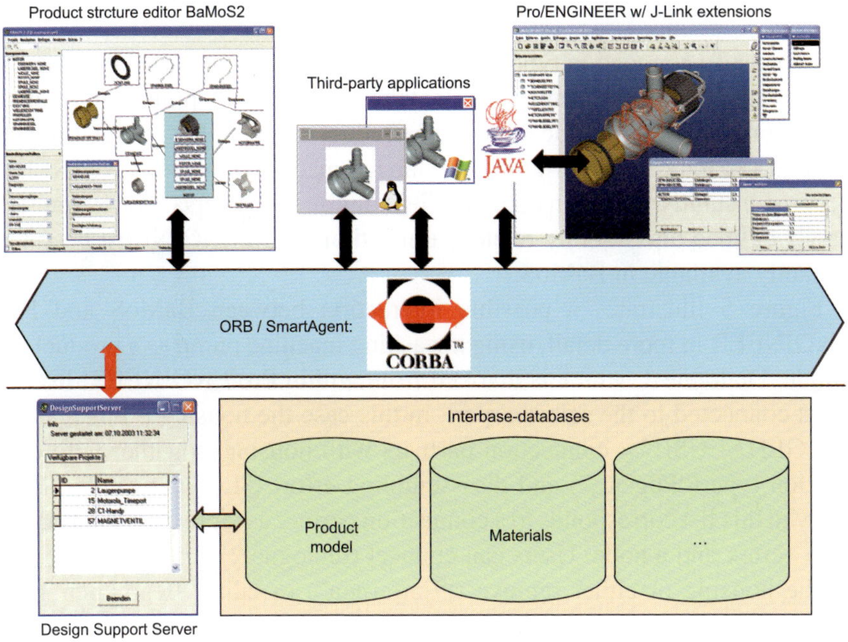

Fig. 4: Architecture of the Design Support System

Central to the system is the Design Support Server. Its main task is to map the data stored in the tables of the database system to corresponding CORBA-objects and to handle all client requests. These client requests, automatically invoked by the manipulation of a CORBA object, are interpreted by the server. The server being the only application with direct access to the product data, returns the desired result (e.g. the mass or material of a specific part) or executes the according operation (e.g. connecting two parts). Therefore, the client applications of the Design Support System do not directly manipulate any product data, but invisibly communicate with the Design Support Server.

The interaction of the CAD software Pro/ENGINEER with BaMoS 2 is an example for a typical workflow within the system. BaMoS 2 is a modeling/analysis tool for product structures (Baustrukturanalyse- und Modellierungs-System [7]). It delivers a graphical representation of the product structure, basically displaying parts and assemblies as nodes, and connections as vertices. A number of analysis functions provides information about the material composition, connection types, disassembly effort, etc.

Using J-Link, PTC's Java-based programming interface, the user interface of Pro/ENGINEER was extended with functions related to disassembly and recycling. Users can now interactively define connections between parts, specify connection type and solution effort, assign materials from the design support system's materials database, etc. By mapping the internal product model of the application to the product model of the Design Support System (as indicated in Figure 4), changes made within Pro/ENGINEER, become instantly visible in other applications and vice versa. For example, a newly defined disassembly-related connection in Pro/ENGINEER becomes instantly available in BaMoS.

Figure 5 illustrates a possible interaction between BaMoS and Pro/ENGINEER in more detail, using a washing machine pump as a product example. Dialogue 1, which is invoked from within Pro/ENGINEER, lists all parts connected to the selected part, in this case the housing ("Fügepartner mit GEHAEUSE" = connection partners with housing) together with connection type ("Fügeart") and disassembling effort ("Löseaufwand"). Each entry in this list corresponds to a connection partner as represented in BaMoS by a vertex and a node. Users can connect (dialogue 2) a new ("Neu") part to the housing or either remove ("Entfernen") or edit ("Bearbeiten") the connection to the part highlighted in the list. In the example shown, the user has chosen to edit the connection of the housing with the shaft seal ("WELLENDICHTRING"), changing the connection type from force fit ("Einpressen") to inlay ("Einlegen"). Once all changes are confirmed, the product data is automatically updated using the mechanisms described above. A user of BaMoS could now access the properties of the connection between the shaft seal and the case and see the updated information (dialogue 3).

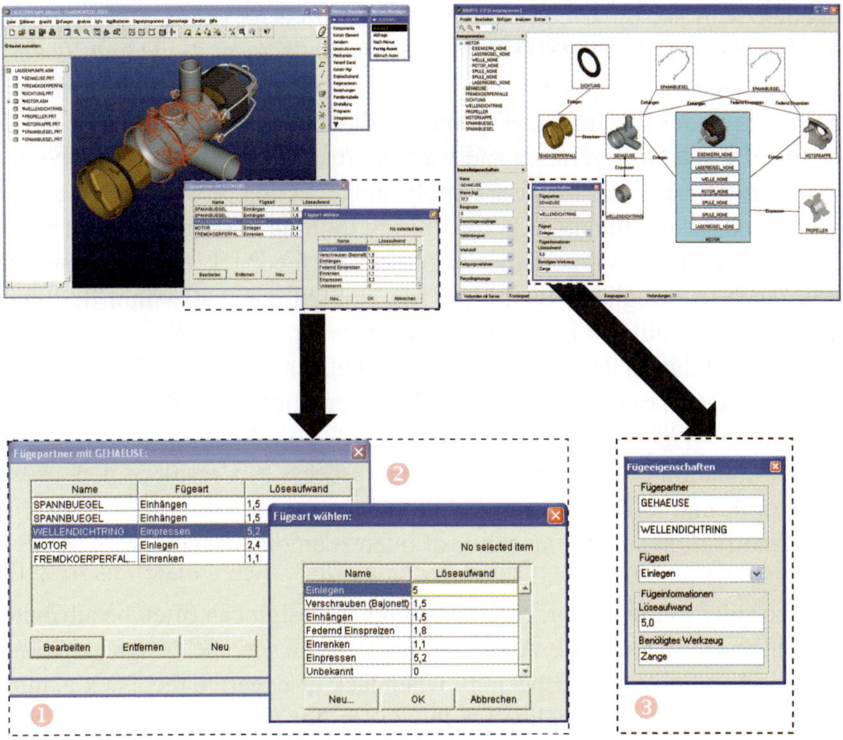

Fig. 5: Real-time interaction between Pro/ENGINEER (top left) and BaMoS 2 (top right)

4.1.3 Feedback of Design-relevant Information

Unlike the recovery of raw materials and parts, which has been the main scope of industrial disassembly by now, the retrieval of information from products at the end of their life cycle is an option that has not been widely considered yet. This information could prove to be a key factor in the development and optimization of sustainable products.

Industry has already picked up this issue. IBM, for instance, operates nine major "Asset Recovery Centers" in six countries over the world that are used as a source of disassembly and recycling related knowledge for product development. Channels for sharing expertise include an annual "Product Stewardship Workshop" where design engineers meet with recyclers. Another example is the 2001 joint venture of Sharp Corporation together with six other companies, who set up a recycling facility with a processing capacity of 600,000 units per year. "Kansai Recycling Systems Co. Ltd." does not only serve as a facility that recycles four home appliance categories under the Japanese Home Appliance Recycling Law (air conditioners, TV sets, refrigerators and washing machines) but is also thought of as a competence centre for disassembly and recycling providing feedback to development departments, e.g. by offering training to designers.

These examples suggest the usefulness of feedback from disassembly experience to designers. Busby [8] has pointed out that within designing organizations feedback in general contributes to:

- the accumulation and retention of knowledge among designers,
- the adaptation of design goals and design practices to a changing environment,
- the evaluation of changes to the design process as a result of new practices or design tools,
- the motivation and maintenance of interest among designers.

According to the author, certain failures in design can often be attributed to problems with feedback. Assumed success with specific design guidelines can breed complacency, especially when lack of negative feedback is interpreted as success. A guideline for the design for reconditioning and reuse, for instance, includes instructions such as "enable cleaning without damaging", "limit wear to specific, easily adjustable or replaceable elements" or "facilitate the assessment of wear" [2].

The difficulty, as with any "Design for X", is that designers are rarely directly involved in any activity that the "X" represents (e.g. disassembly). Those who are involved (e.g. recyclers) might be able to point out that there is a design flaw (e.g. a product being difficult to disassemble), but usually cannot judge if it is caused by

- not following a design guideline,
- inappropriately following a design guideline,
- appropriately following a flawed design guideline, or
- other factors, such as incorrect production or damage during transport or use.

Complicating the abovementioned problem are the temporal, geographical, and organizational distances between development and disassembly and recycling processes (Figure 6).

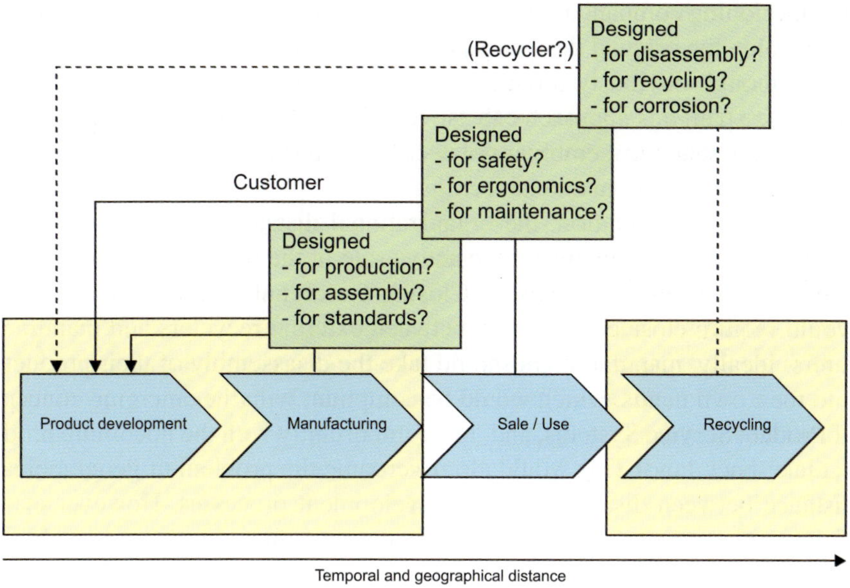

Fig 6: Possible feedback on applied design guidelines at different stages of the product life cycle. The yellow boxes denote organizational boundaries.

Depending on the life span of the product, several years might lie between development and disassembly in quantities necessary for the establishment of an experience base from which useful information can be fed back. In a global economy, it is quite common for new products to travel thousands of miles from manufacturer to customer. Still, in most cases, returning the used products to the origin of their making would render their disassembly highly uneconomical. Consequently, most scenarios involve few centralized manufacturing sites that are outnumbered by many (possibly subcontracted) decentralised disassembly facilities [9] – multiplying the organisational gap.

While the retrieval of feedback can be relatively problematic, it is far from sufficient. The feedback needs to be interpreted and documented to eventually be turned into design knowledge. Just about the most important step, however, is to decide what information is relevant for either the design of a specific product or product design in general. Obviously, the most competent group of persons to define this relevance are the design engineers.

In conclusion, feedback of design-relevant information from industrial disassembly and recycling processes can be a method to support continuous product optimization although the long-term nature of this approach has to be recognized. The temporal distance between the development and the disassembly of a product can be several decades – a period of time after which the developing company might make other products, use other technologies, not employ the original designers anymore, or might not even exist any longer. Although obviously more practical for products with a comparatively short life-span, this approach can as well be useful for long-living products, considering that disassembly is not necessarily an end-of-life option but also part of, for example, maintenance.

In the current situation, the organizational distance could be shortened by assigning the designers a more active role in the retrieval of information from disassembly and recycling. Closing the organizational gap, however, would require close co-operation between external recyclers and manufacturers. Ideally, manufacturers should take the disassembly of their products into their own hands, which would be compliant with the emerging concept of Product Service Systems, and, more importantly with the upcoming product take-back laws. This would also overcome the problem of geographical distance between disassembly and development processes. However even if the used products are not returned to the manufacturer, the geographical distance can be easily overcome since it is information that has to be exchanged. For that purpose, IT-based tools need to be developed or existing ones, e.g. Product Data Management (PDM) systems, adapted. However, the approach discussed here is in any case meant to supplement, not replace current practice.

References

1 Brandt, R.: Demontageorientierte Produktentwicklung – Ein Beitrag zur Gestaltung umweltgerechter Produkte. Dissertation, TU Berlin, 2000.

2 Pahl, G.; Beitz, W.: Engineering Design – A Systematic Approach. 2nd Ed., Springer-Verlag, Berlin, Heidelberg, New York 1996.

3 VDI guideline 2243: Recycling-oriented Product Development. VDI-Verlag, Düsseldorf 2002.

4 VDI guideline 2225: Technisch-wirtschaftliches Konstruieren. VDI-Verlag, Düsseldorf 1977.

5 Dose, J.; Gries, B.; Fleischer, G.; Blessing, L.: Entwicklung nachhaltiger Produkte unter ökbilanziellen Gesichtspunkten. In: Proceedings of the 14th Symposium "Design for X", Erlangen, 2003.

6 Gries, B.; Blessing, L.: A Software Based System to Support the Design for Disassembly. In: Proceedings of the International Design Conference "Design 2004", Dubrovnik, 2004.

7 Radtke, M.: Bamos-Baustruktureditor. DIN-Loseblattwerk "Umweltgerechte Produktentwicklung". Beuth-Verlag, Berlin 2000.

8 Busby, J. S.: Why Designers don't Learn Effectively from Feedback. In: Proceedings of the 11th International Conference on Engineering Design – ICED 1997, Tampere, 1997.

9 Brandt, R.: Demontagefabriken – Schlüssel zum nachhaltigen Wirtschaften. VDI Technik und Gesellschaft – Schwerpunkt Umwelt, Vol. 2, 1998.

4.2 Connections

Lucienne Blessing, Jan Klett, Berlin, Germany

As mentioned in the chapter before, beside the materials and the product structure, connections determine the disassembly properties of a product. Disconnecting-supporting connections still exist but are rarely used. The outcome of a survey about the utilization of disconnecting-supporting connections and of the experiments to investigate the static and dynamic strength of a specific disconnecting-supporting fastener (quarter-turn fastener) is described. To link the (dis)connecting properties with the connection design parameters, the (dis)connecting process is classified in subprocesses. These subprocesses then are allocated to the connection's design parameters. The combination of design parameters in turn results in connection principles. To focus on the disconnecting topic, disconnecting methods and disconnecting supporting connections of different connection principles are regarded and their design parameters are analyzed. To support the design of disconnecting supporting connections in the early design phases, a scheme is developed which links the connection design parameters with the necessary disconnecting procedures and allows a combination according to a desired disconnecting process.

4.2.1 Disconnecting-Supporting Connections

If connections are designed such that they support disconnecting, they are defined as disconnecting-supporting connections. The large number of existing disconnecting-supporting connections and the lack of fundamental knowledge and quantitative properties which are necessary for dimensioning and selecting connections complicate the application of these connections in the design process [1]. An investigation of German standards and guidelines [2] confirmed that no real standardisation of disconnecting-supporting connections exists. Only geometrical data is available, provided by the manufacturer. Values of endurable stresses or the necessary (dis)connecting forces and torques are rarely given. The authors assume that because of these circumstances designers are not familiar with disconnecting-supporting connections. To test this hypothesis, a postal survey was carried out within German manufacturing industry [3].

Utilization

The reasons for designers not using disconnecting-supporting connections were found to be the following (n = 196, boxes with the predefined reasons are to be ticked, multiple answers possible): missing demand (75 %), missing familiarity (41.8 %), unavailable information (38.3 %), nonexistent knowledge (31.6 %), inadequate strength characteristics (15.8 %), high costs (15.3 %), inadequate safety (14.3 %), nonexistent standardisation (8.2 %), inadequate disassembly characteristics (7.7 %) and miscellaneous (5.1 %). When asked about the criteria that were used for the selection of connections, the answers were as follows (n = 196, boxes with the predefined reasons are to be ticked, multiple answers possible): endurable stress (94.9 %), function (87.2 %), assembly (81.1 %), disassembly (79.6 %), material (79.1 %), geometry (78.6 %), costs (77.6 %), knowledge (67.3 %), recycling (25 %) and miscellaneous (1.5 %).

From the outcome it was concluded that the fulfilment of requirements under utilization of disconnecting-supporting connections seems hardly practicable because of the limited access to relevant data such as the above mentioned values of endurable stresses and fears of inadequate strength characteristics.

To investigate missing data and to check if fears of inadequate strength are justified, some experimental investigations with quarter turn fasteners were undertaken and comparison with standard bolt connections was drawn. Further investigation is done to increase the demand and familiarity with disconnecting-supporting connections.

Strength Characteristics of Quarter-Turn Fasteners

This section describes some of the findings of the investigated experiments. For more details see [2]. The main criteria for the selection of connections are the endurable stresses. As one representative for a disconnecting-supporting connection, the quarter-turn fastener connection was tested in three design variants (Figure 1):

- *Variant 1*:
  The spring element is separate (not integrated in the receptacle) and the lead is integrated in the receptacle.
- *Variant 2*:
  The spring element is integrated in the receptacle; the lead is integrated in the stud.
- *Variant 3*:
  Both the spring element and the lead are integrated in the receptacle.

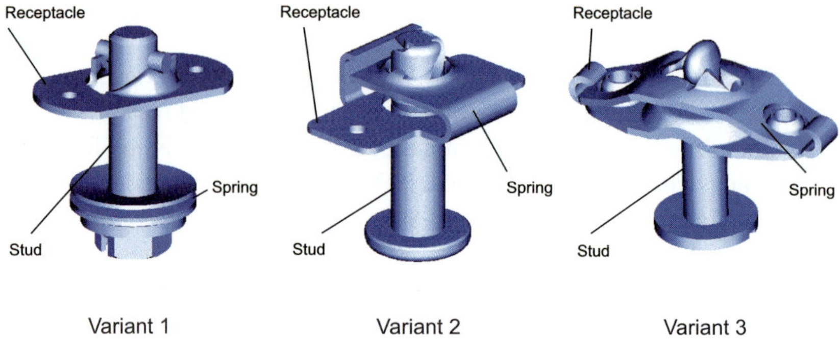

Fig. 1: Variants of quarter-turn fasteners

Static and dynamic tension testing was performed. Typical force-elongation graphs resulting from the static tension test are shown in Figure 2. Types 1 to 3 correspond to the variants 1 to 3, types 4 to 6 to variant 1. To enable comparison with frequently utilised connections, the graphs of standard bolt connections in different sizes (M4, M6, and M8) are added.

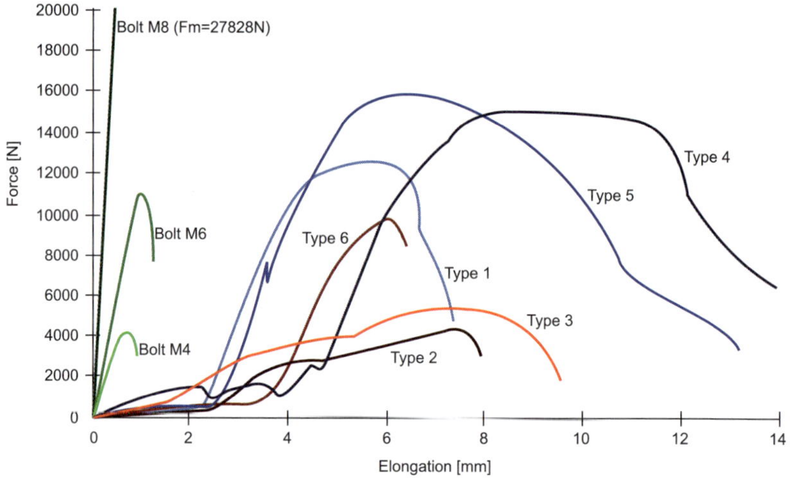

Fig. 2: Typical force elongation graphs of the tested quarter-turn fastener types and of standard bolt connections in different sizes (M4, M6, and M8)

Inspection of the tested connections shows that the weak point of most of the tested quarter-turn fasteners regarding static and dynamic load is the receptacle. Analysing the cause of damage showed that the strength of most of the tested quarter-turn fasteners can be increased through realising the design principles of uniform strength and direct and short force transmission [4]. Common quarter turn fasteners are mostly used for uncritical connections with low stresses. However, if designed accordingly quarter-turn fasteners can be utilised also for critical connections which is a prerequisite for increasing their utilization and in turn a step towards shorter disconnecting times compared with bolt connections.

4.2.2 Parameters Determining (Dis)connecting

The demand and the familiarity with disconnecting-supporting connections can be increased through developing them systematically according to their requirements. Connections are then specified and will not only be selected from the existing mass of standardized solutions. As a prerequisite for the systematic design of disconnecting-supporting connections, the design parameters which are determining the dis(connecting) process must be investigated. Therefore, the process of connecting is divided in three subprocesses, one for bringing the components together (joining subprocess), one for adding additional components (fastener(s)) which are necessary for connecting (fastening subprocess) and one for restricting the remaining movements between all components (locking subprocess). The disconnecting process is divided accordingly in the unlocking, unfastening and separating subprocesses. Through unlocking, the previously restricted movements are possible again. Through unfastening, the fastener(s) are removed. Through separating, the components are detached [5].

The detailed procedures which are necessary for performing the determined subprocesses (locking, unlocking, fastening, unfastening, joining and separating) depend on the connection design parameters for locking, fastening and joining. In the following, the existing kinds of these parameters are defined and classified and the necessary procedures of their subprocesses are described.

Kinds of Locking / Locking and Unlocking Procedures

In literature locking is described and classified differently as the following references show. LeBacq et al. [6] differentiate between mechanical fastening, welding and adhesives, but do not consider soldering and brazing. Brandon and Kaplan differentiate between mechanical, chemical and physical locking [7]. These categories are overlapping because mechanical locking consists of physical effects. The German literature [4, 8, 9] groups connections in material, form, and force connections. Material connections are locked through specific chemical or physical processes, e.g. welding, soldering, brazing, and adhesives. Form connections are locked through geometrical interaction, e.g. snap fits. Force connections are locked by forces. These forces can be transmitted either constantly through the geometry of the components, e.g. bolted joint, or derived from a force field, e.g. magnetic connection. The classification makes no difference between these two sources of the force although there are different procedures necessary for applying or removing these forces. For planning the procedure of the (un)locking subprocess a differentiation between the force origins has to be made. If the force is applied through geometry then the (un)locking procedure is similar to that of form connections. If the force derives from a force field the (un)locking procedure is different from that of form connections. To account for this, the following classification of kinds of locking has been developed [10]:

Force field locking (ff): the existence of a force field (magnetism, gravitation) results in the attraction of the components. For locking, the components are exposed to a force field and a force field is activated, respectively. For unlocking, the components have to be removed and the force field has to be deactivated, respectively.

Material locking (mat): chemical and physical reaction at the components' interfaces results in adhesion or cohesion. For locking, the interfaces have to be treated according to the utilized technology (welding, soldering, brazing, gluing). Furthermore the additional necessary materials, e.g. solder, braze, and adhesive, have to be distributed and processed. Depending on the utilized technology, the unlocking process can be realised by heating or cutting the connection area or by supplying specific substances, e.g. acid.

Mechanical locking (mech): geometrical interaction restricts the components' movements. The following three cases exist: The movements are restricted through: A. geometrical interaction only, B. normal force which is applied through geometry and which results in friction force, C. both, geometrical interaction and normal force which results in friction force. For (un)locking, the components have to be moved or deformed.

Fastening and Unfastening Procedures (Only with Mechanical Locking)

Only at mechanical locking the need for additional components (fasteners) like bolts and nuts can emerge. The additional necessary procedures caused by fasteners (distribution, removal) are defined as (un)fastening subprocess.

Kinds of Joining / Joining and Separating Procedures

The joining and separating procedures depend on the kind of joining. These in turn depend on the geometry of the components' interfaces which are classified in planar, convex, and concave (Figure 3).

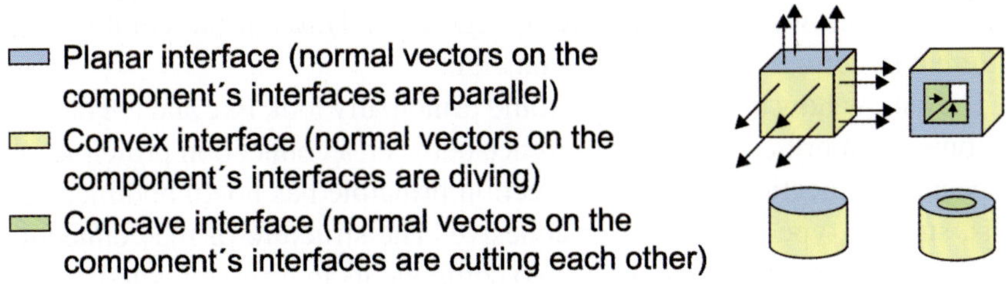

Fig. 3: Classification of interface types

With these above mentioned three interface types two kinds of joining are possible (Figure 4):

- *Planar joining (p):* The components can be joined and separated in five different translatory directions of the 3D orthogonal system. For joining, the movements 1-5 (Figure 4A) are possible, for separating the movements 2-6 (Figure 4B).
- *Convex/concave joining (cc)*: The components can be joined and separated in one up to four translatory directions depending on the interface geometry. Figure 4C-F shows the separating directions.

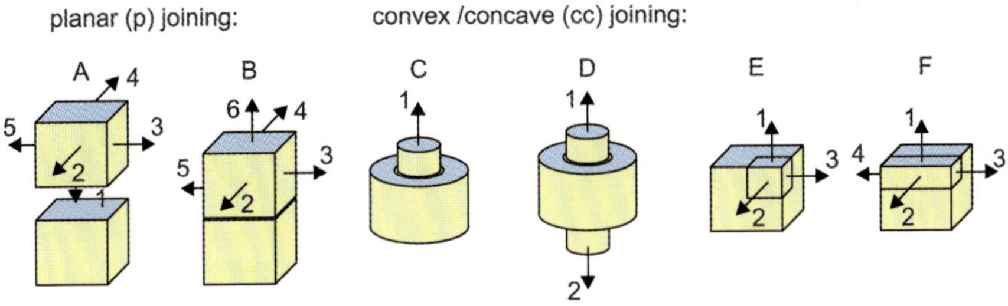

Fig. 4: Kinds of joining, joining and separating procedures

With planar joining, where only one movement is restricted, the separation of the components mostly emerges without additional effort so that a separating subprocess is not necessary and can be neglected. The more movement directions are restricted at convex/concave joining the more a separating process is necessary.

4.2.3 Classification Scheme of Connection Principles

The variety of possible connections results from combining the investigated connection design parameters for joining and locking. To visualize the variety of connections some, through not all, combinations are displayed in the classification schemes (Figure 5, Figure 7). Through the combination, different structures of connections emerge.

They show how the connection could principally look like and are defined as connection principles. The difference between a connection principle and a specific connection is that a connection principle has no embodiment details, such as type of fastener or adhesive. The structure of the connection principles is shown as sections of two components in different colours (blue and green) with the working surfaces of the locking displayed in red. If the locking is initiated through fastener(s) then these are presented in red.

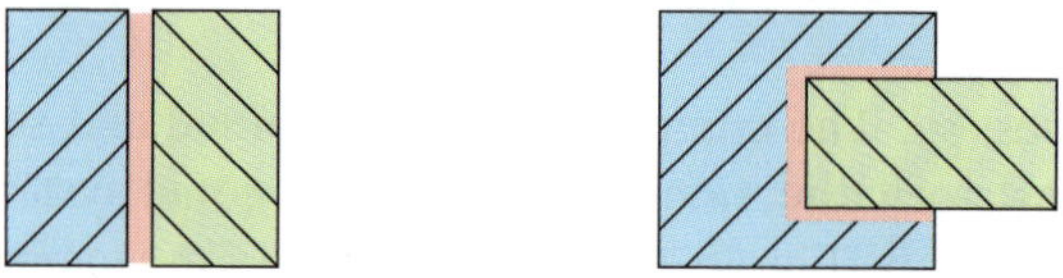

Fig. 5: Classification scheme of connection principles with force field and material locking (blue and green: components to be connected, red: working surfaces of the locking)

Force field locking (ff) and material locking (mat) directly affect the components' interfaces and each result in two different connection principles (Figure 5).

p-mech-		Joining/separating: in 5 different translatory directions
1		Locking is initiated at the interfaces. A fastener is not necessary. Example: suction cup, velcro fastener
2	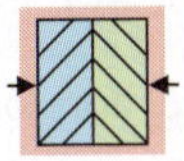	Locking is initiated at the surfaces which are parallel to the interfaces and which are situated outside of the components' contour. The fastener entangles the components completely. Example: cable tie
3	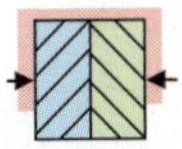	Locking is initiated at the surfaces which are parallel to the interfaces and which are situated outside of the components' contour. The fastener entangles the components partly. Example: vice
4	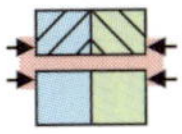	Locking is initiated at the outside surfaces which are parallel to the interfaces and which are situated outside of the components' contour. A notch in the components leads the fastener through the components. Example: fast clamper
5	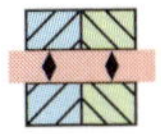	Locking is initiated at the surfaces which are perpendicular to the interfaces and which are situated inside the components' contour. The fastener penetrates the components through a hole. Example: press fit
6	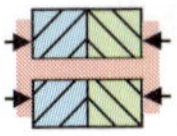	Locking is initiated at the surfaces which are parallel to the interfaces and which are situated outside of the components' contour. The fastener penetrates the components through a hole. Example: bolted connection

Fig. 6: Classification scheme of connection principles with planar joining and mechanical locking; (blue and green: components to be connected, red: working surfaces of the locking or fastener, applied normal force is displayed as black arrow)

Mechanical locking is initiated mostly outside of the components' interfaces and results in a higher number of connection principles (Figure 6, Figure 7). The different connection principles are categorised according to their kind of joining (p = planar, cc = convex/concave). Six planar (p-mech-1 – 6) and seven convex/concave (cc-mech-1 – 7) connection principles are shown. Beside the structure also the necessity of additional components (fasteners) is shown schematically. Some examples of specific connections are given for each connection principle. Depending on the need, the schemes may be extended.

cc-mech-		**Joining/separating in 1 up to 4 different translatory directions**
1	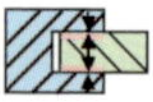	Locking is initiated at the horizontal interface(s). A fastener is not necessary. Example: press fit
2	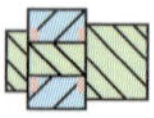	Locking is initiated at the vertical interfaces. A fastener is not necessary. Example: snap fit
3	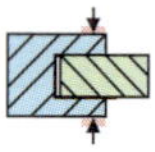	Locking is initiated through one component via the horizontal surface(s) which is situated outside of the components' contour. The fastener entangles the components completely. Example: hose clamp
4	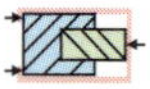	Locking is initiated at the vertical surfaces which are situated outside of the components' contour. The fastener entangles the components partly. Example: draw latch
5	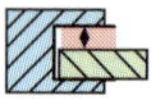	At one component the locking is initiated at a horizontal surface which is situated inside and at the other component at a horizontal surface which is situated outside of the components' contour. The fastener is positioned between the components. Example: wedge
6	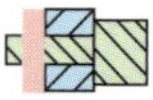	At one component the locking is initiated at a vertical surface which is situated inside and at the other component at a vertical surface which is situated outside of the components' contour. The fastener penetrates one comoponent. Example: splint pin
7	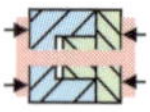	Locking is initiated at the vertical surfaces which are situated outside of the components' contour. The fastener penetrates the components. Example: bolted joint

Fig. 7: Classification scheme of connection principles with convex/concave joining and mechanical locking (explananation of sketches see Figure 6)

Disconnecting Characteristics of Connections

In most cases disconnecting is performed through making a physical contact with the connection. However, having disconnecting processes which do not need a physical contact supports to treat connections which are difficult to reach. The following disconnecting methods can be distinguished [10]:

Contacted disconnecting (manual or automated): For both, the manual and the automated disconnecting, the number of tools should be as low as possible. While complex kinematics can easily be performed manually, they are difficult to perform automated. While high disconnecting forces can easily be applied in automated processes it is difficult to apply them manually. If disconnecting should be performed automated, the procedure and the position where to perform must be known and the states unlocked and unfastened must clearly be identified by the robot. If different products should be disconnected automated, the products must be identified before and the tools have to be controlled.

Contactless disconnecting: The unlocking sub-process is activated without physical contact with the connection. Therefore a trigger is necessary which unlocks the connection, e.g. light or heat.

Some disconnecting-supporting connections are mentioned and allocated to their connection principle. Their support for the disconnecting process is described.

p-mat: With magnetic particles embedded into an adhesive, contactless unlocking can be realized through an electromagnetic field. Thereby a minimum amount of heat is transferred to the components [11].

p-mech-4: With the fast clamper (Figure 8A), unlocking is performed through turning a lever which enlarges the distance “a” and thus reduces the normal and resulting friction force. The fastener can be unfastened in one part because it does not penetrate the components but is lead in a components’ notch.

p-mech-6: With the quarter-turn fastener (Figure 8B), unlocking is realised through rotating the stud only about 90° related to the receptacle. With the removable bolt head (Figure 8C), unlocking is realised through moving the bolt head in translatory direction away from the bolt shaft. The bolt head consists of different components which together form a mechanism. If the cap of the bolt head is lifted the mechanism releases the geometrical locking between the shaft and the head of the bolt. With the “Schnelle Mutter” (fast nut) (Giehl Systems, Figure 8D) unlocking is realised through moving the nut in translatory direction apart from the bolt [12]. With the segmented bolt head (CRC 281 development, Figure 8E) unlocking is realised contactless through electric current. The segments of the bolt head are placed in a groove which runs around the end of the bolt shaft and are held in position by a plastic band. In this plastic band, a heating wire is integrated. By leading the electric current through the wire, heat is generated; the plastic band melts and the connection is unlocked.

cc-mech-2: With the connection system for hoses (Figure 8F), unlocking and separating can be performed through a single translatory movement. With shape memory material (triggered through temperature), unlocking is performed through heat due to geometrical changes of the components [13]. Nokia conducted some disconnecting tests with this technology [14].

cc-mech-3: With the plastic band hose clip (CRC 281 development, Figure 8G) unlocking is realised contact-less through electric current. The plastic band is fitted with a resistant wire. Applying an electric current the plastic band melts and the connection is unlocked [15].

<table>
<tr>
<td rowspan="2">A: p-mech-3/-4: fast clamper with different component geometries (A: accords to p-mech-3, B + C: accord to p-mech-4)
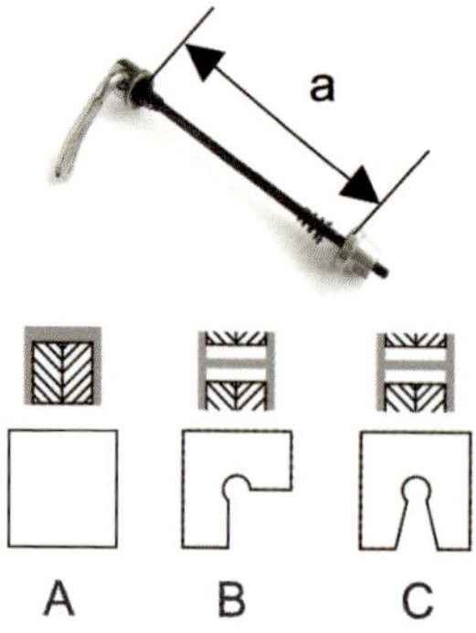
</td>
<td>B: p-mech-6/cc-mech-7: Quarter-turn fastener
</td>
</tr>
<tr>
<td>C: p-mech-6/cc-mech-7: removable bolt head
</td>
</tr>
<tr>
<td>D: p-mech-6/cc-mech-7: “fast nut” of Giehl Systems
</td>
<td>E: p-mech-6/cc-mech-7: unlocking by electric current
</td>
</tr>
<tr>
<td>F: cc-mech-2: connection system for flexible pipes

unlocking and unfastening in one direction</td>
<td>G: cc-mech-3: hose clip with heating wire
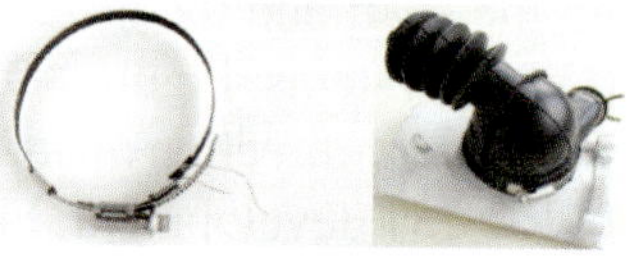</td>
</tr>
</table>

Fig. 8: Disconnecting-supporting connections/fasteners

The examples show that both, the connection principle and the embodiment design of a connection determine the possible disconnecting procedures.

4.2.4 Systematic Design of Connections

To consider disconnecting properties from the early design phases, a scheme (Table 1) was developed which consists of the defined connection design parameters and which describes for each the general procedures of the disconnecting subprocess. Proposals for realising these procedures contacted and contact-less are given. In the column "contactless disconnecting", the constraining issues are mentioned in brackets for highlighting where investigation can lead to advantages for the support of the disconnecting process.

Tab. 1: Scheme for selecting connection design parameters according to the desired disconnecting process

Disc. sub-proc.	Kind of locking and joining	Necessary disconnecting procedure	Contacted disconnecting	Contactless disconnecting
Unlocking	force field (ff)	remove components/overcome force field	move components apart	remove, change, deflect the force field
	material (mat)	undo cohesion/adhesion	heat the locking area; cut the locking area	trigger undoes cohsion/adhesion (locking technologies, materials, triggers)
	mechanical (mech)	change components' geometry; move components	deform; move; cut components	trigger changes components' shape or position; cut components (materials, triggers)
	Unfastening	remove fastener(s)	move fastener(s)	trigger changes fastener(s)' position (materials, triggers)
Separating	planar (p)	Planar joined components do not need to be separated.		
	convex /concave (cc)	move components apart	move components apart	trigger changes components' position (materials, triggers)

Through combining the connection design parameters, all connection principles can be realised and their general disconnecting process can be determined. If a specific disconnecting procedure is required the scheme supports the selection of accordingly suitable design parameters. If, e.g. connections are required which can be disconnected by a single movement, the design parameters have to be selected and combined so that each resulting disconnecting subprocess can be performed through movement. Later, these movements must be unified. The same applies for the contactless disconnecting. If disconnecting should be performed by one trigger then the design parameters are to be selected and combined such that the necessary procedures for the disconnecting subprocesses can be realised with one trigger.

4.2.5 Conclusion

A survey showed that fears concerning the non secure strength characteristics of disconnecting-supporting connections exist. These fears are justified due to experimental testing of quarter-turn fasteners. However, potential exists to reinforce quarter-turn fasteners and to use them also under higher loads as a competitive alternative to conventional fasteners. To achive that disconnecting-supporting connections will be accepted in the future, detailed testing of connections according to standardised test programs should be performed and documented. Also, comparison with common fasteners should be undertaken to provide a reference and application scenarios.

The constant emergence of new disconnecting supporting connections shows that the space of solutions is not open completely yet. To open up this space, connections should be developed systematically according to disconnecting requirements. With the presented schemes, an approach for the design of disconnecting-supporting connections can be derived. The steps of the approach are allocated to the first three phases of the product design process defined by [4]. During planning and clarifying the task, requirements for the disconnecting process have to be defined. During conceptual design, useful kinds of joining and locking have to be selected and combined to connection concepts so, that the disconnecting process requirements can be fulfilled. This is supported by the schemes developed into Figure 5, Figure 6, Figure 7, and Table 1. During embodiment design, the best connection concept has to be selected and adapted to the product.

With the approach, the product and the disconnecting properties, which are determined through the connection, are considered and designed at the same time. Although the whole solution space is regarded, the designer does not have to choose connections from an unmanageable number of possible solutions. Instead, he is lead through a process through which the number of solutions is limited step by step.

References

1 Schmidt-Kretschmer, M.: Untersuchungen an recyclingunterstützenden Bauteilverbindungen. Schriftenreihe Konstruktionstechnik, Nr. 26, Berlin, 1994.

2 Wünsche, T.; Meyer-Eschenbach, A.; Blessing, L.: Testing Connections and Fasteners to Determine Strength Characteristics. In: Proceedings of the International Design Conference – Design 2006, Dubrovnik, Croatia, May 15 – 18, 2006.

3 Wünsche, T.; Meyer-Eschenbach, A.; Blessing, L.: Analysis of Connections and Fasteners to Determine Disassembly and Strength Characteristics. In: Proceedings of the 15th International Conference on Engineering Design, Melbourne/Australia, August 15 – 18, 2005.

4 Pahl, G.; Beitz, W.: Engineering Design – A Systematic Approach. 2nd Ed., Springer-Verlag, Berlin, Heidelberg, New York 1996.

5 Klett, J.; Wünsche, T.; Blessing, L.: Process-oriented View of Connections. In: Proceedings of the 2002 International CIRP Design Seminar, Hong Kong/China, May 16 – 18, 2002.

6 LeBacq, C.; Brechet, Y.; Sherckliff, H.R.; Jeggy, T.; Salvo, L.: Selection of Joining Methods in Mechanical Design. Materials and Design 23 (2002), pp. 405-416.

7 Brandon, D. G.; Kaplan, D.: Joining Processes – An Introduction. John Wiley & sons Ltd. Chichester 1997.

8 Roth, K.: Konstruieren mit Konstruktionskatalogen, Band 3, Verbindungen und Verschlüsse. Springer-Verlag, Berlin, Heidelberg, New York 1996.

9 VDI 2232: Methodische Auswahl fester Verbindungen. Beuth-Verlag, Berlin 2004.

10 Klett, J.; Blessing, L.: Systematic Design of Connections. In: Proceedings of the 13th CIRP International Conference on Life Cycle Engineering, Leuven 2006, pp. 417.

11 Pridöhl, M.: Zimmermann, G.; Hartwig, A.; Lühring, A.: Kleben mit Magnetfeld. Zeitschrift Automobil Produktion, Landsberg, Februar 2005.

12 Giehl Systems: http://www.schnellemutter.de (Access: November 2006).

13 Chiodo, J.D.; Jones, N.; Billet, E.H.; Harrison, D.J.: Shape Memory Alloy Actuator for Active Disassembly Using 'smart' Materials of Consumers Electronic Products. Materials and Design 23 (2002), pp. 471-478.

14 Tanskanen, P.; Takala, R.: Build-in Disassembly Mechanisms for Portable Electronics. Konstruktion (2003) 1/2, pp. 57-60.

15 Sfb 281: Ergebnisbericht 2001-2003, Teilrojekt D1. TU Berlin, 2003.

4.3 Simulation Tools for Disassembly Design

Frank-Lothar Krause, Andreas Romahn, Uwe Rothenburg, Berlin, Germany

Today's product development has to regard an increasing number of requirements that follow from customer needs and restrictive directives of governments as for example the Waste Electrical and Electronic Equipment Directive (WEEE) [1]. One important subject is disassembly-friendly product design. During a product's life processes such as servicing, maintenance and recycling of materials and product components depend on their operating efficiency and, consequently, their cost effectiveness.

In the last years the result of virtualizing the development process resulted in a Digital Mock-Up (DMU) which is defined as "a computer-internal model for spatial and functional analysis of the structure of a product model, its assembly und parts respectively" [5]. The DMU aims to optimize the development process by evaluating a product's virtual prototype and by avoiding cost intensive Physical Mock-Ups (PMU). Current DMU implementations focus on packaging problems and are mainly based on geometries. This approach is insufficient for the simulation of disassembly of a product in use because it does not regard the fact that the component properties change during its life. Product use and its influence on form and function require the simulation of usage in order to investigate their impact on the product and therefore on the disassembly process. This calls for an extension of the DMU, named Functional Mock-Up (FMU) which additionally facilitates the digital representation and investigation, compared to the geometric based DMU. The main objective of the FMU is the analysis and test of a product's function and behavior [4]. Thereby, the availability of appropriate product data models plays a crucial role for the applicability of computer aided methods in product development tasks [8]. Furthermore the increasing diversity of applied tools leads to a fast-growing number of data exchange processes which are time consuming and have a possible negative impact on data integrity. Thus, the development of methods and tools supporting a complete and lossless exchange of product data is of increasing significance [3].

The objective of simulation tools for disassembly-friendly product design is to define modeling methods that provide the representation of product states by integrating the spatial description of the product components, its geometry, physical connections and the impact of usage processes on the properties of the components and their connections. Therefore, a graph-based approach was developed that provides the representation product states including the changes on the product properties. This generic graph structure flexibly expresses the properties of components, the physical connections between the parts, the dependencies between product states, applied wear process simulations and also the modeled and simulated disassembly sequences. The result is a strongly connected structure that tightly couples product and process related information [6].

Product states represented in this way are used to automatically generate feasible disassembly processes and sequences which are used in a Virtual Reality based disassembly simulation. The outcome is information about the disassemble ability of components based on the solvability of connections and the accessibility of components which altogether represent the disassembly friendliness.

4.3.1 Product Condition Model

The simulation based product condition model represents the properties of product components at certain points in time. Each product state is given by the component properties that refer to the same point in time so that those properties together represent a virtual "snapshot" of the product.

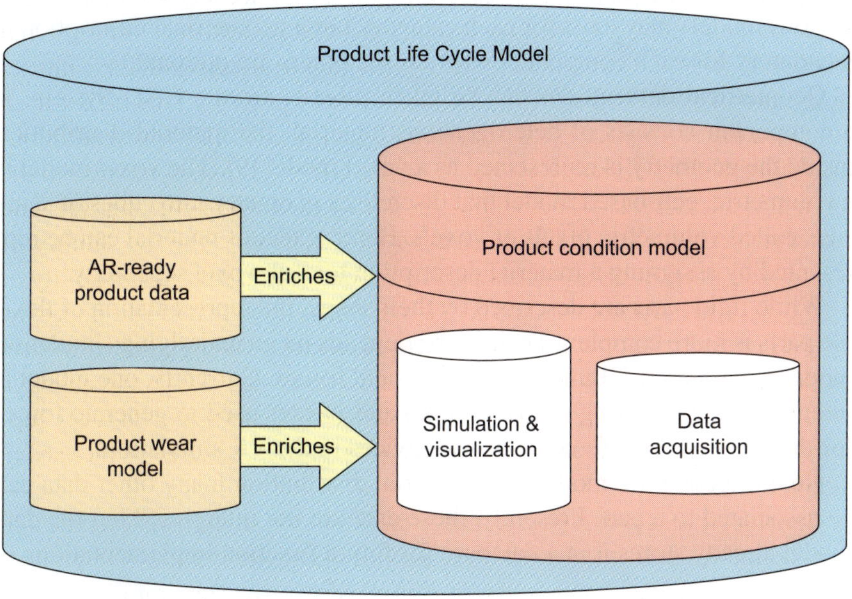

Fig. 1: Product condition model

The focus of the project "Disassembly Oriented Information Technological Infrastructure" is on the acquisition of information about the condition of a real product whereas this project aims to simulate and visualize a product in explicitly modeled states. In the first case information about the impact of usage are tracked and stored whereas in the second case the virtual product condition is created by modeling and simulation (Figure 1).

In the product condition model product states are modeled as a graph structure that contains the physical parts as nodes and the physical connection between them as links between the nodes. Both, nodes and links, are abstract representatives whose ascertained properties are stored in associated attributes. In order to extend the system by additional properties, a generic attribute model was developed that provides a flexible assignment of attributes to graph elements. Properties can be of different types like geometry, the spatial transformation, material or weight. All properties together describe a component from different views and aim to provide a virtual description of as many aspects of a real component as possible in order to enrich the virtual product description and to support the downstream disassembly simulation.

The different representations of a component can be separated into the four categories geometry, material distribution, simulation models and as the last group all other kinds of descriptions like center of gravity or weight. Various models may exist for each category but a geometrical description is mandatory for each component whereas the others are optional.

Geometrical descriptions can be taken directly from a CAD-System. If a component consists of heterogeneous material, the material distribution inside the geometry is represented as a voxel model [9]. The voxel model is a volumetric, cell-based model that discretises geometry into cubes of same size, called volumetric pixels or voxels. Heterogeneous material can be represented by assigning a material description to each voxel separately.

While rigid parts are described by their shape, the representation of flexible parts is more complex. Their shape depends on an underlying simulation model that regards parameters like time and forces. Currently, one model is integrated. It is a spring damper model and is later used to generate forces for the integration of force feedback devices in the VR simulation.

Next to geometric models and material distribution many other data can be associated to a part. Presently, these data are not interpreted but the data model already stores it in a database for future function implementations.

Modeling of Product Wear

Product usage changes the properties of product components depending on the time passed, the environment it operates in and how skilled and intensively it is used. These influencing factors referring to the impact of usage are summed up in the following in the term usage profile. The developed system models the product usage and its effect on the product properties by coupling a product state with processes that express different effects of usage and are called wear processes. By applying wear processes to a product state the outcome is a new product state.

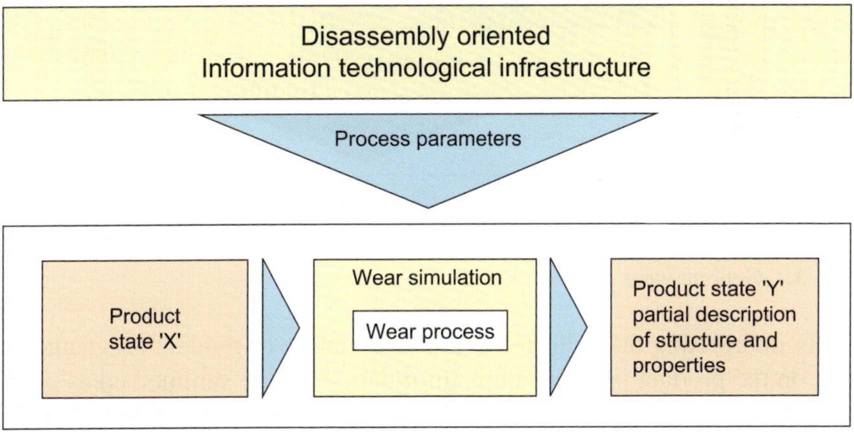

Fig. 2: Product state modeling

All wear processes together are called wear simulation (Figure 2) and include the modeled impact of a usage profile on a product. In order to provide the investigation of certain effects and components wear processes can be selectively applied to a subset of components instead of generating a complete new product state. The wear processes of the wear simulation model are implemented as functions or simulations whose parameters are either explicitly modeled (Figure 3) or are imported from the product condition model that was developed in the project “Disassembly Oriented Information Technological Infrastructure”. This product condition model stores information about the properties of a product in use. The information is either collected automatically with sensor devices or is entered manually by service workers.

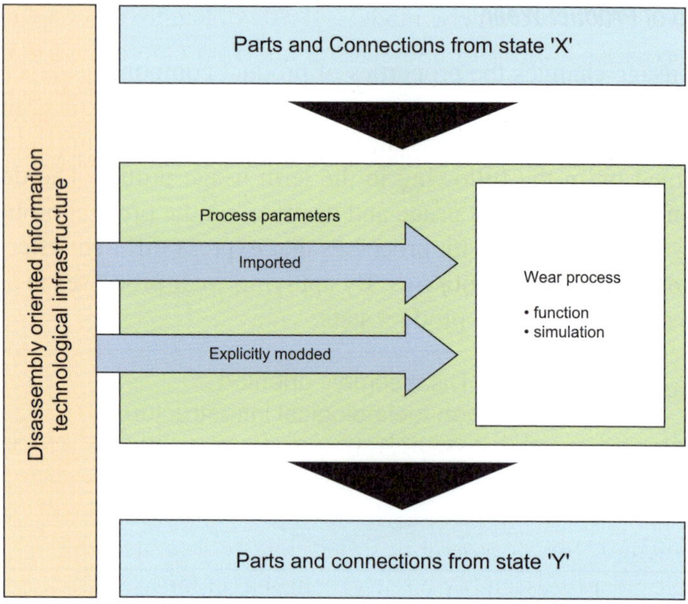

Fig. 3: Generic wear process model

By interpreting the collected data as a coupling of product usage and effect on the product properties this information can be summed up as usage profiles. Usage profiles can be used to configure process parameters in the product usage modeling stage. The generic wear process model (Figure 3), as currently implemented, is providing processes for simulating change of shape, change of material distribution and contamination (Figure 4).

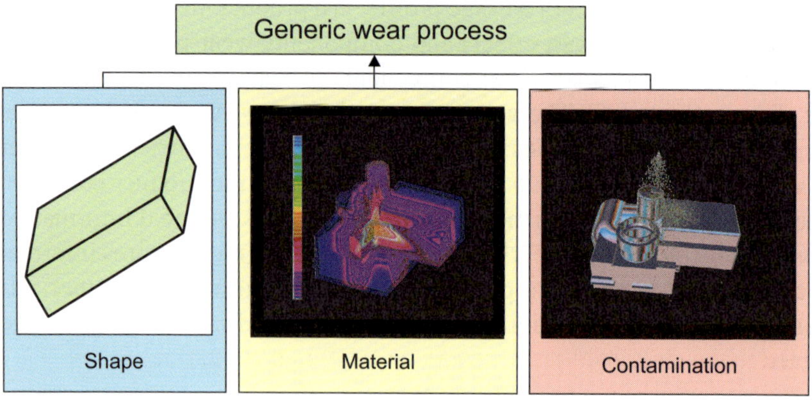

Fig. 4: Wear processes

An advantage of the generic model is the flexible integration of additional wear processes. In order to integrate results from external simulations, as they may come from commercial systems, the modeled wear process represents a document that holds information about parameters to be applied, which elements are to be regarded in a simulation and where to write the result in the targeted product state.

4.3.2 Virtual Disassembly

Sequence Modeling

The developed virtual disassembly system uses the modeled product states to automatically compute all disassembly processes that can be derived from the underlying contact graph expressed in the previously described graph structure. For disassembly simulation the product state structure is extendable by a disassembly environment that comprises all tools that are used during a disassembly process.

The physical contacts between the environment and the product affect the possible disassembly work orders. Thus the contacts have impact on the generated process graph that is the superposition of all possible sequences (Figure 5). The process graph generation algorithm takes the geometry of each component into account which includes the product components as well as the disassembly environment.

The modeling of disassembly sequences is performed by choosing a path of successive processes in the graph. Each sequence represents a feasible work order to disassemble components. In order to evaluate the cost of a disassembly process, two different types of evaluation functions can be applied. The first type assigns disassembly costs to a process based on various functions. An example is assignment of the effort for cleaning processes that are currently not available in an interactive simulation. The second type of evaluation function is a VR-based disassembly simulation that provides a VR-interface in order to perform a disassembly simulation based on the properties of the product and the disassembly environment. The result of the simulation, like the motion paths of the geometries and used tools, are assigned to the simulated process. Part and tool movements can be replayed in an animation or can be used for further evaluations.

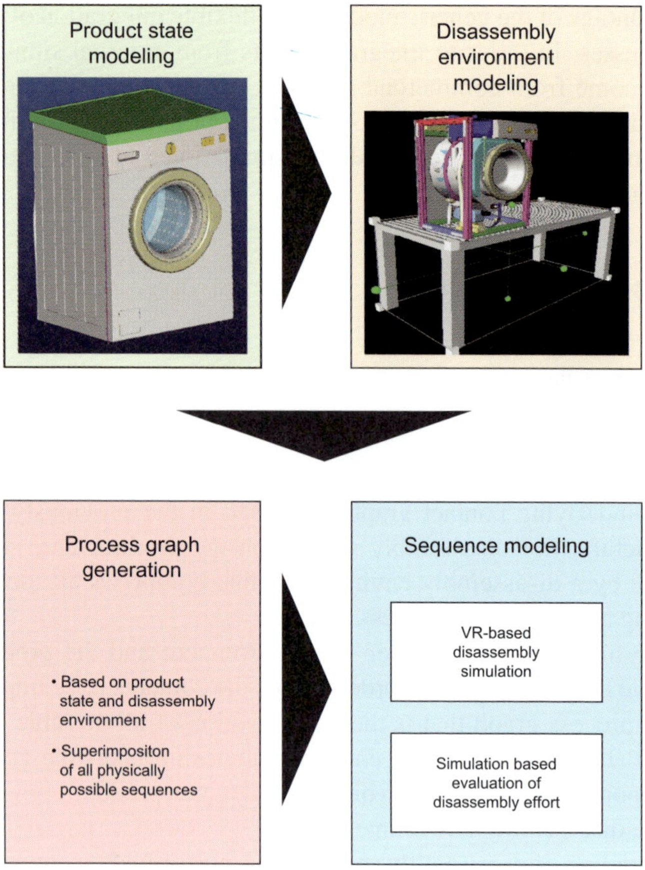

Fig. 5: Sequence modeling

Simulation

The disassembly system provides the VR-based simulation of the previously described generated processes. Two different VR simulations are available and it depends on the process task which one is to be used.

The first simulation type provides functions to disassemble rigid components from an assembly. The user can interact with the scene with the available integrated input devices, as for example mouse and space mouse. Thereby does the process define the elements of the virtual scene like the assembly, the tooling environment, the component to disassemble and which tool to use. Each of these elements has a geometric representation. The geometry of the assembly and the component that has to be removed reflects the current product state that is stored the product condition model. The movements performed with the input devices directly move the components to be disassembled where the geometric shape and collision detection functions restrict the feasible transformations. All movements are recorded and stored as a result of simulation.

The second type of VR simulation focuses on representing flexible parts. As an example the disassembly of flexible hoses was integrated into the system. In case of flexible parts a certain simulation model is required. Because of the enormous computational requirements of the model even for simulating one flexible part makes it not feasible to provide a combined simulation model for flexible and rigid parts. The task has an assigned description of the properties of the flexible part that is required by the simulation model. The VR-simulation integrates force-feedback devices that act as a three dimensional input device and can display forces that are generated in the simulation.

The results of both simulations are stored in the processes and are then available for evaluation functions. In the first case the motion path of the geometries that was recorded during the simulation is stored and in the second case the required forces to disassembly the hose are stored to the disassembly process.

4.3.3 Enhancement of Service and Disposal

The information representing a product state and the simulated disassembly sequences created during the product development can be re-used and enriched for downstream processes like maintenance, repair, dismantling and training. For that purpose two augmented reality applications have been developed that support those tasks by providing computer-generated information in order to support those tasks and improve their execution performance.

Augmented Reality

The Term “Augmented Reality” (AR) entitles the research area of a new form of human-machine interaction. Augmenting the reality means to enrich it with artificially generated information that is context sensitive and can be noticed with the human senses. Today’s typical systems focus on the visualization of virtual objects in the user’s field of view of the real environment through a Head Mounted Display (HMD) [2].

Augmented Reality encompasses a part of the Reality-Virtuality Continuum [10] that is contained in the more general concept called Mixed Reality (MR). This implicates that virtual objects should behave with respect to physical laws to act as it is expected from objects of the real environment. The integration of virtual objects in the real environment requires the update of their representation according to the user's point of view.

Augmented Reality Environment

The MARE Augmented Reality Environment (MARE) was developed as a framework for augmented reality applications by providing implementations for required fundamental functions. Those functions are organized in the five modules tracking, visualization, Graphical User Interface (GUI), interaction and browser. The tracking module provides functions to estimate the spectator's pose with a marker-based approach, which is realized by the integration of the ARToolKit [11], which was developed at the Human Interface Technology Laboratory (HITLab) at the University of Washington. With the pose virtual geometries can be integrated into the real world. Input functions for AR-scenes are provided by virtual interaction components that have been developed and put into the module GUI. Those components are themselves virtual objects and can be placed in a virtual scene in order to design two and three dimensional user interfaces. The interaction module provides interaction functions to activate the GUI components. Finally, the browser module provides a component that facilitates displaying Hypertext Markup Language (HTML) pages which is useful for many applications.

Application for Maintenance, Repair, Dismantling and Training

Product life cycle tasks like maintenance, repair, dismantling and training benefit from task-supporting information that is supplied by an AR-system. The increasing variety of different products and a growing variety of product variants and shorter product lifetimes add up to more often changing work content. Further benefits follow for the performance of complex tasks that require extensive quantity of information. Using AR helps to accomplish task execution more efficiently and to employ workers more flexibly. Replacing paper based instructions in those tasks can be advantageous for at least two reasons. The first reason is the avoidance of interruption of the work procedure that occurs during the alternation between the search for and examination of instructions in a handbook and their accomplishment at the real object. The second advantage is the extended range of different media type that can be offered by an AR-system such as texts, pictures but also audio files and videos. Furthermore does the interconnection between such data provide the facility to investigate those information in a non-linear way as it is well known from hypertext documents.

In order to support the tasks repair, maintenance, dismantling and training an authoring tool and two AR-applications have been developed.

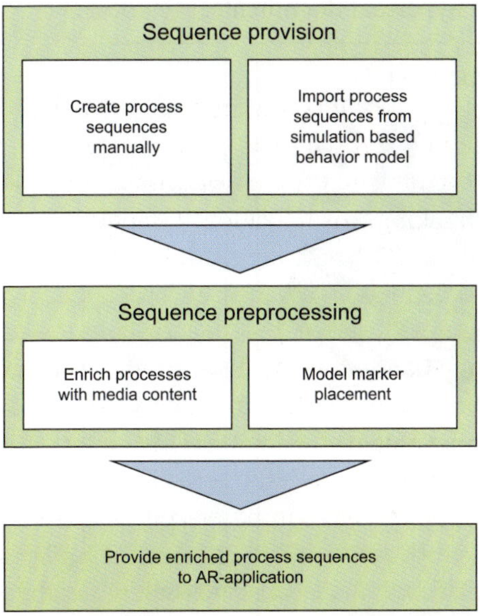

Fig. 6: Authoring tool

Core function of the authoring tool is to create work orders and to prepare them for being usable in the developed AR-system [7]. Two different ways for defining work orders are available (Figure 6). The first one aims to reduce preparation costs by reusing information already created. Sequences that have been created with the virtual disassembly simulation can be imported in order to use information created during the development. The second way of defining work orders is provided directly in the tool and facilitates the definition of process sequences by modeling directed graphs. After the sequences have been modeled, they must be prepared in order to be used with marker-based tracking. This is done by defining the placement of markers in the virtual scene that is assigned to a process so that a correct alignment of the virtual objects and the real world can be computed and displayed. Parallel to the marker placement task the processes can be enriched with various different media content. Almost always virtual objects are assigned to processes in order to visually emphasize a component in the real world or add a non existing component depending on the current process. If a process sequence is imported from the virtual disassembly system, the information created during the simulation may suffice but it can also be enriched. Since the simulation focuses on the development phase, supplementary information required for AR-system supported tasks can be assigned additionally. Until now supported media content types are images, movies, audio files, plain text files and hypertext documents.

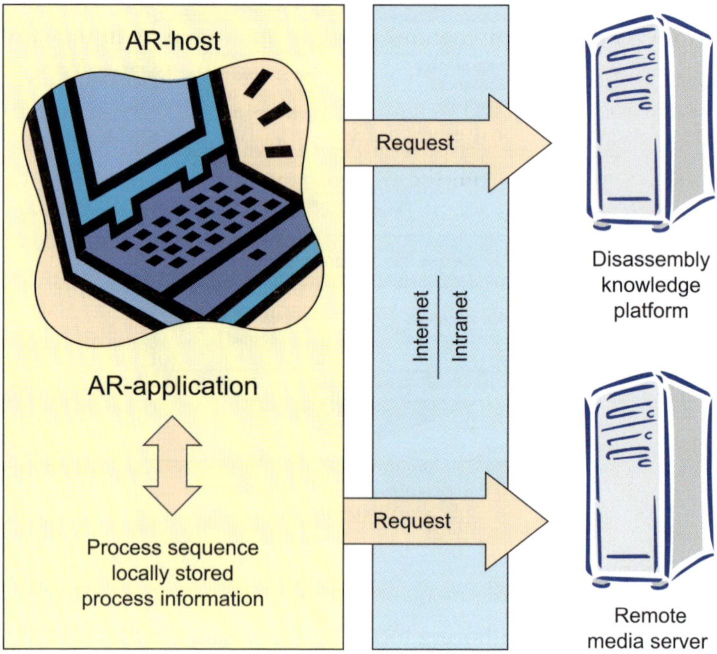

Fig. 7: Process sequence media content

Such information can be either stored locally, on the host where the AR-application runs, or it can be put on a remote server (Figure 7). By supporting references to remote servers the authoring process is decoupled from the media content creation and provision process. The locally stored information for a process is not subject to any changes after it has been created as it is the case with the marker placement and the virtual object geometries. Opposite to that, all other information may become outdated so that a local copy would be a drawback. Putting them on remote servers eases the processes of reworking, updating and sharing.

Based on the concept of the process sequences, two AR-applications have been developed. Both are for displaying the interactive content that was created in the authoring tool but each aims at different tasks and users. The first focuses on tasks like repair, maintenance and dismantling and has a user interface that is designed to support those tasks, especially to display the information created in the virtual disassembly system such as disassembly environment, cleaning processes, required tools and disassembly paths for removed components that can also be played in an animation [7]. The second application has a rather generic user interface and is designed to support training and learning tasks. While there is no specific expected information, the handling is more complex and is not as restricted as in the case of the first application. So the user can select certain subsets of information that are assigned to a process. Training tasks based on work orders that have been created with the authoring tool can be practiced. By the integration of the browser component from the MARE framework an interface to the information that is stored in the disassembly knowledge platform is provided so that an integration of training tasks and the collected knowledge from the SFB project can be linked with the authoring tool during task creation and enrichment.

References

1 Directive 2002/96/EC of the European Parlament and of the Council, 2002.
2 Azuma, R.: A Survey on Augmented Reality, Presence: Teleoperators and Virtual Environments 6 (1997) 4, pp. 355-385.
3 Baumann, R.: Structure-Oriented Exchange of Product Model Data. Dissertation, TU Berlin, 2004.
4 Krause, F.-L.; von Lukas, U.; Rothenburg, U.; Wöhler, T.; Romahn, A.; Anderl, R.; Melk, K.; Awiszus, B.; Ufer, R.; Binotsch, C.; Schützer, K.; Gemignani, B.H.: Engineering 2010: Setting the Course for Innovative Engineering.VMM Wirtschaftsverlag, Augsburg 2005.
5 Krause, F.-L.; Lüddemann, J.; Neumann, J.; Rothenburg, U.: Modellierungsmethoden und Systemgestaltung – DMU für verteilte kooperative Produktentwicklungen. In: Proceedings GI-Fachtagung CAD'98 Tele-CAD – Produktentwicklung in Netzwerken, 1998, pp.2 02-211.
6 Krause F.-L.; Rothenburg, U.: New Application Areas of Digital Mock-Up within Product Development. In: Proceedings of the International CIRP Design Seminar „Methods and Tools for Co-operative and Integrated Design", Grenoble, France, 2003, pp. 303-314.
7 Krause, F.-L.; Romahn, A.: Integrated Application of Augmented Reality in the Product Life Cycle. In: Proceedings of CIRP Life Cycle Engineering Seminar, Paper DIS2 on the LCE-05 CD, Laboratoire 3S, Grenoble, France, 2005.
8 Leemhuis, H.: Funktionsgetriebene Konstruktion als Grundlage verbesserter Produktentwicklung. Dissertation, TU Berlin, 2005.
9 Lüddemann, J.: Virtuelle Tonmodellierung zur skizzierenden Formgestaltung im Industriedesign. Dissertation, TU Berlin, 1996.
10 Milgram, P.; Takemura, H.; Utsumi, A.; Kishino, F.: Augmented Reality: A class of Displays on the Reality-Virtuality Continuum. In: SPIE, Vol. 2351, Telemanipulator and Telepresence Technologies, 1994, pp. 282-292.
11 http://www.hitl.washington.edu/artoolkit/ (Access: November 2006).

4.4 Product Life Cycle Planning and Tracking

Frank-Lothar Krause, Holger Jungk, Christian Kind, Berlin, Germany

Early product life cycle concepts evolved from a business management background, looking for costs and revenues in different phases. Nowadays there exist a multitude of definitions and interpretations of the product life cycle, in which environmental aspects are getting more and more in the focus of interest in public and research. For many products, the majority of environmental impacts are caused by its use or operation. In the automotive industry for example, more than 80 percent of environmental burden results from actual driving operation, against about 10 percent from production and less than 10 percent from the recycling overhead [1].

It gets clear that ecological improvements, as well as further cost reduction and increase in profit, can only be achieved by the contemplation of the entire life cycle. Currently life cycle management has its focus on product data management and is accepted as a powerful means of managing descriptions, properties and workflows of a product within a company. The shift from using the term Product Data Management towards Product Lifecycle Management already expresses the need of a more holistic point of view, however does not imply yet the management of the actual life cycle but only the management of product data beyond product creation.

The research on "disassembly oriented IT infrastructures" followed the steps necessary to be able to support an optimized planning and tracking of product life cycles. Different aspects, both methodical and technical, need to be addressed:

- An appropriate IT infrastructure is needed in order to overcome spatial and chronological differences in a disassembly oriented environment.
- Interfaces and data models must be adapted to be able to monitor individual products and create a seamless information flow along the product life cycle.
- With this prerequisites installed, life cycle scenarios can be planned and tracked aiming at longer and/or more use phases. During the tracking, decisions and actions are taken based on current product condition data.
- The linking of knowledge supports decisions that are often very complex because of functional and organizational interdependencies. It also provides continuous learning and improvement of life cycle processes.

The chapter describes the concepts and software prototypes resulting from this manifold approach, emphasizing the current research.

4.4.1 Information Technological Infrastructure

Life cycle oriented product design as well as planning and operation of ecological and economical optimal product life cycles require phase spanning communication and cooperation of all participants in the product life cycle. According to this, the superior target aimed by the subprojects of the Collaborative Research Center – increase of the life cycle productivity of resources – can not be achieved through isolated applications, but only through continuous data exchange between systems in fields as design, maintenance, disassembly planning and recycling. Therefore it is necessary to cope with the heterogeneity of these systems on the one hand and the spatial and chronological differences of the life cycle phases on the other.

The integration of distributed, heterogeneous systems can be achieved by component-based middleware architectures. Numerous scientific and industrial projects deal with middleware concepts and standards, which differ in their grade of abstraction and their application focus. Concepts for disassembly oriented IT infrastructures have been investigated [2, 3]. A simulation system has been developed that allows strategic planning of IT-infrastructures based on a reference model and activities [4].

An integration platform for enabling continuity in data processing and networking spanning various classes of business arose within the framework of the iViP-project "Integrated Virtual Product Creation" [5] and defined the basis for the conception of a disassembly oriented IT infrastructure within the project. The iViP-architecture is based on the integration of various services which are accessible from a common user interface - the iViP-client. Individual services can run on different computers transparently, so the user does not need to know where the services are actually running. Originally the iViP-platform was geared towards development processes by emphasising the development-oriented product data management. Therefore, additional services had been added in order to manage and acquire product information on an individual product level.

Efforts on achieving a continuous computer aided product development aim at a completely computerized representation of the product model up to a virtual product [6]. The product model is specified during the development process and serves as a common data base for different software applications, for example design, calculation and simulation programs. During the manufacturing process a physical product is created according to the product model.

However, the existing individual product has properties and parameters whose values deviate from the nominal values. It has also characteristics which are not regarded at all in the digital model. This is caused by unpredictable variations of the production processes and material composition, and by modifications of the product through wear and external influences during the use phase. The representation of these deviations and modifications in a computerized model is a function of the Product Condition Model (PCM).

The PCM essentially consists of geometric and technological data from the product development and incorporates product and time related use and wear information, as well as external changes on the product, for example due to maintenance or adaptation processes. Therewith, it can be seen as an extension of the product model beyond the product creation process (Figure 1).

Since the attribute values of a product in use change continuously, the PCM has to be updated, requiring an information flow from the physical product to the model. A way of capturing and recording product condition data automatically is provided by the Life Cycle Unit (LCU), consisting of sensors and memory components amongst other. The IT infrastructure can connect to the LCU in manifold ways in order to read out the data and store it in the PCM.

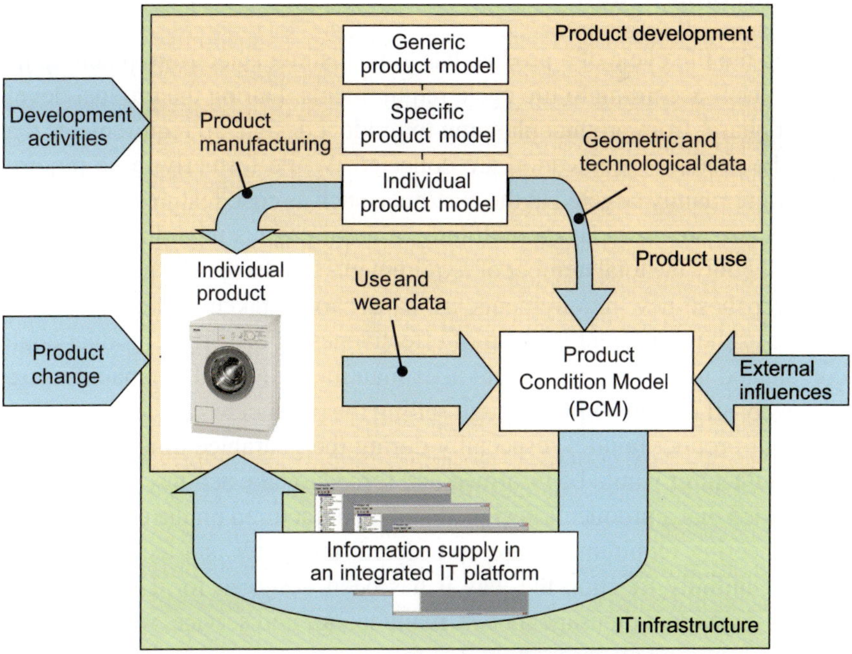

Fig. 1: Product condition model for the representation of product changes

The matching between real product and digital model is referred to as synchronization. Synchronization is done within certain intervals, for which frequency and regularity depend on the planned application. Between two synchronization points the divergence of 'real' and 'virtual' condition increases. During synchronization the divergence is reduced to an extent the PCM has been designed for. The acceptable tolerance essentially complies with the requirements of the application. Therefore, the PCM exclusively represents product attributes which are relevant for the monitoring task. Since the application requirements can change during the life cycle, the model is flexible by means of data types and composition of attributes, so that a large spectrum of possible parameters can be represented. For the configuration and administration of the Product Condition Model software tool, the PCM-Administrator, has been developed.

The model is able to acquire and represent product condition information in a way it can be processed throughout the product life cycle. Along with consistent interfaces to the IT-platform and to the product itself it serves as an information carrier that reduces interfaces and connects systems [7, 8].

4.4.2 Implementation

Planning the life cycle of a product has to be distinguished from planning the product itself according to life cycle requirements. During the product development phase, from product planning to product design, all requirements that affect the product specification are collected. While formerly these requirements were mainly targeted to the customer, such as cost, features and product design, there are nowadays a multitude of requirements regarding ecological goals. Not only the total number of requirements that have to be met increased, moreover these new requirements are harder to be taken into consideration because of their different horizon of effectiveness. Formerly a product could be evaluated to a large extent right after manufacturing and the manufacturer gave away all responsibility through selling the product to the customer. Only in case of product failures, especially during the guarantee time by contract, the manufacturer would be informed and forced to take actions.

Provided that a product is designed and manufactured on the condition that all engineering requirements are considered, the result is a technical object that will initially function the way it has been designed for. However, it is strongly dependent on usage modes, maintenance and service, whether it will meet the requirements throughout its life.

Planning of product life cycles aims at achieving the most utility possible, related to consumption of resources. This can be separated into two goals: The first is using the products longer, and the second is achieving different ways of use after the products failed to meet the original requirements.

The question arises, how can the life cycle be measured, and how can it be influenced to achieve the planning goals. A process model illustrates inputs and outputs and controllable factors [9]. In a product centric view the product itself follows the process (Figure 2). In each process step, the product function is initiated by its operation. Requirements define the expectation on the product behaviour. The actual output of the process step has to meet these requirements and makes up the utility of the product.

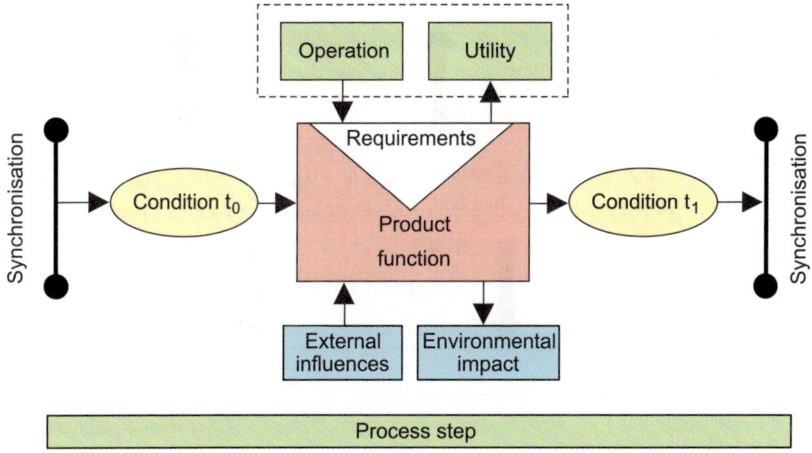

Fig. 2: Product centric life cycle process

The product itself changes by its function and this can be measured by tracking the product condition before and after a process step.

External influences are at the one hand those external factors that cannot be controlled, such as environmental situations, but more important these are also controlled actions aiming at maintaining, changing or improving the product function, so that a particular use can be achieved. The so called adaptation processes include actions such as repair, change of components, modernization, and expansion [10].

Each use phase is defined by its specific set of requirements (Figure 3). Because of product deterioration at the one hand, or increase of requirements at the other, the product utility will diminish during the use phase, up to a point at which use is ceded. In order to avoid this, the product use has to be kept up by means of adaptation. The trend will be that the need for adaptation increases with the duration of use.

The life cycle planning approach consists of first defining the life cycle requirements, and then specifying the necessary adaptation processes to maintain fulfilment of these requirements. However, since the life cycle is a non-deterministic process, it cannot be foreseen when exactly an action has to be executed. Therefore, rules are required that can be checked during the use phase, which is referred to as tracking. Setting up of these rules and checkpoints is an essential planning task which also prepares the tracking infrastructure.

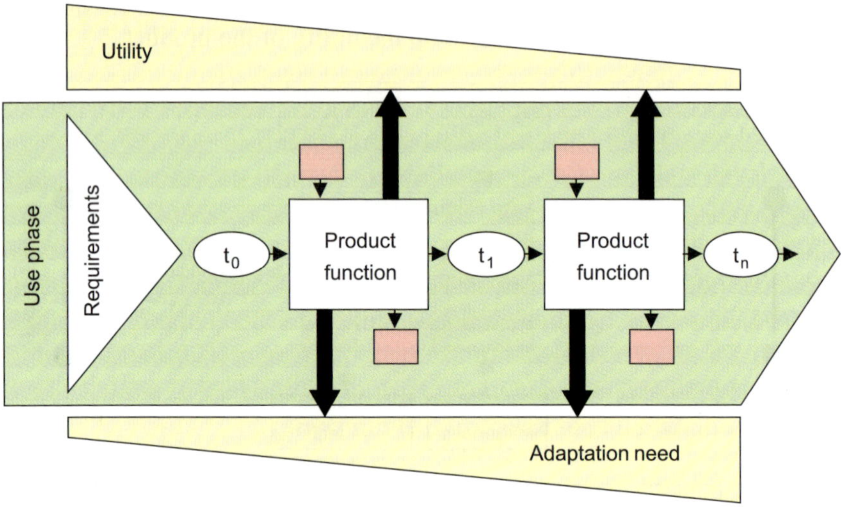

Fig. 3: Definition of the use phase

A software tool supporting the planning approach is the Life Cycle Modeller. It allows the definition of use phases and its requirement profiles. Each requirement can be correlated with product condition information that will be the basis for the tracking. The Life Cycle Modeller therewith provides the context of product conditions, their influences and how to interpret them. The definition of possible actions is supported by a repository that contains basic adaptation processes.

The planning of the life cycle should start in an early phase of the product development because the identification of life cycle requirements and critical product components can be an important source of information for life cycle oriented product design. Regarding methods and tools of Quality Management, which aim at avoiding the occurrence of failures in advance rather than have them removed after their discovery, there are analogies which put Life Cycle Planning into the Quality Management context (Table 1):

Tab. 1: Analogies

	Quality Management	Life Cycle Planning
Aim	Avoidance of product failures	Better use productivity by longer and more use phases
Target	Product design Product creation process	Total product life cycle
Feedback	Service protocols Replacement parts orders (Passive feedback)	Product accompanying information systems (Active feedback)
Countermeasure	Repair	Adaptation

Product failures generally force the customer to act. There may be direct feedback if service is called, but also if the customer repairs a product himself information about failed product components could be gathered by the number of sold replacement parts [11]. Only in case the customer replaces a product by a product of another manufacturer there will be no feedback, also meaning the loss of the customer.

Life Cycle Planning does not only aim at product failures but a set of requirements that need to be fulfilled. Any deviation of a requirement is not as obvious as a product failure. It can only be detected by monitoring the product condition and the requirements. Therefore an adequate effort needs to be put into the information feedback, not only to analyze, learn and eventually improve but moreover to be able to successfully drive the process.

Product Life Cycle Tracking

During the tracking, requirements and product use, represented by product condition, are compared regularly. If there is any deviation or may elevate any, appropriate actions are initiated.

Since the planning refers to the specific product level, the planning model needs to be instantiated for each individual product. This means that all checkpoints and events will become properties with absolute time values or references to individual product condition models, defined in the so called tracking model (Figure 4). The tracking model also includes the stored history of each action.

Tracking itself is controlled by the checkpoints that have been defined in the Life Cycle Modeller. Each checkpoint refers to a specific requirement and organizes the gathering of necessary product information. Checkpoints that have the same execution date are grouped into a checklist, which is presented in the Life Cycle Protocol.

The Life Cycle Protocol is a dynamic document which structures relevant product information along with the life cycle requirements. Only the information is displayed which is critical or at least relevant in that current stage. It is therefore a filtered and structured form of displaying product condition data. An expert will be enabled to take decisions and initiate actions, and have them documented. It is a bidirectional document, that is to say it also allows data input, for example update of requirements or results of taken decisions.

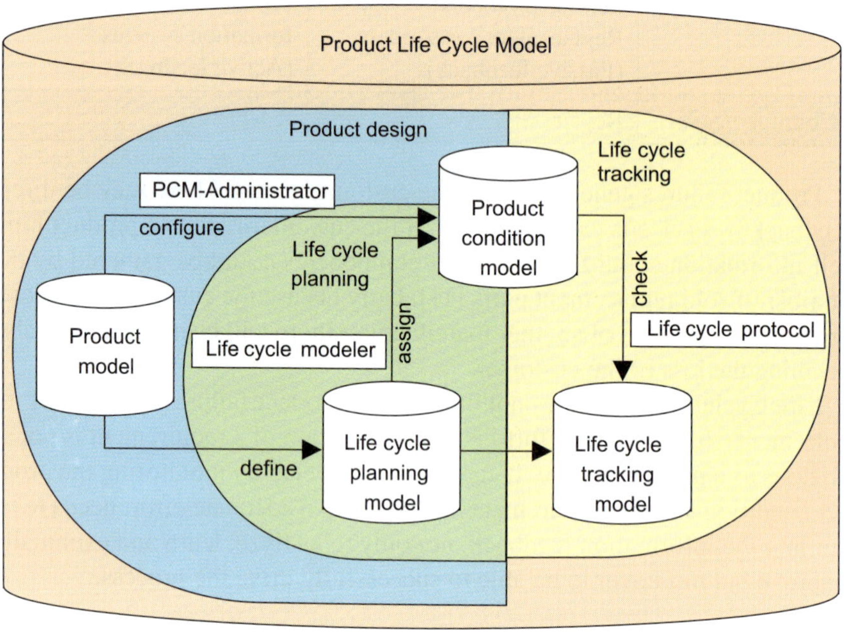

Fig. 4: Partial models and associated software tools of the Product Life Cycle Model

4.4.3 Knowledge as a Process Driver

The actions made during the tracking can partly be supported by knowledge processing systems, provided that the relationship between product condition and requirements is known and unambiguously formalized. In many cases however, an expert needs to judge the situation based on experience und personal skills, for example to decide whether an adaptation process is reasonable or another use phase should be targeted.

Knowledge management is needed in order to share the expertise of all partners involved in the product life cycle, and to reveal the dependencies of different knowledge domains. Therefore, a knowledge platform has been created that contains research results, case studies, multimedia and publications created during the research periods of the Collaborative Research Center (Figure 5).

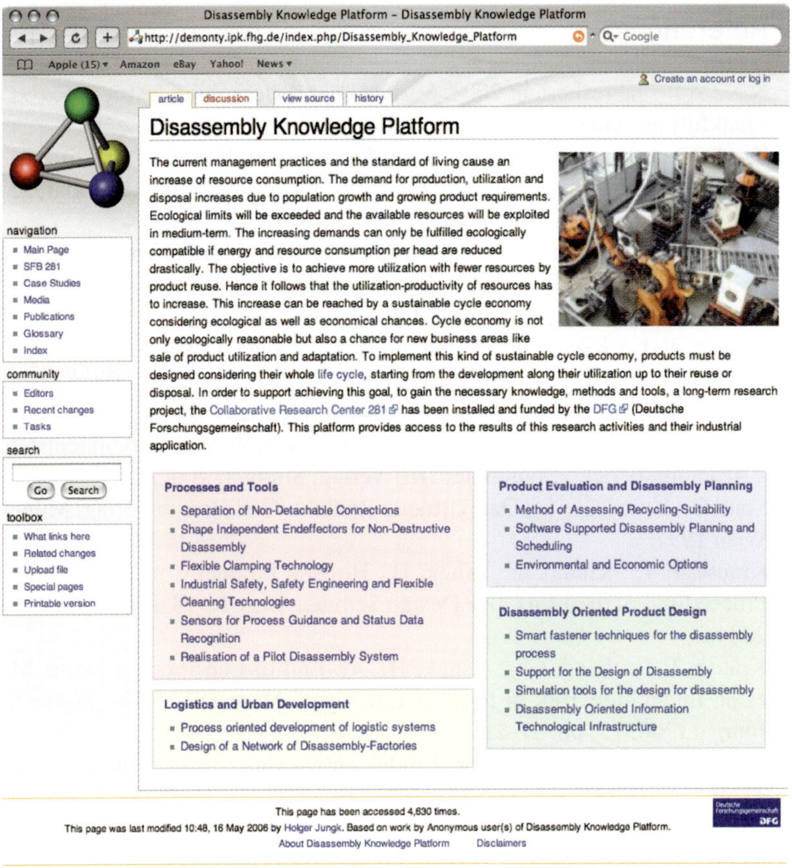

Fig. 5: Disassembly Knowledge Platform

That way it serves as a common information pool related to all aspects of a closed loop recycling economy. Due to its open concept, by which the users can edit and create articles and links, it will dynamically be updated and grow [12].

Repository elements of the Life Cycle Modeller as well as the Life Cycle Protocol are assigned with a knowledge link that leads to context specific knowledge within the platform. An expert is therefore enabled to take decisions based on facts and knowledge. The integration of the different approaches of knowledge management, process modelling and information systems can lead to a broad and holistic overview over processes, methods and techniques necessary and helpful for making the right decision at the right time of the product life cycle.

References

1 N.N.: Auto Jahresbericht 2004. Verband der Automobilindustrie e.V. (VDA), Frankfurt am Main 2004.

2 Kind, Chr.: Demontageorientierte informationstechnische Infrastruktur. In: Tagungsband zum Kolloquium zur Kreislaufwirtschaft und Demontage des Sfb 281, Berlin, Januar 2000.

3 Jungk, H.; Kind, Chr.; Rothenburg, U.: Demontageorientierte informationstechnische Infrastruktur und Werkzeuge. Futur 01/2002, 4. Jahrgang, S. 16-17.

4 Krause, F.-L.; Kind, Chr.: Simulation of Product Development Processes. In: Proceedings of the ProSTEP iViP Science Days 2005, Cross Domain Engineering, Sept. 28/29, 2005, Darmstadt, Germany 2005.

5 Krause, F.-L.; Tang, T.; Ahle, U. (Hrsg.): Integrierte Virtuelle Produktentstehung – Abschlussbericht. Fraunhofer IRB Verlag, Stuttgart 2002.

6 Spur, G.; Krause, F.-L.: Das Virtuelle Produkt. Carl Hanser Verlag, München, Wien 1997.

7 Krause, F.-L.; Kind, Chr.; Jungk, H.: Holistic Product Condition Model. In: Proceedings of the 14th CIRP Design Seminar, May 16-18, 2004, Cairo, Egypt 2004.

8 Krause, F.-L.; Kind, Chr.; Jungk, H.: An End-to-End Product Status Model as an Integration Tool in Product Life Cycle Management. In: ProductData Journal (2003) 2, pp. 57-61.

9 Raupach, C.: Simulation von Produktentwicklungsprozessen. Dissertation, TU Berlin, 1999.

10 Grudzien, W.: Beitrag zur Steigerung der Nutzenproduktivität von Ressourcen durch eine Life Cycle Unit. Dissertation, TU Berlin, 2002.

11 Edler, A.: Nutzung von Felddaten in der qualitätsgetriebenen Produktentwicklung und im Service. Dissertation, TU Berlin, 2001.

12 Jungk, H.: Wissensplattform für Demontage und Kreislaufwirtschaft. Futur 1/2006, 8. Jahrgang, S. 26-27.

4.5 Reference Model for Managing Product Development

Henrique Rozenfeld, São Carlos, Brazil

The Product Development Process (PDP) is considered a critical business process for improving the competitiveness of companies, mainly because it leads to greater product diversity and continuous reduction of product life cycles. A world-class PDP enables a company to create new and more competitive products, shorten the time-to-market and sustainability meet its strategies, marketing requirements and technological innovations.

Market uncertainty and PDP unpredictability have traditionally hindered any initiative aimed at systematizing PDP activities. Discipline was seen as an obstacle to flexibility, creativity and innovation. Management models are recognized today as important tools for process reliability, repetitiveness, performance and success. If the model includes best practices and lessons learned, a company's PDP and its learning capacity can both be continuously improved.

Many sustainable manufacturing initiatives aim to improve ongoing productive processes and usually focus on the environmental optimization of existing products. If the requirements of sustainability were taken into account during product development, more sustainable manufacturing would be a natural consequence.

Sustainable Product Development (PD) could be defined as a new discipline, but this new approach and its related practices must be combined with proven best practices of other knowledge areas in order to enhance the PD body of knowledge. A real application must be built upon a company's maturity level in PD, aiming at the harmonious evolution of its process.

4.5.1 Sustainability

Although the notion of sustainable development may not be clear for some researchers [1], since many definitions have been put forward for this term [2], we can state that its generally accepted meaning is "… sustainable development can be defined as development that satisfies today's needs without compromising the ability of future generations to meet their needs"[3].

Sustainable development embraces approaches like green design, which was replaced by ecological design (eco-design) [4]; Design for the Environment (DfE), Design for Recycling (DfR), Design for Sustainability (DfS), and Life Cycle Analysis (LCA). We do not intend, here, to discuss the particularities of many definitions, since they depend on a temporal perspective and have different connotations when used in Europe or America, for instance. Sustainable development can represent more than the abstract, important and generally accepted meaning mentioned above. It is a framework that encompasses concepts, strategies, systematic processes, guidelines, checklists, rules, methods and tools that indicate a shift in attitudes on environmental issues in PD [1, 4].

Some knowledge areas in engineering involve PD that focuses on physical, chemical, and biological phenomena or on a combination of such phenomena. They are so called technological areas and their main interest is to create analytical and numerical solutions or experimental solutions verified by experiments to simulate those phenomena. Thus, the behavior of products and processes can be predicted in order to improve their performance.

The focus of design and ergonomics is to define a better interface between human and artifacts to fulfill user requirements, including the cognitive. Functional requirements must be defined and deployed based on market analysis, focal group studies and other techniques that are studied by professionals of the marketing and quality areas. In this context, some knowledge areas are concerned with the use of statistical tools such as Design of Experiments (DOE) and Robust Design (RD) to define, verify and manage product characteristics like reliability and maintainability. Broader approaches for this are Critical Parameter Management (CPM) and Design for Six-Sigma (DFSS). In the quality area, for instance, there are tools for Quality Function Deployment (QFD) and for managing and avoiding product failure, such as Failure Tree Analysis (FTA), Failure Mode and Effect Analysis (FMEA), and Design Review Based on Failure Mode analysis (DRBFM).

We can integrate other tools in the PDP, such as value engineering, and tools for supporting creativity, such as TRIZ. The contributions from the financial area for PD involve managing target cost and monitoring the financial analysis indicators of a new product development. In addition, there are techniques from technical areas such as industrial engineering, which deal, for instance, with product architecture definition and product modularity.

There are works on organizational and human-related aspects of product development for the definition of better organizational structures, the improvement of work team performance, leadership in PD etc. Many of these aspects are integrated to the areas of knowledge management and learning organization.

Another category of works involves the development and application of information systems to support most of the aforementioned methods and tools, from specific systems integrated to CAD/CAE/CAM environments (also called Virtual Product Development – VPN) to systems that integrate the entire life cycle, such as Product Life-cycle Management (PLM).

Issues of a more strategic nature are portfolio management, technological surveillance and innovation management. Some works deal with performance measurement and its integration to business strategy definition through balanced scorecards.

Simultaneous engineering, which is still an up-to-date buzzword, is an integral part of most of the methods and tools listed here, but it is currently part of the scope of the Integrated Product and Process Development (IPPD) approach. Lean design reused most of the aforementioned concepts and added some new methods, e.g. set-based concurrent engineering.

During a PD, factory-planning methods must be integrated when new plants or facilities have to be designed for a new product. Process planning and the broader concept of Manufacturing Resource Management (MRM) as part of virtual manufacturing, in which the manufacturing systems are simulated, are also knowledge areas of PD.

This brief overview shows many of the possible perspectives of product development. PD management intends to create a unique integrated view of those concepts, techniques, methods and tools to allow companies to systematically apply them in order to continuously improve their PDP.

Sustainable development should then also be integrated in a complete PD management strategy as a complementary approach, a knowledge area, to ensure a fully integrated and harmonious PDP and avoid waste and conflicts.

Processes versus Projects

Business Processes (BPs) contain a set of activities aimed at producing a product or a service for a specific type of customers (internal or external to the company). BPs can represent repetitive operations that are normally well structured, such as finance management or production. PDP, however, is not so structured, since each product development may be unique.

Projects also represent a group of activities, although a project is a unique and temporary endeavor, with a beginning and an end.

If we systematize a PDP for a company, we define a pattern that can be followed for defining projects for developing products. A team or a manager can adapt the practices described in the pattern according to project's specific requirements. This contributes to the standardization of some practices, to the use of a common language, to the repeatability among projects and to their quality (Figure 1). Customization of the standard for each project considers many characteristics, such as product complexity, reusability of existing solutions, level of innovation and uncertainty or technology.

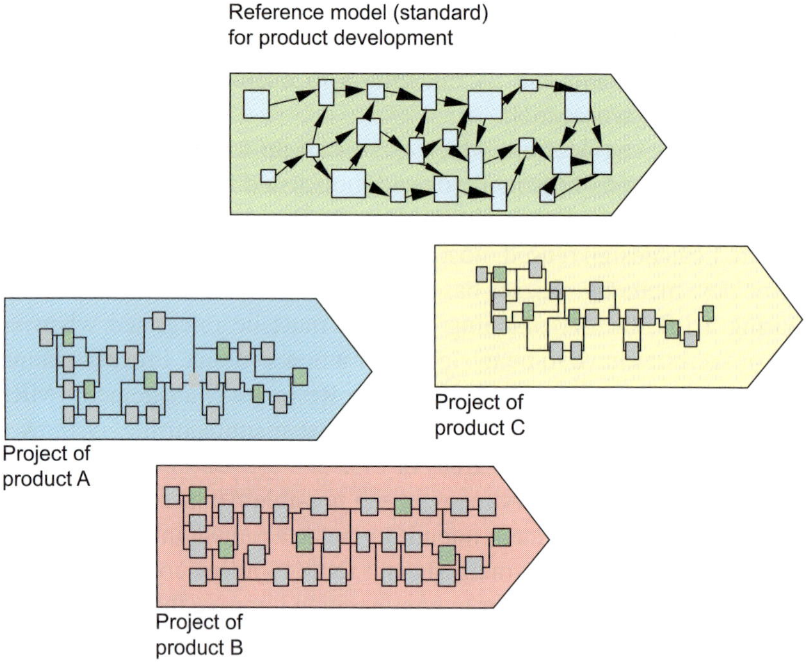

Fig. 1: Definition of different PD projects based on a common PDP standard

4.5.2 Modeling

The standard PDP on which development projects are based is normally represented in a reference model. The literature of PD describes this process as a sequence of steps, phases, activities, i.e., there are, in fact, many reference models, although they are not structured and presented in the same way.

Since a model represents only part of reality, normally the vision of a process that is modeled is what we call a partial model [5]. The most common vision used is the functional one, which encompasses the phases, activities, steps, and tasks of a process. Other visions may be informational; organizational; the resources. Some authors call the representation of activities, information, events and organization in the same diagram a "process vision", like the event Process Chain (ePC) [6].

Many companies have a defined reference model for PD (their standard process). There are many different forms used in real cases for documenting a process. A figure or a chart can show a general view of the process. In other cases, only the standard activities are described. Some models also contain more detailed information, e.g., procedures, methods to be applied, tools to support an activity, concepts, and bibliographical references, to allow someone to learn more about the knowledge needed to perform an activity, etc. The forms found are also very diverse, including figures, charts, graphics, texts, handbooks, and guidelines (Figure 2). Today it is very common to find such information stored in an intranet.

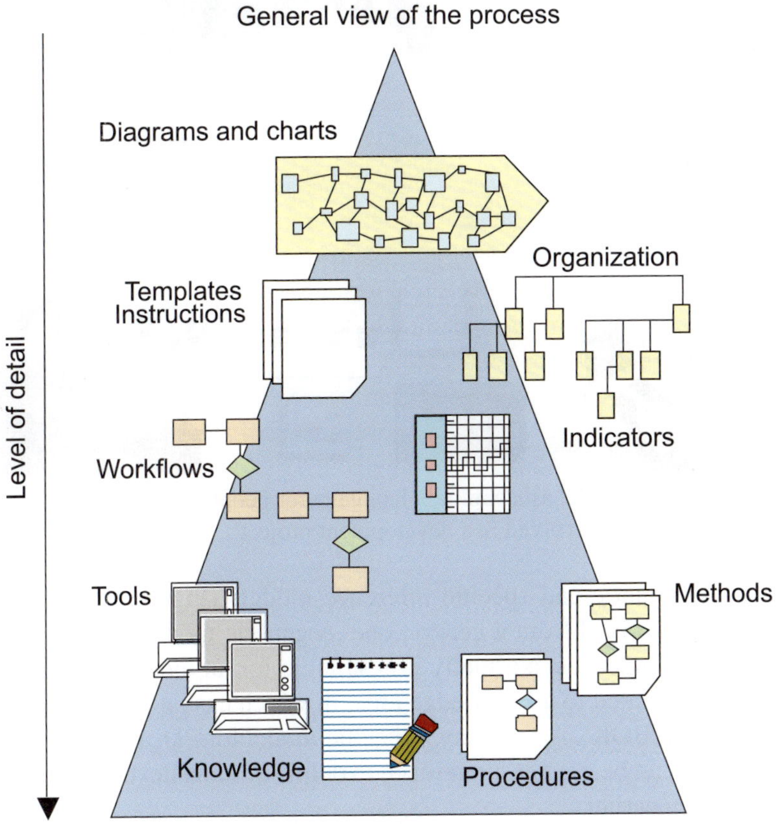

Fig. 2: Different levels and types of information in a reference model

Companies often do not have a unified vision of their PDP. Each company has its own understanding of the process and uses its own vocabulary to describe its PDP according to its own professional background. This limitation leads to problems and inefficiencies in PD due to poor communication and integration among professionals from the marketing, sales, engineering, manufacturing, technical support, customer service and managerial levels. This problem is further aggravated in collaborative developments among different partner companies.

Reference models can be used to minimize these limitations. The PDP is described in the model in order to establish a common language for all the professionals involved in a project (Figure 3).

Fig. 3: Reference model as a common language for people with different backgrounds involved in a development project

There are generic and specific reference models (Figure 4). A specific model can be derived from a generic one. A generic model includes a collection of best practices. Normally it is suitable for an industrial sector but it can also be related to a knowledge area, in which case it is called a Body of Knowledge (BOK). A company's specific model for PD, also called standard process, can be used as a reference to define product development projects, as shown earlier.

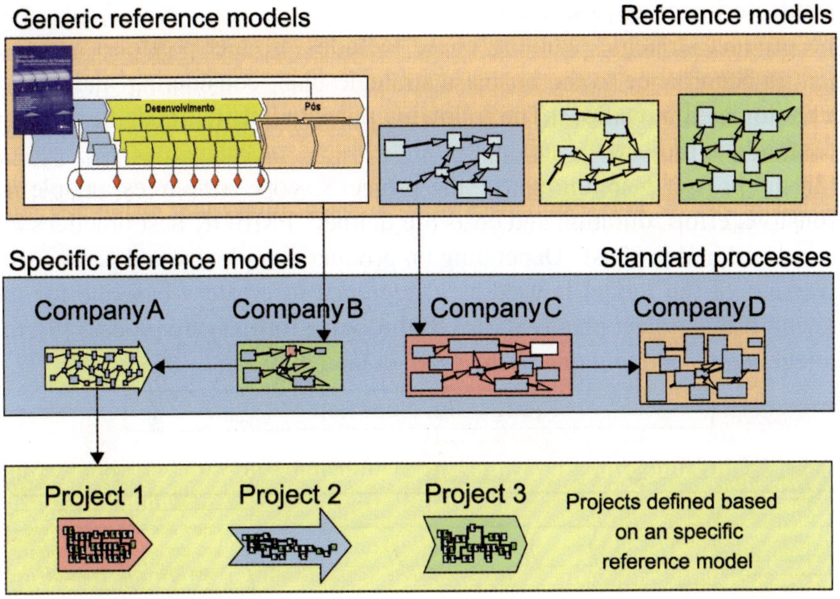

Fig. 4: A company's specific reference models (standard processes) deployed from generic reference models (Body of Knowledge – BOK) and PD projects derived from a standard process

4.5.3 Generic Model

A description is given below of a generic reference model for developing durable and capital-goods. This model resulted from experiences that have been collected and systematized since 1990 and were acquired through consultancy, research projects, coordination of PD projects, case studies, diagnostics carried out at many companies, and analysis and comparison of reference models of several companies [7]. This reference model was created in a joint project of three research institutions based on a community of practice [8] on the internet for knowledge sharing among universities and companies in PD. Standard processes for some companies have been derived from this model, which helped us to improve it [7].

A general view of the reference model divided into macro-phases is presented in Figure 5. The pre- and post-development macro-phases are generic, as the informational design, production preparation and product launch phases of the development macro-phase. The conceptual and detailed design phases consider the technological particularities of products and their corresponding

manufacturing processes.
The product strategic planning phase includes product portfolio management in accordance to the business strategic plan, considering market and technological innovations. The following phases relate to a specific product, i.e., a unique project.

In the project planning phase, the project's scope, resources, people responsible, effort, duration and costs are defined. PMBOK best practices are considered in this phase. Depending on product complexity and parts reuse, a version of the model is used as a reference to create a baseline for the project. If the project plan is approved through a formal gate process [9], the project begins and will end in the product launch phase.

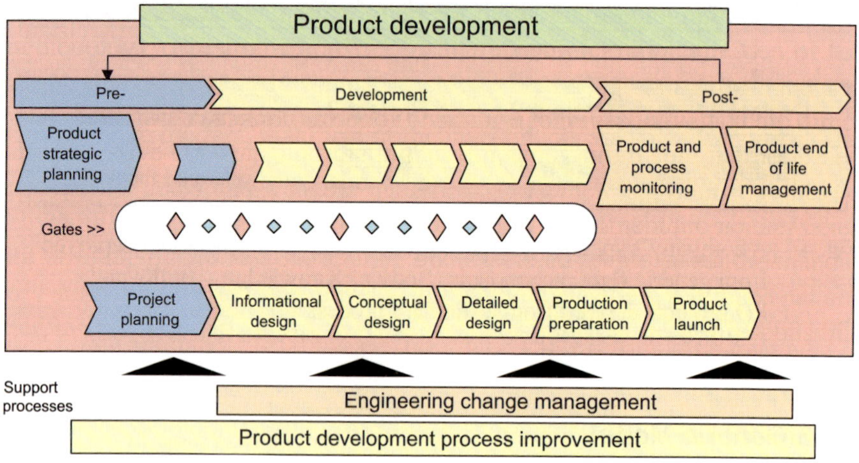

Fig. 5: The unified reference model for PD

The product's life cycle, the stakeholders and their requirements are determined in the informational design phase. The product requirements, which must be quantified in measurable variables with target values are derived from the stakeholder requirements. This is not the first time they are defined, since their definition begins in product strategic planning, when marketing delivers information about the market, which is now detailed in the informational design for a specific product.

The product functions, e.g. physical, quality, and interface, are established in the conceptual design phase to meet the product requirements. The technological solutions and product architecture are also determined at this point. Creativity methods may be applied in this phase. Innovations may emerge based on new technologies developed by the R&D process. Nevertheless, not all projects go through the conceptual design phase, since the product architecture is not normally changed in derivative projects.

The next phase is detailed design, which consists of three integrated cycles: detailing, acquiring and improving cycle. The calculations, simulations, product modeling, drafting, bill of materials, process plans, failure analysis, prototypes, evaluations and tests are carried out in this phase. All the manufacturing resources are specified; even a new factory, when necessary. Product handbooks and instructions for technical assistance are also produced, as well as sales support information systems.

The supply chain is defined at the beginning of a PD, when agreements are made with main strategic partners and co-developers. The latest supplier contracts must be signed in the detailed design phase. Based on the prototypes, the product is then certified.

In the next phase, production preparation, the first new equipment arrives. It is received, tested and, when approved, a pilot production lot is produced to certify the production facilities and products being manufactured with the new resources. In this phase, a new production process can be mapped and established to define, for instance, whether production will be controlled based on orders or Kanban.

The product launch phase takes place in parallel with production preparation. Other BPs are mapped in this phase, such as technical assistance and customer service, when, for instance, a new help desk script for the new product must be created.

In summary, production preparation aims to move the supply chain from the internal standpoint and the product launch phase from the external standpoint (market and customers).

After the product is launched, production and sales begin under the responsibility of other BPs and other areas of the company. The project is concluded, the team is disbanded, and its members are allocated to other projects or returned to their original functional areas. Nevertheless, PLM continues, since efforts must focus on monitoring the product and its manufacturing process. Ongoing customer support and Engineering Change Management (ECM) must be provided to eliminate failures or improve product performance. At this time, configuration management ensures product information integrity throughout the product's life cycle.

Product changes are managed by the ECM process, whereas improvements in PDP are carried out by other supporting processes.

At the end of a product's life, it is discontinued, reused, remanufactured or recycled according to the end-of-life plan, which has been under development from the beginning of the PD. The end-of-life requirements are taken into account in the informational design phase. In the conceptual and detailed phases, sustainable solutions are defined and tested, after which the end-of-life plan is established.

This brief description of the process provides only a functional vision of PDP, since only the main activities have been mentioned. Other complementary visions have not been addressed here.

Methods and Tools in the Business Process Vision

Methods and tools support the execution of PD activities. Both terms can be used as synonyms. Some examples of methods are: Quality Function Deployment (QFD), Design for Manufacturing and Assembly (DFMA), Failure Modes and Effect Analysis (FMEA), Life Cycle Analysis (LCA), or Robust Design (RD). These can also be called PDP supporting tools. The term "tool", however, is more commonly employed for information systems such as Computer-Aided Design (CAD), Computer-Aided Process Planning (CAPP), Computer-Aided Engineering (CAE), or Product Data Management (PDM).

The scope of most of these methods and tools has expanded. Thus, their functionalities overlap each other, and users risk getting lost if they attempt to apply them in an integrated context.

A method often consists of a list of steps. Many PD publications show some of these methods in a PD phase and activity. Since the original method was not conceived based on a process vision of the PDP, it is not easy to connect these steps to a PDP activity. Indeed, some activities are carried out entirely by one method, while others require more than one method (or steps – parts – of different methods), and a method can be part of many activities. In a good reference model, this is transparent to a user. In such a model, the relationships among methods and activities are consistently recorded. The same problem occurs with information systems that support PD. PLM system is a term that indicates a solution which integrates processes, methods and tools.

Knowledge Areas Sustainability

Activities for representing a reference model can be grouped in two ways: by knowledge areas, also known as process areas, or by phases. Each form of representation has advantages and disadvantages. The activities of a BOK model are normally grouped by knowledge areas or process areas. Examples of these are Project Management BOK (PMBOK) and Capability Maturity Model Integrated (CMMI). This is useful for increasing the model's flexibility and when the model is taken as a reference for process certification. Grouping activities by phases is more didactic and suitable for representing the usual sequence of activities.

In the proposed model, an activity can belong to the two kinds of categories. There is the sequential representation we have shown in the previous section, but it is possible to have a cross-vision of the activities based on knowledge areas. Thus, a team member of a company's division, normally functionally structured, knows what his contribution and responsibilities in the process are. Hence, the contribution of a functional area to the process can be understood. The proposed reference model comprises the following knowledge areas: project management, marketing, product engineering, process engineering, production, supply, quality, costs and environment.

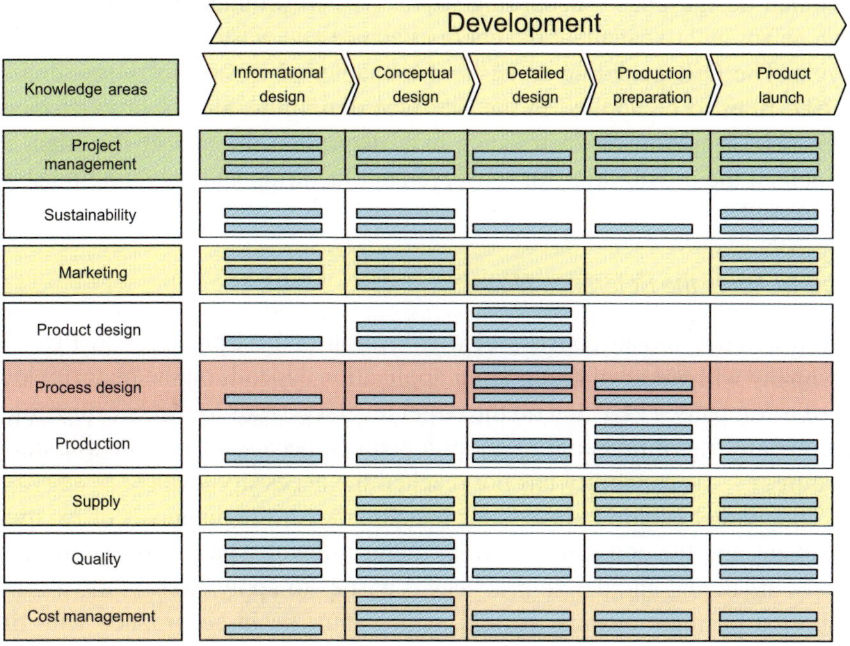

Fig. 6: Cross-vision of knowledge areas and their intensity of use throughout the development macro-phase

The latter knowledge area, environment, comprises the activities involving sustainable development, as well as methods and tools to execute them. We can therefore find all the concepts, methods, guidelines, rules and tools for eco-design. These elements of the model are related with the corresponding phase, according to the tools' scope, from product strategic planning to product end of life. Figure 6 qualitatively illustrates the intensity of the use of a knowledge area (indeed, the knowledge area's methods, tools etc.) during the development macro-phase.

During the information design phase, a team member can consider, for instance, the sustainability requirements that may have been systematized based on a checklist of common requirements. This checklist may grow from each new application and can be reused for the development of a new product for a known market, or if the industrial sector has standard rules based on regulatory agencies.

In the conceptual design, the team can apply DfE to access guidelines for a manufacturing process, as in the case of founding processes. Such guidelines may contain best practices for defining material suitable for recycling. A system for LCA is also available, which can be used in the conceptual or detailed design phases according to the type of product. These guidelines can be applied to calculate the energy balance and waste throughout the life cycle. The DfE guideline consists of a chapter of Design for Disassembly (DfD) rules to be applied in the interface definitions among product modules. The resulting characteristics may determine the end-of-life plan, in which all the possibilities for reuse, remanufacturing and recycling must be addressed.

How to Apply the Reference Model

The reference model consists of a collection of best practices in PD. No company will use all of them. Their application depends on the maturity level of a company's PDP and on the target level it intends to reach. Sometimes a company is not prepared to adopt a practice because it is not sufficiently mature, i.e., its capability has not reached the necessary level.

This model should be used as a benchmark for the diagnosis of existing practices and the company's current maturity level. The company can then study the model and define practices it wishes to apply. To do this, it must follow a business change method, which must be in accordance with the model. Such a method must take into account other business dimensions that may influence the improvement of this PDP.

However, the description of this change method lies outside the scope of this chapter.

4.5.4 Open Knowledge Support to Small Enterprises

We are currently building an internet portal to publish the model and related practices, methods, tools etc. free of charge. The first version can be viewed at the URL [10], but is now in Portuguese, because it is intended for use by SMEs in Brazil. Any company interested in it can download the model's content, for free, to define its specific model. It can also download a tool to build up an intranet with its specific content, in order for instance to organize it like an ISO 9001 procedure manual. The reference model must, of course, be adapted to specific requirements. It must also be possible to integrate current practices to the model.

In this environment, communities of practices involving a specific theme could be organized in accordance with the knowledge areas. The information could be published for free access like open access knowledge. New experiences and lessons learned could be added in order to share them among community members. One case would involve sustainable development. Users could access e-learning courses and tutorials to enable them to apply the guidelines and tools, which could also be downloaded free of charge.

This portal could also include cases to give to visitors a better idea of the potential and real use of these techniques. Information and contacts with specialists like researchers and consultants could also be related to the subjects of a knowledge area to allow for further connections.

The current version of the model is not structured for Engineering to Order (ETO) products, but there is another version for that kind of production strategy, in which bidding activities are inserted before detailed project planning occurs.

Sustainability must not only cover existing products and processes but also avoid future environmental unfriendly products and processes. Therefore, sustainable development sits squarely on today's agenda. Nevertheless, we must integrate sustainable development in a broader context of PD and PLM to meet the requirements of all knowledge areas relating to a specific product in a balanced way.

The proposed model contributes to the creation of a BOK in sustainable PD, establishing a bridge between technical and environmental areas. The aim is to make these practices applicable to practical cases.

References

1 Baumann, H.; Boons, F.; Bragd, A.: Mapping the Green Product Development Field: Engineering, Policy and Business Perspectives. Journal of Cleaner Production (2002) 10, pp. 409-425.

2 Bhander, G. S.; Hauschild, M.; Mcaloone, T.: Implementing Life Cycle Assessment in Product Development. Environment Process, Vol. 22, No. 4, Dec. 2003.

3 World Commission on Environment and Development: Our Common Future, Oxford University Press, Oxford, U.K. 1987.

4 Karlsson, R.; Luttropp, C.: EcoDesign: What's Happening? An Overview of the Subject Area of EcoDesign and of the Papers in this Special Issue. Journal of Cleaner Production (2006) 14, pp. 1291-1298.

5 Vernadat, F. B.: Enterprise modelling and integration: principles and application. Chapman & Hall, London 1996.

6 Scheer, A.-W.: ARIS Modellierungsmethoden Metamodelle.Anwendungen. Springer-Verlag, Berlin, Heidelberg 1998.

7 Rozenfeld, H.; Amaral, D. C.; Forcellini, F. A.; Toledo, J. C.; da Silva, S. L.; Alliprandini D. H.; Scalice, R. K.: Gestão de Desenvolvimento de Produtos: uma referência para a melhoria do processo. Saraiva, São Paulo, Brazil 2006.

8 Rozenfeld, H.; Eversheim, W.: An architecture for Management of Explicit Knowledge Applied to Product Development Processes. CIRP, Vol. 51, No. 1, pp. 413-416, Spain 2002.

9 Valieri, S. G.; Rozenfeld, H.: Improving the Flexibility of New Product Development (NPD) through a New Quality Gate Approach. Journal of Integrated Design and Process Science 8 (2004) 3, pp. 17-36.

10 http://www.pdp.org.br (Access: November 2006).

4.6 Cleaner Production

João Fernando Gomes Oliveira, Aldo Roberto Ometto,
Américo Guelere Filho, São Carlos, Brazil

The 1980s marked a worldwide intensification of environmental awareness, as well as a considerable change in the behavior of companies with respect to the management of environmental impacts caused by the wastes generated through their activities. In response to the population's increased environmental awareness, environmental laws became more stringent, rendering the technologies that supported the end-of-pipe approach in use insufficient to meet the stricter emission standards. In those days, the predominating vision among companies was that being an environmentally aware enterprise negatively affected its economic performance. Taking care of the environment would require heavier investments in products and processes. Then, starting in the mid-80s, leading companies began to envision investments in environmental protection not primarily as costs, but rather as investments in the future and, paradoxically, as a competitive advantage [1].

In addition to the legal aspects, this change in attitude also came about from the realization of the actual costs of the traditional end-of-pipe approach. Behind the normal costs of treatment and disposal are related costs (not usually recorded), such as loss of raw material, water, electric energy, legal and normative nonconformities, and company image-related costs [2]. From this economic standpoint, according to data published by the United Nations Environmental Program (UNEP), typically, for each dollar spent on waste treatment or disposal there are two or three additional dollars that are "hidden" or simply ignored, a finding that applies even to large, well-managed companies [3]. It is in this context that some industries began to adopt environmental management approaches aimed at reducing the quantity of wastes generated through techniques of reuse, recycling (internal or external to the process) and recovery of material [4].

Although they do contribute to minimize the generation of wastes, these techniques do not constitute definitive solutions, since they are based on the assumption that the waste exists and also because they generate subproducts and involve energy and other operational costs. In view of these facts, starting in the late 80s and early 90s, companies began to reassess their environmental management approaches, adopting a position favoring the prevention of waste generation and a balance between environmental and economic gains. This was the context in which the concept of cleaner production emerged.

The concept of Cleaner Production was introduced by UNEP's Industry and Environment arm in 1989. Cleaner production is the continuous application of an integrated preventive environmental strategy applied to processes, products, and services to increase eco-efficiency and reduce risks for humans and to the environment. It applies to:

- *Production processes:*
  conserving raw materials and energy, eliminating toxic raw materials and reducing the quantity and toxicity of all emissions and wastes,
- *Products:*
  reducing negative impacts along the life cycle of a product, from raw materials extraction to their ultimate disposal, and
- *Services:*
  incorporating environmental concerns into design and delivery services.

Cleaner production requires changing attitudes, responsible environmental management, creating conducive national policy environments, and evaluating technology options [5].

4.6.1 Methodology

The methodological structure of cleaner production can be divided, according to [5], into four stages:

- planning and organization,
- opportunity identification,
- availability analysis, and
- implementation/monitoring.

The main causes of the sources of environmental aspects listed by [5] are technology, technical status of equipment, process and/or product specifications, raw materials, production planning and management, employee skills and motivation, process efficiency, and focus on wastes, effluents and emissions.

The opportunities for cleaner production are generated based on the identification of the main causes of the sources of environmental aspects. The main opportunities indicated in [5] are technological changes, equipment modifications, process and/or product modifications, raw material modifications or replacement, improved process control, housekeeping, reuse in the process, and recycling outside the process.

The opportunities for cleaner production can be grouped on three levels (Figure 1), with level 1 – reduction at the source – being the focus of the actions of the cleaner production program.

The main opportunities for reduction at the source by means of process modification are housekeeping, the replacement of raw materials, and technology modifications. Housekeeping actions involve organization, cleanliness, safety, maintenance, training, and the process specification control operations.

Raw material replacement is aimed at finding material with better technical and environmental qualities, reducing the environmental impact associated with their use, the consumption of materials and energy, and the generation of wastes.

The change to a cleaner technology requires stringent studies and technical analyses according to the characteristics and criteria of each productive process. For the metal-mechanical manufacturing sector, some of the opportunities for cleaner technology studied by our groups are [6]:

- biodegradable vegetal oils,
- Minimum Quantity Lubrication (MQL) and
- dry machining.

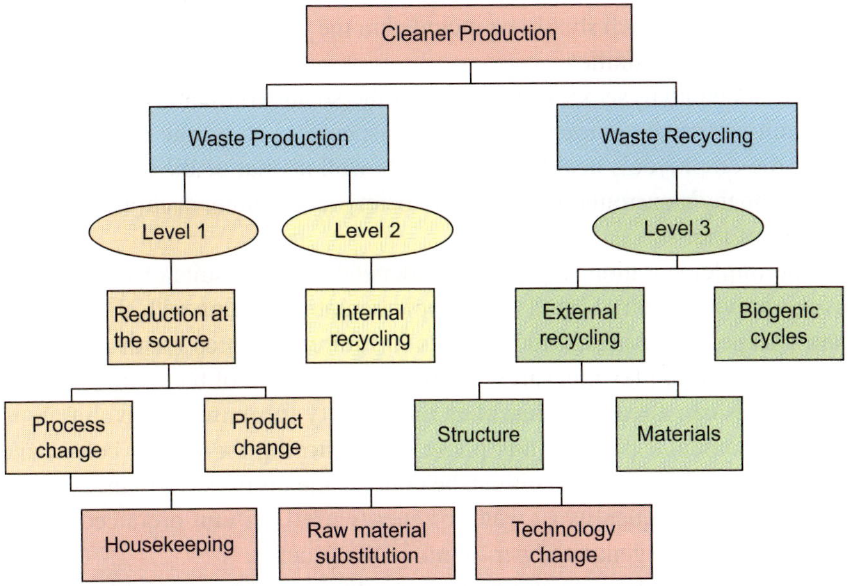

Fig. 1: Levels of opportunities for cleaner production

After the opportunities for cleaner production have been identified, each one must be evaluated from the technical, organizational, environmental, economic and social standpoints in order to prioritize the implementation of the most sustainable ones. The technical analysis involves aspects relating to the production process and to the product quality.

The organizational analysis of the measures serves to verify how easy (or difficult) it would be to implement the measure vis-à-vis the company's manufacturing and administrative procedures.

The environmental analysis identifies the opportunities that will bring the company the greatest environmental benefits. To this end, environmental impact assessments are carried out based on the environmental aspects identified earlier. The methodologies for environmental impact analyses include the impact matrix, interaction networks, modeling, control listing as well as specific impact analysis, such as global warming, environmental acidification, human toxicity, ecotoxicity, among others, which are given in the EDIP (Environmental Design of Industrial Products) method described by Wenzel et al. [8].

The economic analysis involves from the valuing of environmental aspects and the hidden costs of wastes, losses and waste to the financial analysis of each opportunity for cleaner production, such as investment return time, image improvement and the capture of new markets. Among the hidden costs of waste are the costs of raw material, energy, labor, administrative costs, etc., which should be included in the financial analyses of cleaner production opportunities.

The social analysis seeks to ascertain the social consequences of the opportunities for the company's internal aspects, such as the generation of jobs, the employee's health and safety as well as that of the community in the company's surroundings and for society as a whole, such as improved quality of life.

The implementation sequence will depend on the results of the analyses of the previous item. Usually, the implementation begins with the simplest and cheapest measures, which results in future resources for the more expensive and complex measures. A simple segregation of the more problematic wastes can already represent an opportunity for reuse or recycling.

A fundamental point in this phase is monitoring the results before and after the implementation. This can be done by means of performance indicators, such as the quantity of water or energy used per unit produced, or the quantity of waste generated per month in the process.

Monitoring of the indicator by documental or electronic means is important in order to record and check if the foreseen goals and objectives have been met.Thus, the cleaner production program comprises a PDCA (Plan, Do, Check, Act) cycle, and checking to correct the implementations depends basically on a good monitoring system.

4.6.2 Industrial Cases

The Sao Paulo Estate Environmental Agency (CETESB) has fostered the cleaner production through an organization called "Sao Paulo State Round Table on Cleaner Production". The results include several real industrial cases of implemented cleaner production programs. Some of the most relevant are described below [7].

Industrias Romi S.A.

The casting manufacturing process uses large amounts of sand for the production of molds and dies. This sand is usually extracted from pits or rivers, with a varied granulometry.

Based on experiments and studies developed by the company, the following measures were implemented:

- replacement of phenolic resin by furanic resin in the molding system and core shop, by cold-curing with an organic-base catalyzer,
- recovery of the used sand (containing furanic acid) by a mechanical process at room temperature. Currently, the company reuses 98 % of recovered sand with 2 % of new sand in the production of dies and molds,
- in the cold box process, 85 % recovered sand and 15 % new sand is used.

The company's investments to attain the above mentioned improvements amounted to about USD 25,000, which were spent mainly on training, the purchase of dosers for mixing new sand with recovered sand, and for the preparation of anatomical molding boxes for the items with larger production volumes.

With the implementation of these improvements, the consumption rate of new sand dropped from 800 kg to 200 kg for every 1,000 kg of castings produced. This gain represented a reduction of 80 % in the consumption of new sand and of 80 % in the disposal of foundry sand. For the sake of comparison, note that a benchmark performance index is 500 kg of new sand for every 1,000 kg of castings produced, indicating that the index achieved is very good indeed.

The reduction in the volume of discarded sand, in addition to representing an enormous environmental gain, allowed for annual savings of USD 500,000 as a result of the reduction in purchases of new sand and, especially, in the costs of disposing of the waste in an industrial landfill.

The company also intends to develop a form of using the foundry sand as a construction material, thereby completely avoiding its disposal in landfills.

Ford Motor Company Brasil Ltda. – Taubat® Industrial Unit

The manufacture of engines, transmissions and chassis components involve metal parts machining processes that generate an average total of 375 tons/months of metal chips. These chips, originating from different types of process and impregnated with different types of cutting oils and oily emulsions, were mixed together and passed on to the effluent treatment station.

The cutting oils/emulsions used in each stage of the processes were re-evaluated and divided into eight families of oils. Selective collection of the metal chips impregnated with oils/emulsions was implemented, based on these eight oil families.

Today, the oils/emulsions are separated and stored in containers. They are physicochemically analyzed to check and possibly chemically adjust the technical conditions. After this determination, they are reused in the cutting process from which they originated or passed on for use in equipment in the company, whose operations allow for the use of cutting oil/emulsions with less stringent characteristics.

The results attained was a reduction of cutting oil purchases in the order of 1,400 liters/month, representing savings of USD 5,500/month, an average reduction of 1 % in water consumption at the industrial unit, representing 20,000 liters of water per month and savings of about USD 39/month, an 88 % reduction in the amount of oil sent to the effluent treatment station, with a 3 % reduction of the overall volume of liquid effluents to be treated, generating savings of 3 % in the consumption of chemical products used in the effluent treatment station, or USD 60/month.

In the medium term, the company plans to install a cutting oil regeneration facility in order to extend its reuse of regenerated oils.

BSH Continental Eletrodomesticos Ltda.

This company's household appliance manufacturing processes consume large quantities of water, estimated at 5,500 m^3/month or 66,000 m^3/year, mostly in the treatment of the surfaces of the products' metal components.

To reduce water consumption, preserving natural resources while simultaneously reducing water-related costs, a system was implemented for the reuse of treated effluent, which consists of a filtration system with an output of 10 m^3/h, adapted to the effluent treatment system.

The filtration material is composed of iron-free sand, which retains contaminants and suspended particles. After the effluent is filtered, it goes to a $5m^3$ water storage tank equipped with two pumps that pump the water to the industrial process and toilets. In the industrial process, the water is utilized for washing floors, removing powdered paint from damaged parts before they undergo the firing process, and for degreasing parts. It is estimated that about 60 % of the water is reused in the rest rooms.

The implementation of this measure has resulted in a 30 % reduction in the plant's consumption of water from the public water service, dropping on average from 86,400 to 60,480 m^3/year. The adoption of this measure has led to an estimated saving of USD 68,000/year.

Studies are also in progress to implement a water treatment system by reverse osmosis or water demineralization, which will allow for the reuse of 100 % of the treated effluent in processes that require better quality reused water, e.g., in the preparation of water-based paints.

Eaton Ltda. – Transmission Division

The vehicle transmission manufacturing process uses large quantities of mineral oil (petroleum-derived). These oils receive additives (animal fats, preservatives, protective agents, etc.), originating whole cutting oils, lubricants and hydraulic fluids with differentiated formulations and applications. Whole cutting oils are used for lubrication and cooling and to ensure the desired finishing on the surface of workpieces during machining.

The oil-recycling program involved the treatment and recycling of cutting, lubrication and hydraulic oils. To implement these measures, the company provided in-house training courses and awareness raising campaigns for its manufacturing staff to assure them that using recycled oil in the machines would not affect the quality of the productive process or the service life of the cutting machinery.

In 2002, the company estimated it had saved USD 124,200 in cutting oil and USD 65,125 in lubrication and hydraulic oils, resulting in total savings of USD189,325. Thus, its estimated investment return time was only 1.5 month.

From the environmental standpoint, it should be noted that the use of recycled oil contributes toward the preservation of natural resources, since oils are petroleum-derived nonrenewable resources.

MAHLE METAL LEVE S.A. – Valve Train Unit

The assembly of a camshaft drive requires the fixing of cams, which are the parts that move the valves along the shaft. In the assembly process, the cams are heated to about 400° C, dilating their measurements, after which they are cooled by injecting pressurized carbon dioxide (CO_2).

Experiments carried out in-house indicated that the CO_2 used for cooling the parts could be replaced successfully by compressed air. This replacement was put into effect without requiring any equipment or process changes.

The substitution of CO_2 by compressed air brought benefits by eliminating emissions of approximately 160 tons/year of CO_2 (a gas that produces the greenhouse effect) and cost reductions by eliminating the purchase of CO_2, on which the company had heretofore spent USD 150,000/year.

4.6.3 Trends

The concept of eco-efficiency was first coined in 1992 by the World Business Council for Sustainable Development (WBCSD) and further defined 1993 as being "reached by the delivery of competitively priced goods and services that satisfy human needs and bring quality of life, while progressively reducing ecological impacts and resource intensity throughout the life cycle, to a level at least in line with the earth's estimated carrying capacity". The WBCSD has identified seven success factors for eco-efficiency [8]:
- reduce the material intensity of goods and services,
- reduce the energy intensity of goods and services,
- reduce toxic dispersion,
- enhance material recyclability,
- maximize sustainable use of renewable resources,
- reduce material durability and
- increase the service intensity of goods and services.

Eco-efficiency embraces cleaner production concepts such as efficient use of raw materials, pollution prevention, source reduction, waste minimization, and internal recycling and reuse. It captures the idea of pollution reduction through process change as opposed to the earlier end-of-pipe approaches.

Today, UNEP and WBCSD consider that both eco-efficiency and cleaner production are "integral parts of the macro-vision of Sustainable Production and Consumption (SP&C), which encompasses the entire commercial system and its interrelations", and that eco-efficiency starts from issues of economic efficiency which have positive environmental benefits, while cleaner production starts from issues of environmental efficiency which have positive economic benefits [8].

Although they have played a major role in placing environmental management on the corporate agenda, the environmental model commended by the ISO 14001 standard do not provide companies with explicit tools to measure the generation of solid wastes, gas emissions and liquid effluents or to identify options for preventing the generation of pollutants and the resulting environmental impacts [9]. Instead this market-oriented environmental standard focuses on achieving the legal compliance based on waste treatments and disposal.

Because the "core" of cleaner production is based on the systematic description of mass and energy balances, thereby evaluating the efficient use of materials, water and energy, it could be applied to fill the existing gaps in the ISO 14001 environmental management system so as to render it, too, an effective management tool from the standpoint of prevention [9, 10].

References

1 Callenbach, E. et al.: Gerenciamento ecológico: guia do Instituto Elmwood de auditoria ecológica e negócios sustentáveis. Cultrix, São Paulo 1993.

2 Bierma, T.J.; Waterstaraat, F.L.; Ostrosky, J.: 1998. Chapter 13: Shared Savings and Environmental Management Accounting, from The Green Bottom Line. Greenleaf Publishing, England 1998.

3 United Nations Environment Programme – Division of Technology, Industry and Economics: Financing Cleaner Production (2004). Profiting from Cleaner Production: Resource Kit for Training Checklists for Action. Available at: http://financingcp.org/trainung/training.html (Accessed on 23 April 2004).

4 Bishop, P.L.: Pollution Prevention: Fundamentals and Practice. McGraw-Hill, Singapore 2000.

5 United Nations Environment Programme – Division of Technology, Industry and Economics: Cleaner Production (2001). Available at: <http://www.unepie.org> (Accessed on 23 April 2004).

6 Oliveira, J. F. G.; Alves, Sale Martins: Development of Environmentally Friendly Fluid for Vitrified CBN Grinding. Cirp Annals Manufacturing Technology, Vol. 55, Nr. 1, 2006.

7 CETESB: Companhia de Technologia de Saneamenta Ambiental. *Meio ambiente*: produção mais limpa. Available at:<http://www.cetesb.sp.gov.br/Ambiente/producao_limpa/casos.asp> (Acesso em: 11 Jun. 2006).

8 United Nations Environment Programme: Eco-Efficiency and Cleaner Production: Charting the Course to Sustainability. The World Business Council for Sustainable Development. United Nations Development Program. Available at: http://www.iisd.CA/consume/unep.html (Accessed on 17 Jun 2004).

9 Fresner, J.: Small and Medium Sized Enterprises and Experiences with Environmental Management. Journal of Cleaner Production 12 (2004) 6, pp. 545-547.

10 Fresner, J.: Cleaner Production as a Means for Effective Environmental Management. Journal of Cleaner Production 6 (1998) 3/4, pp.171-179.

5 Processes and Tools for Disassembly

Disassembly within remanufacturing is a main element of a sustainable cycle economy. Disassembly is needed in maintenance and repair during product usage phases and for adaptation of products and components between usage phases. Thus disassembly enables for an increased use productivity of resources.

Disassembly in the following chapters is considered from different perspectives:

- Materials are regained by developing processes and equipment for separating non-detachable connections.
- Components are conditioned for further usage phases by non-destructive separation with respective tools.
- Due to the huge variety of used products, flexible clamping devices are required for manual and automated disassembly processes.
- Innovative cleaning processes help compensating wear effects and thus setting up components for further usage phases.
- Hybrid disassembly systems integrating manual and automated workstations for a wide range of products to be dismantled provide the testbed for experimentally checking integrated functionality and reliability of equipment for remanufacturing, clamping and handling, identification and measurement, transport and storage.

5.1 Non-Detachable Connections

Eckart Uhlmann, Nayim Bayat, Berlin, Germany

Numerous products or their components are designed with many detachable and non-detachable joints and connections. The range of complexity to disassemble them varies from components made of sheets to more complex workpieces or devices. Thereby, the principles for the connections are rivets, screws, welds or forming processes. Ideally, to disassemble a consumer product all joints have to be dismantled. In most cases this is not possible. This may result from the connection itself as with welded joints, or from external influences on it such as corrosion or damage. Therefore, destructive processes and tools must be used to obtain fast access to reusable parts or components in order to reuse them or recycle the materials. Connections which cannot be taken apart have to be destroyed and are called non-detachable. Also it might prove more effective not to dismantle connected parts but partly to destroy one of it to obtain fast access to downstream parts that may be reusable. The development of disassembly processes and an appropriate reprocessing technology are necessary for the reuse and recycling of used products and for their transfer in the product and material cycles. Innovative disassembly technologies are needed to quickly and efficiently separate a product into its components. The separation methods and processes need to comply with general criteria of operating and economic efficiency.

Accordingly, two approaches exist: One is the classification of components and joints. The other one is the classification of existing manufacturing technologies. Destructive disassembly processes can be pooled in a corresponding process selection system. The aim of the system is the definition of appropriate methods for specific disassembly cases. Generally, it is possible to use the existing separating manufacturing technologies for the disassembly operations. Depending on the recycling purpose of the component that has to be dismantled two groups of processes are available. To extract reusable parts, non-destructive technologies such as unscrewing, dismantling, and evacuating are required. Destructive processes and tools are used to separate material or to provide easy access to subsequent reusable components. They can be divided into processes that cause secondary damages and those without. The use of processes which are secondary damage free, such as shearing, splitting or ripping apart, have the advantage that subsequent post treatment like cleaning or replenishing is not necessary. Disadvantages are high emerging process forces, slow progress, and resulting demolitions of the processed part as well as back stroke impacts on tools [1].

Quality criteria for destructive disassembly processes with secondary influences are short processing times, economical operation and minimized secondary influences of reusable parts. Secondary influences range from minor contamination on surfaces to the destruction of whole parts or from harmless gaseous particles to hazardous aerosols. They can consist of sparks, scrap material, slag, chippings, splinters, dust, steam and vapor. The mentioned alternative methods and processes for the disassembly of connections and joints make clear that an universal classification cannot be derived. The description of available technical methods is an objective basic for a suited classification and fulfills most requirements. The destructive manufacturing processes were categorized into the groups dividing, chipping, jet technologies, melting and deforming. The efficient technical methods used for disassembling are drilling, knife cutting, strike-cutting, splitting, shearing, abrasive cutting, plasma jet cutting, cryotechnical disassembly, and hybrid tools. These technical methods can be subdivided into conventional and non-conventional tools. Basically, it is possible to combine two methods to a hybrid disassembling technology. The efficiency of the technical methods used for disassembling will be described in detail.

5.1.1 Drilling

Screw joints represent a major group of joint technology and are in the focus of disassembly. The application of technical methods is based on two requirements. The first requirement is a maximum flexibility in regard to the disassembly device or tool. In regard to unscrewing this means the independency of the shape and dimensions of the screw head. The second requirement is a level of robustness of the tool towards changes of the joints which could appear in form of diversification, modifications or alterations during the life cycle of the product. Exemplary changes are modifications of the lock type or shape of the joint elements which could be caused by corrosion or deformation.

It is possible to fulfill the aforementioned flexibility requirements by destroying the screw by drilling. Usually, both the screw and the tap hole are destroyed by this, so that in order to further use the product, the created drill hole needs to be opened and a new, bigger thread tapped. This approach is not always feasible due to limiting part dimensions or its shape. Moreover, additional work is required to create the new drill hole. On the basis of these results the objective is to destroy the screw while leaving the thread intact. This way, the functioning of the component parts is maintained, and they can be reused. The general feasibility of destroying the screw without also destroying the thread can be demonstrated with joints with hexagon head. Spiral drills were used for the destruction of nearly 200 screws by drilling. The joints consisted of two metal sheets, of which one had a tapped through hole.

The clamping length amounted to 3 mm. The parameters clamping torque for the screw and feed and revolution speed for the drilling process can be varied. The feed force, the drilling torque, the surface quality and the geometry of the tap hole serve as quality criteria. The results are the determination of the characteristic force or torque curves during destruction of screw joints (Figure 1).

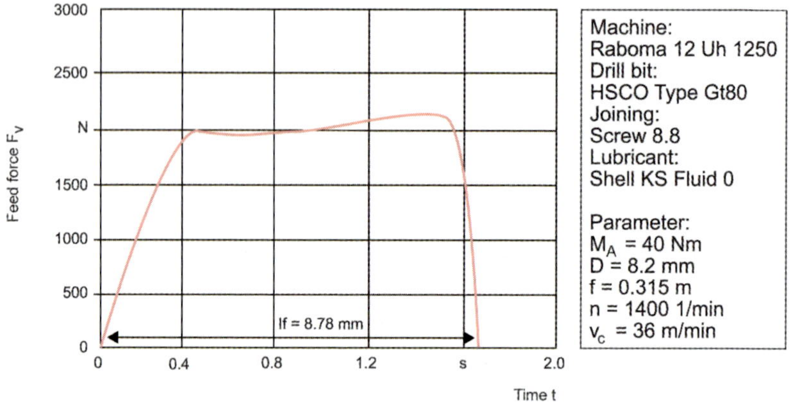

Fig. 1: Characteristic run of feed force during destroying a screw joint

Principally, feed force and torque show similar runs and differ from the conventional drilling process into full material in the area of the tear-off of the screw [2]. This area is characterized by a sudden drop in the curves. The force increases during the breakthrough from the screw head into the screw core with its reduced cross sectional surface. This significant increase of the force leads to a non-ductile fracture of the screw head. The screw shaft is driven out by the drill bit during the following unscrewing process of the remaining screw thread. The feed force and the torque can be disregarded. The width of fracture corresponds with the feed stroke of the drill bit. The selection of the correct drill bit diameter is the main prerequisite for the success of the disassembly process. At too small diameters of the drill bit, the sectional surface gradient of the screw core is not sufficient to initiate the screw fracture. Too big diameters of the drill bit will lead to a damage of the thread. A screw fracture rate of nearly 60 % could be achieved for the metric screw M10 and a drill bit diameter of 8.1 mm. The rate of fracture was nearly 90 % for adequate bigger drill bit diameters. The operating safety, the maximum screw clamping torque and the deviation of the drill axis with relation to the screw axis are the main quality criteria for the evaluation of the tap hole functions. This deviation is nearly independent from the drill bit diameter and is not relevant for the selection of a suitable drill bit. The optimum range of the stiffness for a screw joint is at a stripping security of $D/d_1 = 0.75 - 0.85$. Generally, the maximum clamping torque of a screw is defined for a value at 90 % of the limiting yield strength $R_{p0.2}$. This torque is nearly 48 Nm for a screw M10 in the property class 8.8. A significant higher clamping torque of 78 Nm could be measured within the experimental investigations. This can be explained by the potential influence of the resisting friction during the joining process. Under the assumption that a smaller width of screw head fracture correlates better with a reliable disassembly process, the influence of the misalignment of the drill axis and screw axis on the success of the disassembly process is depicted in Figure 2.

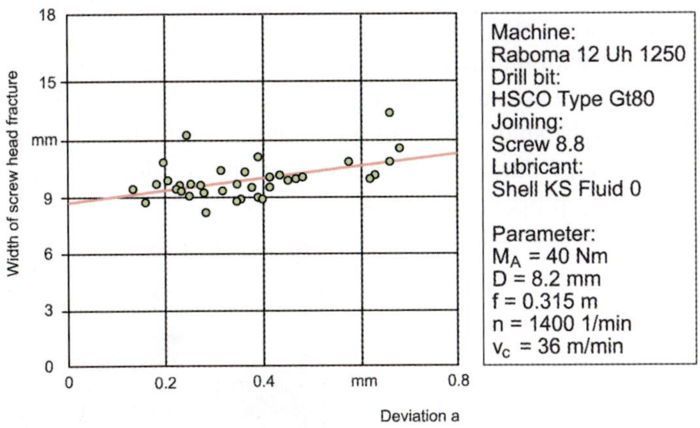

Fig. 2: Width of screw head fracture

The deviation of the failed disassembly trials was more than 0.6 mm. Based on the optimization of the revolution speed and of the feed it could be exposed that a high rotational speed and feed should be selected with regard to economical aspects. It should be considered that a risk of tap hole damages exist for high feeds by resulting forming processes during the drill out. Generally, the disassembly of screws which are also corroded or damaged is possible without any restrictions. The rate of success within the scope of carried out experimental investigations was nearly 98 %.

5.1.2 Shearing and Splitting

The process group "dividing" includes the listed processes shearing, splitting, rupture and breaking. A major advantage of dividing processes to other processes is the lack of secondary influences on other parts. Additional subsequent treatment is not necessary, which reduces the costs for cleaning or reconditioning. Analytical investigations have found that these processes can only be applied in special cases for disassembly. The principle of knife cutting has been used for the destruction of dampers and springs of the washing machines that are disassembled in the pilot disassembly system, where a pair of hydraulic shears was integrated. The process of shearing is suitable to disassemble mainly bars, wires, cables and tubes [3]. The rupture process can be used to tear off screw heads by tightening the screw with a tool until the head comes off. However, the implementation of this process requires an universal gripping mechanism which can grip screw heads with different geometry and is able to transmit the necessary torque. Thus, due to this disadvantage the process of splitting is better suited for this case. By driving a wedge with multiple pulse-like strokes into the joint between the screw head and the base surface, a shearing force is created which splits the screw head from the shaft. The tool with modified effectors and the experimentally determined duration for the disassembly of different screw types and corresponding property classes are described in Figure 3.

		Property class		
		4,6	8,8	10,9
Screw type	M3	0,2	0,4	0,6
	M4	0,5	0,9	1,1
	M6	1,0	1,8	2,0
	M8	1,9	2,5	3,1

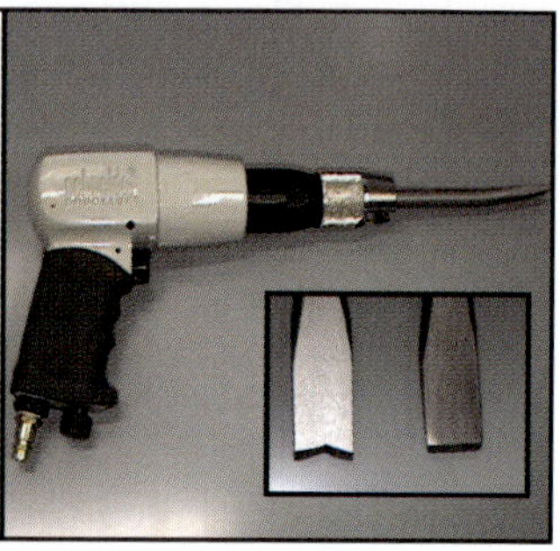

Fig. 3: Duration in seconds for the disassembly of screws

The splitting trials were carried out with a commercially available chisel hammer which operates at a maximum pressure of 7 bar and a power of 0.46 kW with a stroke frequency of 3200 per minute. The shape and geometrical dimensions of the end-effectors have been optimized. The chisels which existed for splitting have been modified by bending them in order to obtain a better suited split angle. Also, the active surface has been changed into a V-shape, so that the end-effector doesn't slip from the screw. With these changes it is possible to split screws up to a diameter of 8 mm within a maximum of 3 seconds. For diameters to 4 mm splitting times of less than a second could be reached. With a cryogenic pre-treatment the application range could be extended to screws with larger diameters, and the splitting times could be reduced further.

Hit shearing tools are suited to disassemble point forming connections by punching out the connection point. Thereby, the shape and geometrical dimensions of the end-effectors have to be optimized for the specific application.

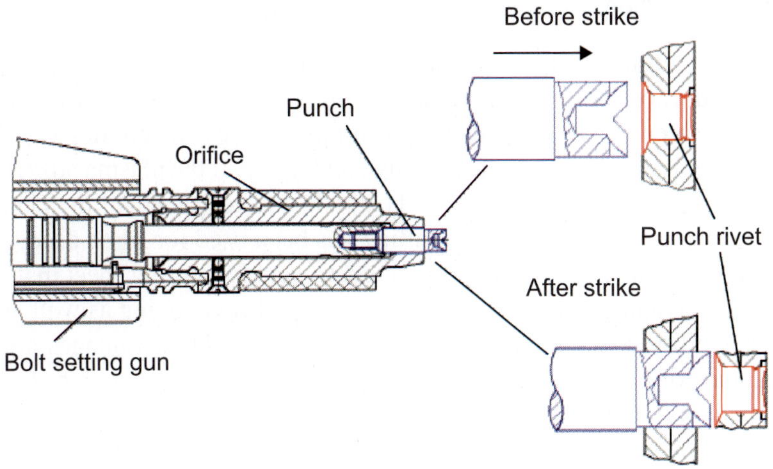

Fig. 4: Principle of a bolt setting gun (left) and strike process (right)

Furthermore, a wide range or various energy carriers can be preset as parameter. Even with such high levels of energy which are generated by an explosion, it is only possible to separate parts with a material thickness of 2 mm.

5.1.3 Jet Techniques

The opening of the case, the removal of the hatch, the cutting of the springs and dampers of the damping system and the disassembly of tubes and cables were identified as application areas for the destructive processes. The jet techniques like water jet, laser and plasma arc are suited cutting processes in cases of sheet materials, high cutting widths or high cutting lengths. Principally, the properties of the material to be cut and the risk of secondary damages have to be considered for the selection of a suited jet technique. Since there is no connection between tool and work piece, jetting processes have the advantages of a nearly force free mode of action, an omni-directional machining, and the free choice of the start and end point. Besides cut-off grinding, also the jet cutting processes are suited to open the washing machine case. In comparison with abrasive water injector jet cutting and laser beam cutting, plasma arc cutting has advantages in regard to cutting costs and speed.

Plasma Arc Cutting

Plasma arc cutting is a thermal jetting technology. It uses the heat of plasma to melt, burn, or sublimate material. Plasma is the so called fourth state of matter. It is generated when adding even more energy to certain gases. The gases de-ionize and are thus electrically conductive. If the plasma nozzle and the work piece are used as anode and cathode, an electric arc of high energy can be generated [4]. Plasma arc cutting has been used for some decades. It finds application in the ship building industry and for cutting sheet material. The main advantages are the fast and economical functioning as well as the high flexibility concerning the place of application. Disadvantages are the appearance of hazardous gases, the low cutting quality, and a high thermal influence on the processed material [5]. Except for the emitted gases these points do not count as disadvantages in disassembly (Figure 5).

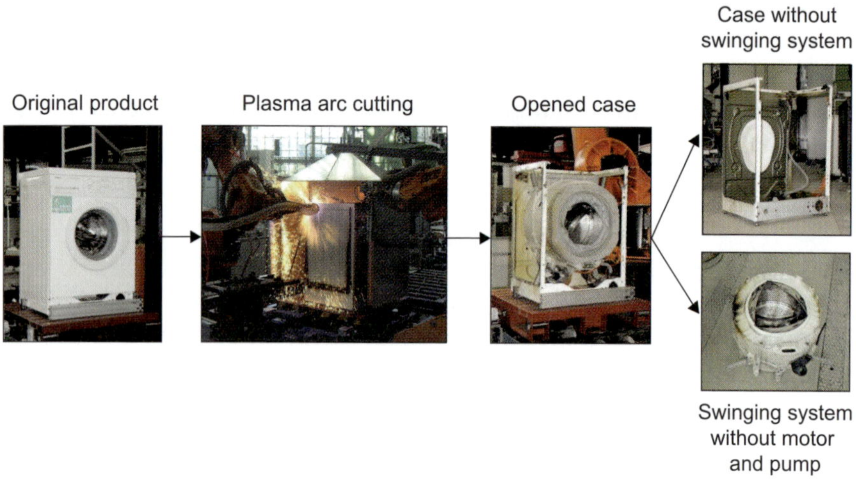

Fig. 5: Plasma arc cutting of washing machine housing

Within disassembly planning the opening of washing machine housings was identified as the first possible automated disassembly step. For this purpose, the application of plasma arc cutting was examined. With it, the panels of washing machines are separated in such a way that only a frame remains which holds the inner components. For different washing machine cases type-specific disassembly sequences were elaborated. Inflammable materials like insulators or cable loops had to be avoided. The aim of the first disassembly step is to reach inner components and assemblies as fast and cost-effective as possible. A robotic plasma arc cutting device with a transmitted plasma arc at cutting currents of 40 A to 130 A, and a cutting voltage of 160

V was used. The technical examinations contained a qualitative gas analysis as well as a process optimization with regard to maximum cutting speed, minimum mechanical and thermal component damage, and minimum removed particle mass [6]. The cutting gases tested were air, Ar/H2, and Ar/H2/N2. The maximum cutting speeds could be obtained with air. However there is a high emission of removal particles. With increasing cutting amperage the maximum cutting speed increases, too. At a cutting amperage of 125 A a maximum cutting speed of 12 m/min can be achieved. The cutting amperage and the cutting gas with a minimum removal particle mass were determined. Figure 6 displays the removal particle mass as a function of the cutting amperage in the case of plasma arc cutting of a work piece similar to the washing machine case material.

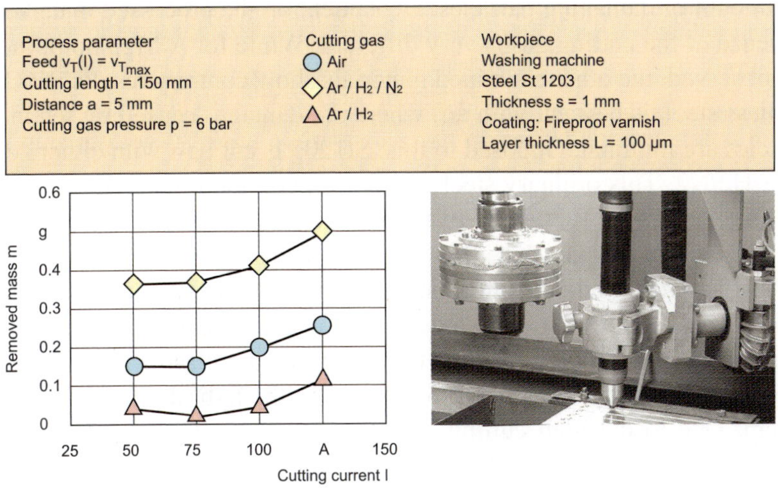

Fig. 6: Removal particle mass of sheet metals with plasma arc cutting

As reference materials AlMg3, St1203 and V2A with a thickness between 1 mm and 3 mm were chosen for the experiments. Only uncoated sheets were investigated in the first step. In the final step, the reference parts were coated with the commonly used varnishes and enamels to reproduce the emissions and byproducts of the real disassembly product. Powder varnish, enamel and 2-component PUR varnish was used in layers of 60 μm, 80 μm and 250 μm. The chosen materials, thicknesses and coatings represent more than 95 % of the market for common washing machines. After an optimization the particle emission during plasma arc cutting of a washing machine case could be reduced by 95 % compared to the initial setting which was a cutting amperage of 125 A and the cutting gas air. The conclusion is that the removal particle mass is almost irrespective of the cutting speed in the case of a defined separating cut and that it is only minimal with the cutting gas Ar/H2 and a cutting amperage of 75 A. A disassembly time of three minutes is needed for the industrial disassembly of a complete washing machine case if optimized parameters are applied during the separation of the four lateral walls of the washing machine case. After that, the assemblies and components can be disassembled in a destruction-free manner [7, 8]. A special test stand was built to determine secondary influence on close components. During the process of cutting, a part of the removed material evaporates or melts. The melted parts consist of cinder and sparks. They are accelerated by the cutting gas and thus get in contact with other components. To reuse these component parts, additional cleaning and rework processes have to be applied. The cutting gas and processed material have the main influences on the amount of damaging particles. Dependent on the processed material the influence of the cutting gas is very different. While for AlMg3 no influence was observed, the other examined materials show contrary results. X6Cr17, stainless steel, is best cut with air, where the damages caused by secondary particles are minimal. Opposed to this, St1203 is cut best with either Ar/H2 or Ar/H2/N2. This ordinary steel is the most common material of washing machine housings. Taking all examinations into account and considering that air is the cutting gas that allows highest feed speeds, it is difficult to find the best parameter adjustment. The fact that for the most common material Ar/H2 causes the lowest emissions and damages turned the balance to this cutting gas. Including the results, the design of the plasma arc device and the periphery were altered and integrated into the hybrid pilot disassembly system. Due to the high cutting speed, good automation and low cutting costs, plasma arc cutting can be used in the pilot disassembly system to open washing machine cases from various suppliers.

Plasma arc cutting is the most effective disassembly process with secondary influences. Due to the optimization of the process combined with an installed housing with a suction system it was possible to minimize the occurring emissions. Therefore, the potential danger for workers within a hybrid disassembly system was reduced.

Water Jet Cutting

Water jet cutting was introduced to industrial practice in the early 1980s, when pumps with sufficiently high pressure and nozzles with small diameters and long tool lives could be produced for the first time [9, 10, 11]. The performance of cutting with a water jet could be considerably improved by adding small solid abrasive particles. Today, with the development of ever more effective and powerful compressors and pumps, ultra high pressure water jets are available. The range of commercially used high pressure pumps goes up to almost 4,500 bars. Combined with an abrasive medium it is possible to process almost every material. It facilitated the cutting not only of soft materials such as leather, synthetic materials, and paper, but also of harder materials like metals. Water jets are produced by transforming the energy of pressurised and compressed water into the kinetic energy of the jet. This transformation takes place at a nozzle. The diameter of the nozzle determines the size of the jet and, together with the pressure, the application of it. Broad jets at low, middle, or high pressure are used for cleaning or deburring operations. With narrow jets at ultra high pressure it is possible to cut any given material. Jet cutting with water can then be divided into the groups water jet cutting and abrasive water jet cutting, while abrasive water jet cutting can be further subdivided into abrasive water injector and abrasive water suspension jet cutting [12, 13, 14]. The three variants of water jet cutting are illustrated in Figure 7.

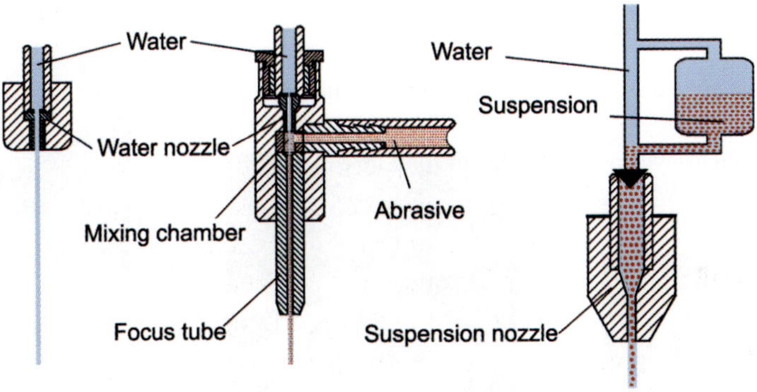

Fig. 7: Pure (left), abrasive injection (middle) and suspension (right), water jet [13, 14]

For all three variants the pressure generation is based on the principle for pure water jets. The highly compressed water runs through a pulsation damper to create a steady flow and leaves the system at the nozzle. Then it meets with the work piece and cuts it. Due to the fact that not all kinetic energy is absorbed in the cutting process, the residual jet leads to secondary damages. The pure water jets are capable of cutting soft materials such as rubber, synthetic foam and insulating compounds. Abrasive water jets are created by adding an abrasive aid to the compressed water in a mixing chamber. The mix has then to go through a focusing tube to assure that particles and water are unidirectional. The injection system is the technologically most relevant process variation. It is used for cutting hard materials such as steal, non-iron metals, or titanium in nearly all fields of industry. The examinations on adjusting water jet cutting to the requirements of disassembly were carried out on a system working with the injection principle. For abrasive suspension water jet cutting an abrasive is added to a compressed reservoir of water. This reservoir has a limited volume and is used as a substream from the main stream. Compared to the injection system the suspension system provides a better mixing of water and abrasive, wastes less of the kinetic energy, and is much more subject to wear. The main disadvantage is the discontinuous process which is caused by the limited volume of the suspension reservoir. Another task of the investigations on water jet cutting is to determine and explain the jet properties, particularly the influence of the secondary jet on inner components and assemblies. If water jet cutting is used for disassembly, damages are caused by the primary and the secondary jet.

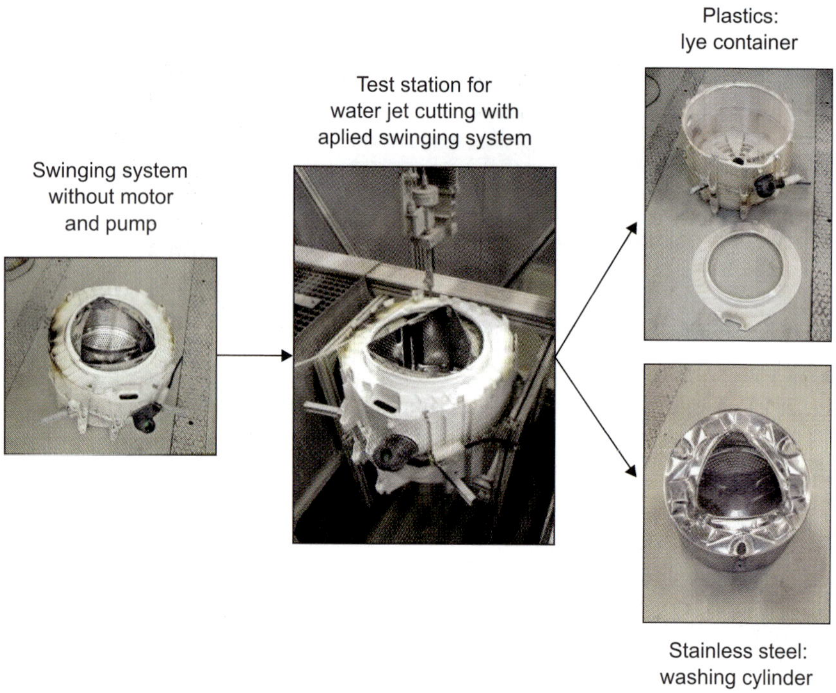

Fig. 8: Water jet cutting for the disassembly of washing machine components

The primary jet is the jet prior to cut and the secondary jet exists behind the work piece that is machined and dies when striking inner components.

With disassembly chances are high that the secondary jet hits an inner component or other assemblies. When the so-called secondary jet distance, the distance between the primary work piece and inner components, is low, reusable parts are at risk of being damaged. Thus, the objective is to minimize the energy of the secondary jet and to predict its influence on other components. The secondary jet distance is measured in a straight line which runs vertically to the primary work piece through the jet exit point. The abrasive water jet was integrated into the disassembly process of standard washing machines to open the plastic lye container and to separate it from the washing cylinder made of stainless steel, which is shown in Figure 8.

The separation of the damping system, which gains access to the lye container and the washing cylinder, is a potential application for water jet cutting. The precondition is that the damping system is stripped from weights, pump and motor before. Therefore, the secondary damages of inner components can be reduced in cases of cutting plastics or soft materials. The use of abrasive particles can be avoided for these soft materials. The secondary damage and the spread of the water jet are obvious at compound materials or hollow shaped components (Figure 9).

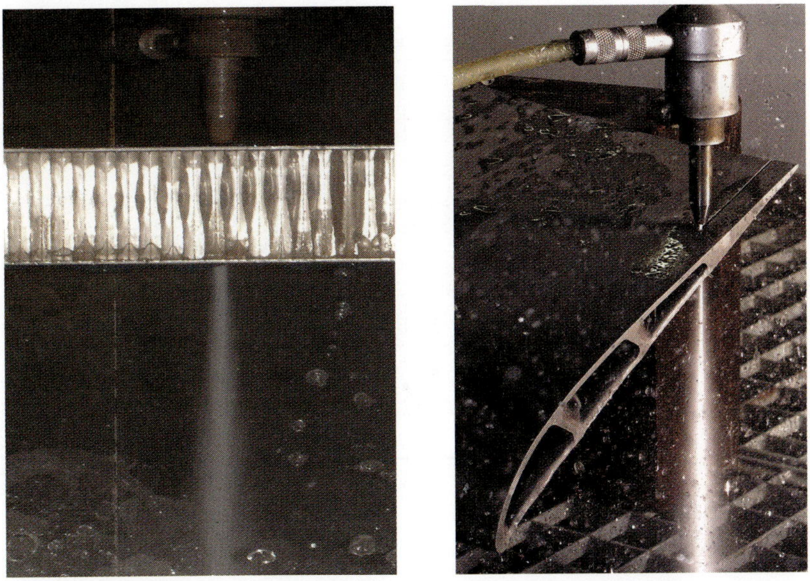

Compound material   Hollow shaped component

Fig. 9: Abrasive water jet cutting of a compound material (left) and a hollow shaped component (right)

These light-weight materials and components are used to reduce the weight of the washing machine and to reduce the acoustic emissions. The challenge for investigations within the field of water jet cutting in disassembly processes is to describe the properties of the jet, e.g. the distribution of the abrasive particles in the jet or their velocity, which is difficult due to high velocities and limited space. The properties of the jet are mostly measured indirectly by measuring the work result, meaning the kerf generated by the abrasive water jet [14]. Examining the kerf however not only helps to draw conclusions about the abrasive water jet, but also to gain information in order to evaluate the quality of the work result for use in manufacturing technology. The kerf profile is directly dependent on the velocity distribution of the jet. The velocities decrease with increasing distance from the center axis of the jet and fall below the minimum velocity necessary for removal [16, 17, 18, 19]. Since the minimum velocity is specific for each material, model and calculation of the coefficient are based only on the aluminum alloy AlMg3. As a result, the damage of an aluminum secondary part can be neglected from a secondary jet distance of 25 mm or above when separating with 90 % of the maximum traverse rate. Figure 10 emphasizes this property with experimentally investigated cutting speed and feed for different hollow shaped profile materials.

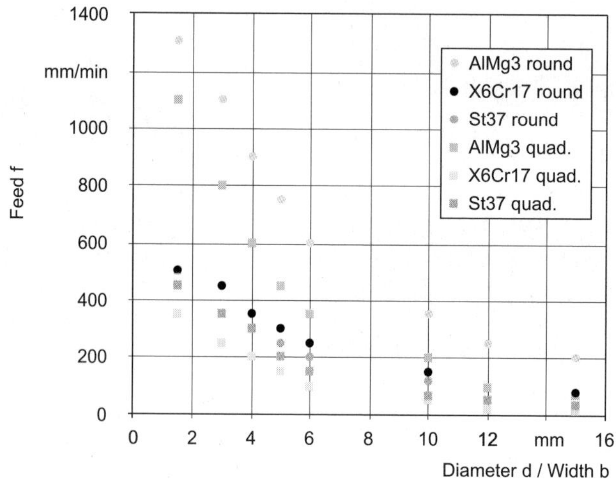

Fig. 10: Maximum feed for different profile materials

For other materials, the coefficients must be determined anew by a few calibration tests. The insights won in this study regarding the potential for damage of both the primary jet exiting from the jet head and the secondary jet form the basis for a three-dimensional application of abrasive water jet cutting. These examinations on water jet cutting show both chances and limitations. It is possible to predict the secondary influences on the reusable part by adjusting the parameters or the cutting direction and thus to avoid them or at least to minimize them. Water jet cutting prove to be economically adequate for various disassembly operations, but only as a stand alone solution. The integration of this process into an automated disassembly line is very complicated. The used water can harm the transporting system and subsequent processes or tools.

5.1.4 Cryotechnical Hybrid Tools

Chipping and jet cutting processes create undesirable by-products like gas, sparks, smoke or molten mass. They pollute or destroy reusable components and put strain on the environment and the machine user. High process forces with the splitting processes can cause deformations of the disassembly object or damages of tools or effectors. Dividing processes with high process forces are used to disassemble joints, which put strain on tools and component parts. So far, a modified pneumatic chisel for the lateral splitting of screw and rivet heads and a hydraulic pair of shears for cutting of cables, wires, bars and tubes have been developed.

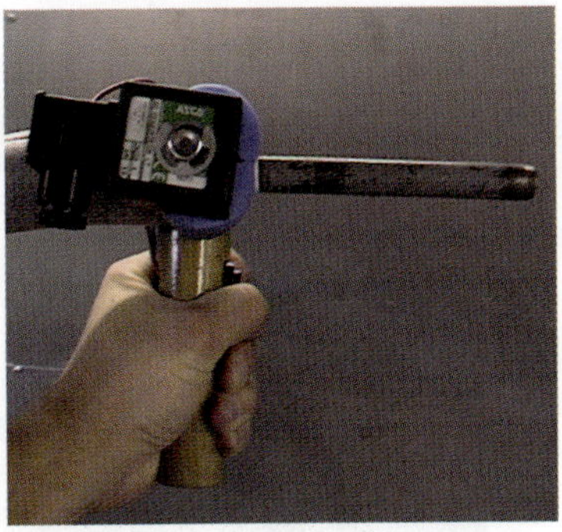

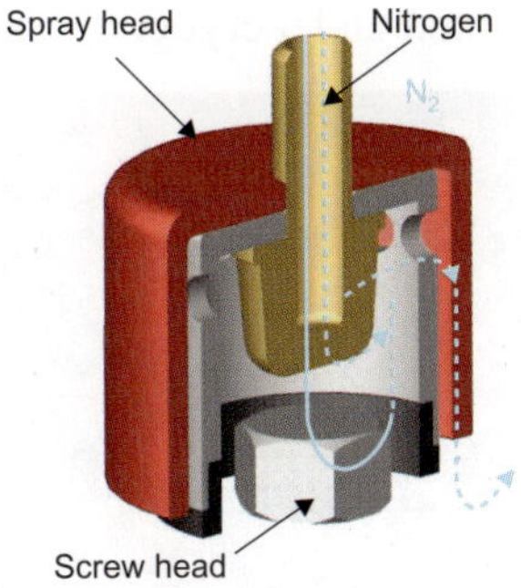

Fig. 11: Cryogen spray pistol (left) and optimized spray head (right)

However, for metallic materials the resulting high process forces limit their application. In order to reduce these forces, the cohesion of the material can be diminished. This can be done by heating or by embrittling with liquid nitrogen in the case of materials with body-centred cubic lattice structure. The processes embrittling with liquid nitrogen and dividing can be combined for cryo-splitting and cryo-shearing. A tool with such a combination was developed to divide screws (Figure 11).

As a first step, the cooling times for metal sheets and screw heads are determined for the combination of the dividing process with the cryogenic pretreatment. Liquid nitrogen is used within a wide application range in the area of refrigeration because of its low temperature of -196 °C and its comparatively simple production, storage and handling. Conventional installations take the nitrogen from the pressure tank with a standpipe. Part of the medium evaporates in the pipe. With a phase separator it is ensured that only the gas phase escapes. For the disassembly applications the use of the liquid phase is of advantage, because a large heat quantity is withdrawn from the material by the evaporation of the liquid and therefore the workpiece cools down more quickly. If the nitrogen is merely poured onto the surface, an insulating gas cushion forms between the liquid and the workpiece, and the cooling time increases. Therefore, a special spray head illustrated in Figure 11 has been developed for the optimal application of the liquid phase. Then, the embrittling was combined with a disassembly tool for the hit shearing of screw heads. The process times and forces could be reduced by up to 50 % and the application range for screws extended from a diameter of 5 mm to 10 mm at a property class of 10.9. The combination of a cryogenic pretreatment with a shearing tool for the disassembly of high-grade steel bars raised the workable diameter from 6 mm to 10 mm. Moreover, the shearing work was reduced by up to 80%. The effect is obvious at the ruptured surface of an untreated and a cryogen treated screw head (Figure 12).

Fig. 12: Ruptured surface of an untreated (left) and pretreated screw head (right)

The hybrid use of the cryogenic pretreatment together with the dividing process leads to halving the process forces, shortening of process times and an extension of the application field. The flexibility of the process is increased and the disassembly more efficient.

The cryotechnical process and the plasma arc cutting represent thermal methods. A further method is represented by the thermal removal of materials with cutting or heating tools at higher temperatures.

5.1.5 Inductive Heating and Thermal Removal

Generally, products with a high complexity consist of different materials like metals, plastics, glass or stoneware. The thermal removal of different materials has to consider the range of melting temperatures. While the melting temperatures of metals are significantly higher than 1000 °C, the plastic materials can be removed with temperatures between 200 °C and 400 °C. The inductive heating is used for alloying, forming, melting or tempering. Its main advantages for disassembly are the non-contact heating and the lack of contamination. Thereby, the inductive heating is used for conductive metallic components by applying an electrically generated alternating magnetic field with high frequencies. Specifically designed induction coils are used as inductor to heat or melt, e.g. metallic bars (Figure 13).

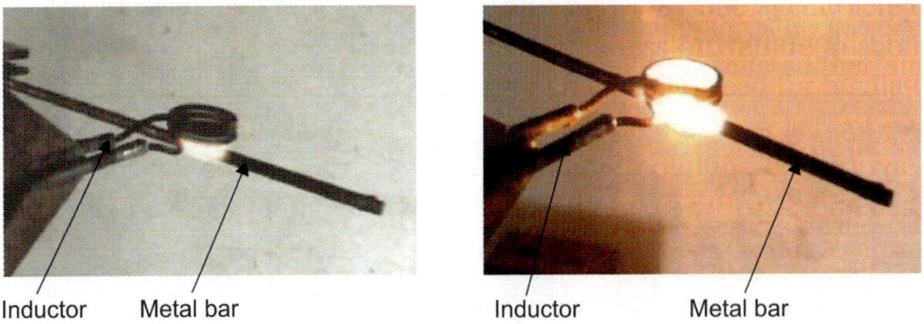

Fig. 13: Inductive heating (left) and melting (right) of a metal bar

Although the maximum temperature and the speed of heating or melting can be varied within a wide range, the main disadvantage is given by the circumstances for the disassembly. The housings of products made of sheet metals can be better cut by plasma arc or laser beam cutting, which have lower negative thermal influences on the ambient components. The restrictions of accessibility to inner components, the minor rate of thick metallic devices, and the damages caused by drops of melted metals have to be considered. Thereby, the combination of heating and traction force are better suited to decrease the effects of the mentioned negative influences. The optimized combination can be better applied for plastic materials, where a heated metallic knife serves as cutting tool (Figure 14).

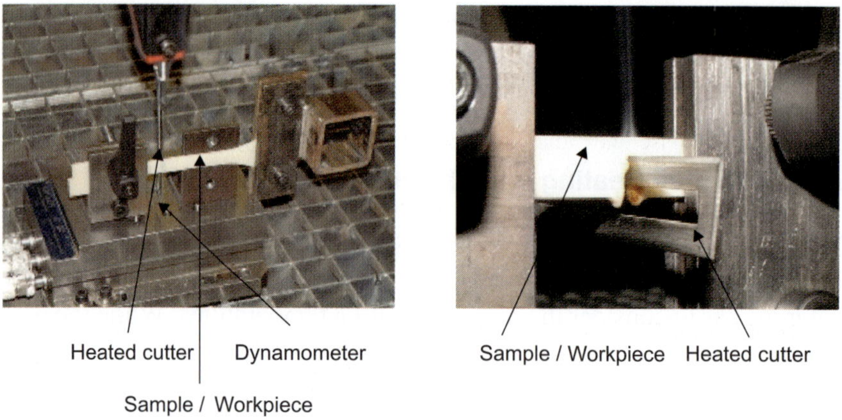

Fig. 14: Horizontal (left) and vertical (right) cutting of plastics with a heated knife

The temperature of the cutter of nearly 600 °C is higher than the melting temperature of plastics. The resistibility force, measured with a dynamometer, emphasizes the range of the cutting force for exemplary plastic materials. An additional traction force enables a further decrease of the mean resistibility force of the cut plastic materials (Figure 15).

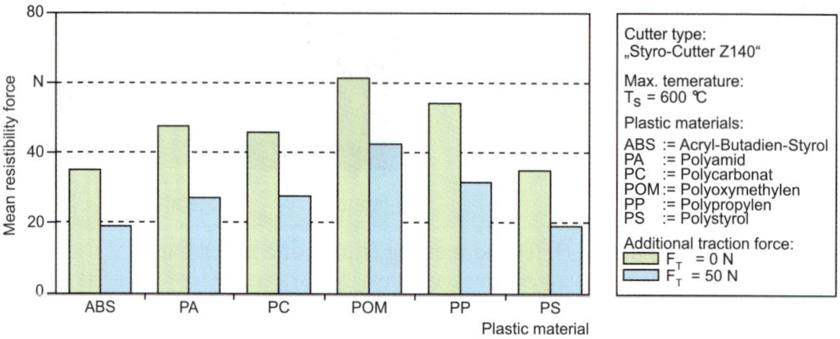

Fig. 15: Influence of an additional force on the mean resistibility force of plastics

The combination of temperature, cutting tool and an additional traction force reduces the risk of damages to further components arranged near to the cutting area. Furthermore, this process can be automized easily with standard parameters for the limited groups of technical plastics in products. In this context the negative effects of toxic or acid smokes can be avoided with adapted exhausts.

References

1 Spur, G.; Axmann, B.; Elbing, F.; Seibt, M.: Fertigungsverfahren der Demontage. Normungsvorschlag zur Begriffsklärung in der Demontage. Institut für Werkzeugmaschinen und Fabrikbetrieb, TU Berlin, 1997.
2 Spur, G.; Stöferle, Th.: Handbuch der Fertigungstechnik, Band 3: Spanen, Handhaben und Montieren. Carl Hanser Verlag München, Wien 1979.
3 Härtwig, J.-P.: Verfahren und Systeme zur Demontage komlexer technischer Gebrauchsgüter. Dissertation, TU Berlin, 2005.
4 Bach, F. W.: Beitrag zum thermischen Schneiden dickwandiger Werkstücke. Habilitation, TU Hannover, 1983.
5 Prismeyer, U.: Thermische Schneidverfahren und Werkstoffreaktionen in Hinblick auf die Entstehung von Staub und Aerosolen. Dissertation, TU Hannover, 1997.
6 Uhlmann, E.; Axmann, B.; Elbing, F.: Fast Disassembly of Point Forming and Welded Connections. In: Proceedings of the 4th World Congress R`99. Recovery, Recycling, Reintegration. Geneva, Switzerland, 1999, P. III. pp. 328-333.
7 Spur, G.; Uhlmann, E.; Elbing, F.; Dittberner, J.; Sundaresan, S.; Thantry P, B.: Flexible Automatic Disassembly for the Recycling of Consumer Goods. In: Advances in manufacturing Technology XIV, Proceedings of the 16th National Conference on Manufacturing Research, London, England 2000, pp. 407-411.
8 Uhlmann, E.; Spur, G.; Elbing, F.; Dittberner, J.: Optimization of Plasma Arc Cutting for the Disassembly of Washing Machines. In: Proceedings of the International Conference on Cutting Technology 2002, Hannover, pp. 61-66.
9 Engel, P. A.: Impact Wear of Materials. Elsevier Scientific Publishing Company, Amsterdam 1976.
10 Wiedemeier, J.: Flüssigkeitsfreistrahlen großer Relativgeschwindigkeit und Bruchkinetik spröder Werkstoffe. Dissertation, TU Hannover, 1981.
11 Oweinah, H.: Leistungssteigerung des Hochdruckwasserstrahlschneidens durch Zugabe von Zusatzstoffen. Dissertation, TH Darmstadt, 1989.
12 Guo, N. S.: Schneidprozeß und Schnittqualität beim Wasserabrasivschneiden. Dissertation, TU Hannover, 1994.
13 Meier-Wiechert, G.: Unterwassereinsatz von Wasserabrasivstrahlen. Dissertation, TU Hannover, 1993.
14 Laurinat, A. H.: Abtrag mit Wasserabrasivinjektorstrahlen. Dissertation, TU Hannover, 1994.
15 Uhlmann, E.; Axmann, B.; Elbing, F.: Model of Kerf and Simulation of Damage in Abrasive Water Jet Cutting. Production Engineering (VI/2), 1999.
16 Blickwedel, H.: Erzeugung und Wirkung von Hochdruck-Abrasivstrahlen. Dissertation, TU Hannover, 1990.
17 Finnie, I.; Levy, A.; McFadden, D. H.: Fundamental Mechanisms of the Erosive Wear of Ductile Metals by Solid Particles. American Society for

Testing and Materials, Philadelphia, USA 1979, pp. 36-58.
18 Uhlmann, E.; Axmann, B.; Elbing, F.: Strahlstoßkraftmessung beim Strahlen mit CO_2-Pellets. ZWF 93 (1998) 6, pp. 240-243.
19 Spur, G.; Axmann, B.; Elbing, F.: Model for Kerf Profile and Jet Velocity in Abrasive Water Jets. In: Proceedings of the 5th Pacific Rim International Conference on Water Jet Technology. New Delhi, India 1998, pp. 407-420.

5.2 Detachable Connections

Günther Seliger, Stefano Consiglio, Berlin, Germany

To enable the reuse of components or products, adaptation processes must be non-destructive. If, however, partly destructive processes are employed, the damages must be compensated. Partly destructive disassembly processes are often more cost effective than non-destructive processes, especially if only the joining elements are damaged [1]. Typical joining elements are often standard parts, e.g. screws or rivets, and therefore they can be easily and inexpensively replaced.

Remanufacturing facilities require flexible tools for cost effective non-destructive or partly destructive disassembly operations due to the large variety of products in small lot sizes [2]. That is, a diverse spectrum of joining techniques or elements must be separated.

Tools which can detach a connection independently of its attributes have a high grade of flexibility. Figure 1 shows different possibilities of the implementation of flexibility in tool design.

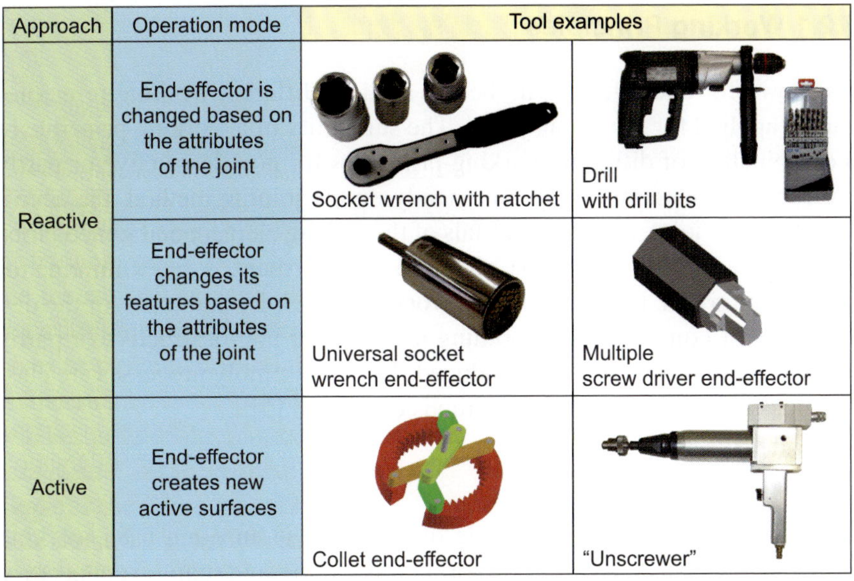

Fig. 1: Active and reactive approaches for flexible disassembly tools

In the first case the tools' end-effector can be quickly changed; thus suitable end-effectors can be attached to the tools depending on the attributes of the joints. The disadvantage of this approach is the long idle time because of the required end-effector changes. In the second and third case the tools have shape-independent end-effectors.

The end-effectors of the second case fit to the shapes of the connectors by adapting their shapes and not by changing the end-effector as in the first case. The universal socket wrench consists of many separately supported pins, which adapt to the contour of the screw head when attaching the end-effector. The multiple screw driver end-effector fits to the active surfaces of slotted head, recessed head, and hexagonal head screws with a metric thread diameter between 4 and 6 mm [3].

Contrary to the first and second case, where the tools or end effectors react to the attributes of the joint, the approach in the third case is an active one. The end-effector generates new active surfaces on the joining element and then uses these surfaces for loosening. The collet end-effector pushes its teeth into the lateral surface of the screw head and can thereby transmit the initial breakaway torque for loosening. The "Unscrewer" generates a notch in the front surface of the screw head and thus even countersunk screws can be unfastened [4].

5.2.1 Working Principles

As shown for end-effectors in the last section, different working principles can be applied for loosening joints. The suitability of a working principle or a combination of different working principles for non-destructive or partly destructive disassembly depends mainly on the joining method, the accessibility of the joint, and the materials of the joining element and joined parts. In some cases also the shape of the joining element is relevant, e.g. for screws and rivets. In the following, working principles and tools for the disassembly of commonly used joining methods for non-destructive or partly destructive disassembly are introduced. The disassembly processes are classified according to the joining methods defined in [5].

Assembling

In the class of assembling, elastic deformation is an often used working principle. Typical connectors are spring rings, clips, and snap-on connectors. Snap-on connectors are often used in parts made of plastic or metal sheet, especially in housings to ease the assembling. Commonly, plastic housings show traces of use after a usage phase, while the technological functionality of the product is still given. Examples of such devices with a high reuse value are mobile phones or computer screens. By deforming the snap-on connector the joint can be disassembled in a non-destructive way. Figure 2 shows a prototypically realized tool for loosening snap-on connections in mobile phones. A partly destructive way to loosen snap-on connectors is to overstrain the connector by pull force, thus the connector breaks or deforms until no form closure is left.

However, snap-on connections are often difficult to access. Therefore a more suitable approach is to integrate the loosening mechanism into the product. Braunschweig discusses different approaches such as snap-on connectors made of shape-memory materials. The connectors deform with heat and thereby relieve the form closure with the opposing part [6].

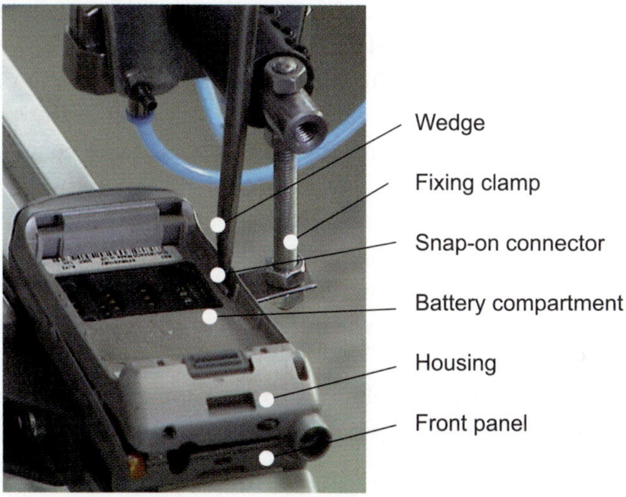

Fig. 2: Disassembly tools for loosening snap-on connections

Joining by Mechanical Means

Most types of joints belong to this class, and screws are the most important connectors within this class. More than 40 % of all joints in machines and cars employ screws [7]. Screws can be unfastened using existing active surfaces such as hexagonal sockets, or by generating and using new active surfaces [4]. When using existing active surfaces for manual unscrewing, the required tool has to be identified, the tool has to be positioned and orientated, and the end-effector has to be engaged with the screw head. The torque required for loosening is usually provided by a mechanized tool. Often the end-effector has magnetic properties, enabling the screw to be retrieved by the tool after a successful unscrewing operation. The control of the whole process is carried out by a technician. The process steps which the technician can execute easily due to his cognitive abilities are challenging for an automated system.

The use of product information for choosing the proper end-effector and for positioning the tool is hardly appropriate for a remanufacturer because the product condition is unknown. Moreover, remanufacturing companies often do not get adequate product data from the product manufacturer [8]. Therefore an unscrewing system must be enabled by sensor systems and intelligent strategies to react to a product and its conditions. The data concerning changes of the product condition during the usage phase as well as the accessibility of the joining elements has to be acquired and processed autonomously. The system has to control and verify the engagement of tool and screw head, the loosening of the screw, and its removal.

For a reliable engagement, the concentricity of the end-effector and the screw head is vital. In order to locate the exact position of the screw and to verify whether the expected screw head is present, a 2D image-processing system is suitable [9]. However, image analysis will not always be capable of identifying and locating the screw head, for example, due to differing ambient light conditions and differing levels of contaminants on the screw head surface. An important factor when using image processing for positioning a tool is that the connection between the tool and end-effector is rigid and free from backlash.

The number of required revolutions for a complete unfastening process can be calculated, if the product data is available regarding the screw type and the length of the engagements. However this approach is not fail-safe because the fastener could be replaced by a shorter or longer one, e.g. during repair work. To ensure a robust process, the torque, the revolutions, and the travelling distance of the screw have to be measured. The removal of the screws might be difficult due to space restrictions, material properties of the screw, and contaminants on the screw head.

For ferromagnetic materials a permanent or electromagnet is suitable. A photoelectric barrier can be used to verify that the screw is properly removed from the join patch and afterwards from the end-effector.

A non-destructive unscrewing tool is shown in Figure 3, where the above discussed issues have been considered. In the developed tool the backlash is compensated by a collet which consists of three jaws, sliding on bevelled guiding actuated by a pneumatic cylinder. The tool has been integrated into a disassembly system and applied to loosen screws of engines.

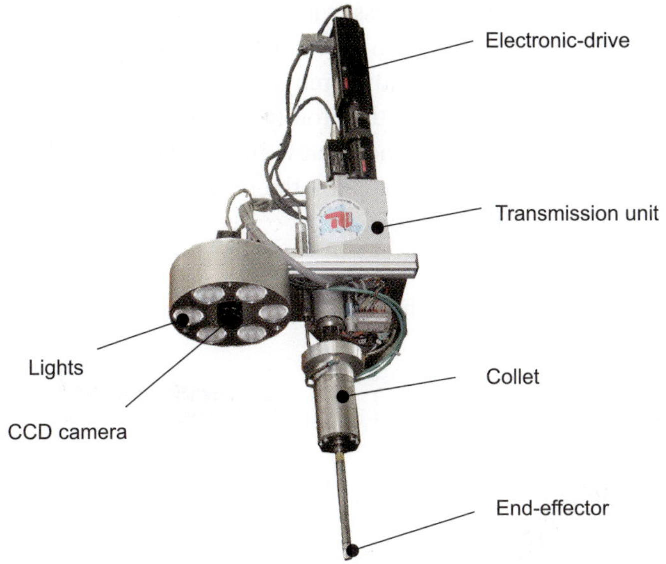

Fig. 3: Non-destructive unscrewing system

For unscrewing independently from the existing active surfaces of the screw head, various working principles are suitable to generate active surfaces: pressing, drilling, milling, sawing, grinding, and notching. Notching the upper surface with a wedge is considered to be the most advantageous working principle, based on a comparison with respect to the following objectives [4]:

- little space and little process time required,
- easy to control and independent of material,
- simplicity, e.g. acting surface-producing component can be used for the transmission of torque, and
- low side effects of the procedure, e.g. little load is transmitted to the other parts.

The wedge strikes in a straight line perpendicularly or diagonally into the surface of the screw head. The active surface is generated by repeatedly striking the rear of the wedge with a moving mass, thus transferring its energy to the wedge. The notch is created very quickly and is relatively large in comparison to the volume of material displaced.

The wedge used to produce the notch is also used for transferring the torque required to loosen the screw. The notching process leads to work hardening in some materials, which favorably affects the torque that can be transferred to the screw.

Further advantage is that process wastes such as shavings, chips, and splinters are avoided, compared to the working principles based on chipping. The applied force is low and thereby the side effects are small compared to pressing. Figure 4 shows the prototypically-implemented hand and automated tool as well as a notched screw head. The wedge is contained within the centering unit which supports the positioning of the tool during the notching and unscrewing process.

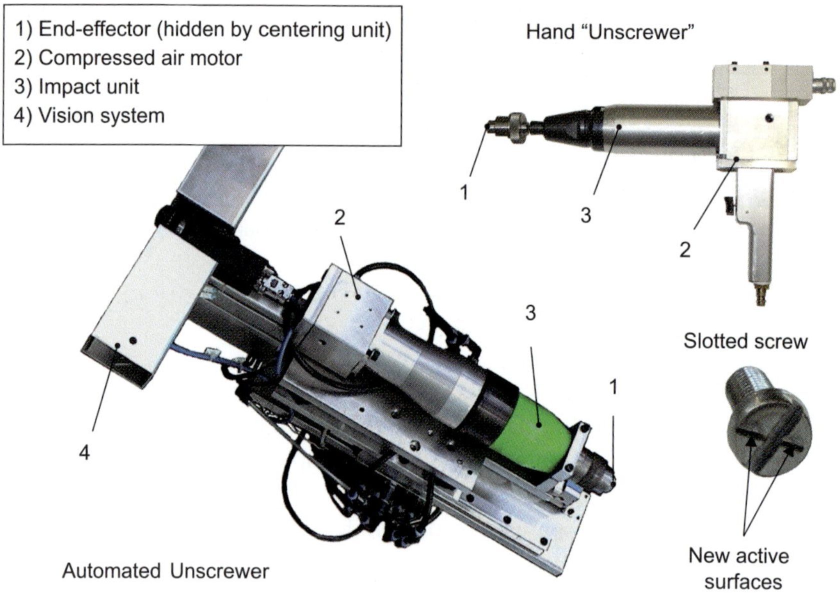

Fig. 4: Partly destructive unscrewing tools

Soldering

Soldering processes can be divided into soft, hard, and high temperature soldering [10]. Electronic components are joined with substrates, usually Printed Circuit Boards (PCB), by soft-soldering processes. These joints can be loosened using thermal and mechanical processes. The removal of soft-soldered components is reasonable for repair purpose, separation of hazardous substances, or if there is a market demand for certain components.

If the components have to be separated from the substrate non-destructively, the spectrum of suitable processes is reduced to those few which employ thermal working principles. Usually, mechanical processes separate the parts from the PCB by cutting or sawing the pins, and thus the pins are too short to be soft-soldered again. In thermal processes where all joints are melted at the same time, temperature-sensitive components can be damaged due to an excessive energy input. The heat sources used for these mass-desoldering processes include ovens, hot plates, or baths of liquefied solder. The thermal impact during the desoldering process can be reduced to a minimum by selective desoldering [11].

Currently, the particular desoldering process to be used on a product is determined on the basis of the operator's experience. The effort to adjust process parameters by experiment is quite high and must be undertaken for new or unknown electrical components. A procedure-independent process model has been developed to determine analytically the minimal required heat input [12]. In the following, the process model and its software implementation is described. The process model takes into account the heat energy and flow in the joint by computing the temperature profiles in different locations of the solder joint. The process model is based on the physics of heat flow and thus it is independent of the specific geometry of the solder joint, the soldered parts, and the soldering principle.

Using specific partial models the machine parameters and the dimensions of the solder joint are translated into process parameters and classified into the classes material, geometry, and energy. For modelling, the solder joint is divided into homogeneous sectors. For example, the pin is divided into three sections, and thereby the heat differential over the part can be taken into account for determining process parameters.

The model is simplified by approximating the differential equations by difference equations and by reducing the three-dimensional equation for heat conduction to a one-dimensional expression. Figure 5 illustrates the simplified geometry for the modelling of a Through Hole Device (THD) and Surface Mounted Device (SMD) solder joint and the energy flow in a THD solder joint. Some heat transfers are not considered due to a negligible influence on the heat flow. Convection and radiation occur around the solder joint, while conduction takes place between the components [13].

The evaluation of the procedure-independent desoldering model is carried out by experiments in an automated desoldering system.

The experiments were conducted using Sn60PbCu2 solder, a standard FR-4 PCB with Cu-pads, and a socket. The energy source in the experimental apparatus is a diode laser. The energy absorbed by the solder joint is determined by a pyrometer. To calibrate the pyrometer an emission factor has to be determined. Due to a certain proportionality of emission factor and temperature, the emission factor can be verified with the help of a thermocouple. An exact indication of the pyrometer is guaranteed when measured temperatures are synchronized by adjusting the emission factor. Figure 6a shows the experimental rig with a laser, a pyrometer, a camera, a power unit, and a fixed PCB.

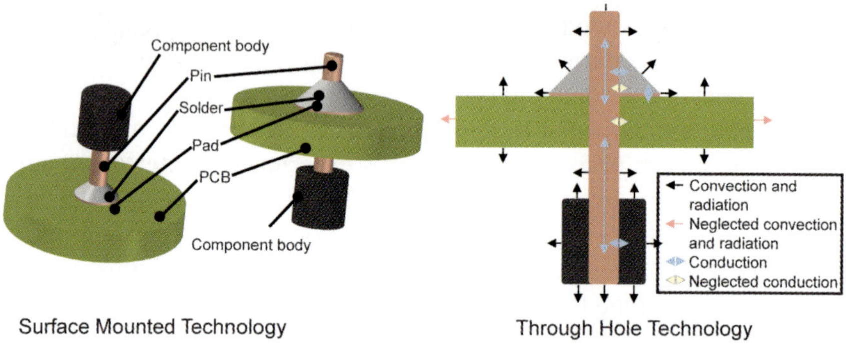

Fig. 5: Elements and energy flow of the process model

The camera focuses on the solder joint and provides a high-resolution greyscale picture to the computer monitor beside the test rig. The laser and pyrometer are installed at opposing angles to the PCB. A lamp illuminates the test rig in order to provide a high-quality camera image. Figure 6b illustrates the simulated run of the temperature curve of the process model implementation (left picture) and a comparison of simulated and measured values (right picture).

The temperature curve can be divided into the heating-up phase where energy is transferred into the solder joint, and the cooling-off phase where the laser is switched off. The temperature of the solder stabilizes briefly while melting and solidifying. Results of the process model such as melting time, solidification time or the required amount of heat are displayed in the software menu.

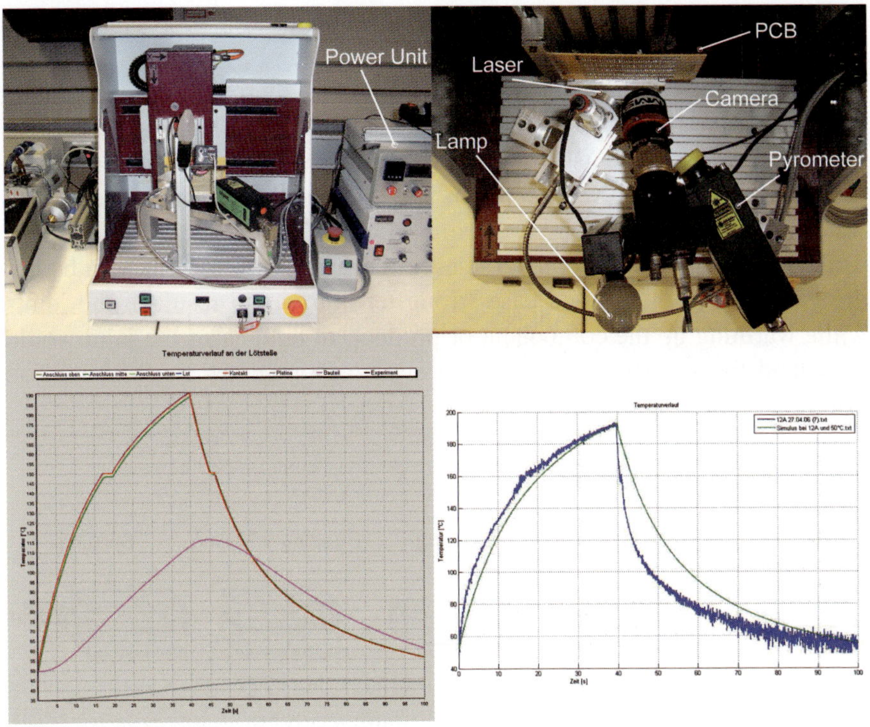

Fig. 6: a) Experimental desoldering apparatus
b) Software implementation and comparison of simulation and experiments

The gradient of the temperature curves between simulation and experiments show slight differences, caused by the dependence of the emission factor on temperature. The emission factor used in the process model is calibrated to account for this difference at the temperature that is most significant, i.e. at phase transition. At other non-critical temperatures, this calibration causes inaccuracy. By this approach the minimal temperature for non-destructive desoldering can be accurately predicted.

Gluing

Gluing can be applied to a large variety of different materials. The endurance of a gluing joint depends on the adhesion between the part and the glue, on the cohesion of the glue, i.e. on the strength of the molecular bonds in the glue, and on the dimension of the glued surface. Glued parts can withstand high strain by pull and shearing because the force is evenly distributed over the whole joined surface. In contrast, the adhesive and cohesive bonds are only partly strained if, e.g. bending moments or splitting forces are applied to the gluing joint (Figure 7a).

Bonded joints can be separated by chemical, thermal, and mechanical processes as well as by a combination of these. Solely chemical or thermal ungluing processes are usually not capable of performing the non- or partly destructive disassembly. Exceptions are for example connections where water sensitive glue was used or connections where the parts have a very different thermal expansion coefficient. In the first case the glue can be hydrolyzed and in the second case strong tension would result in the joints while warming up the component or product. In both cases the joint would withstand far less strain. Mechanical ungluing processes can be based on different working principles.

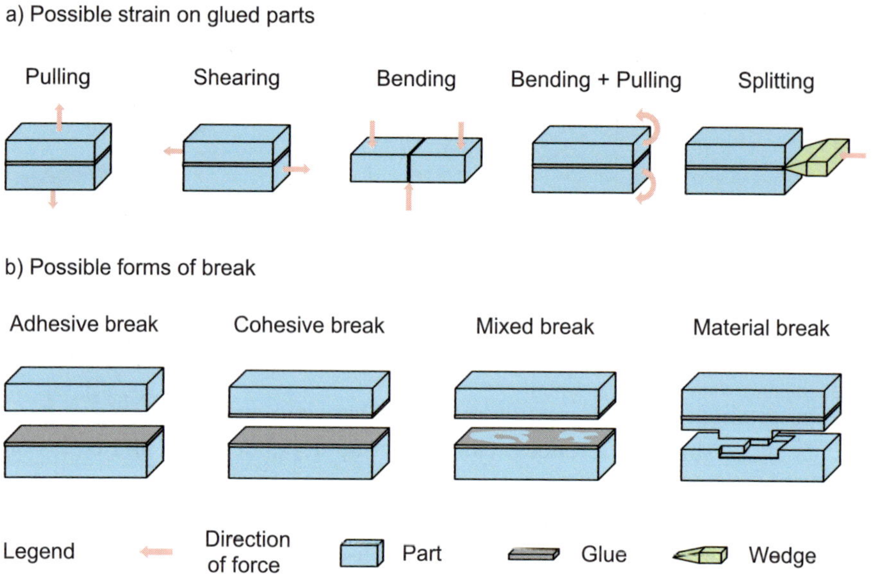

Fig. 7: Strain and break forms of glued parts

The applicability of principles basing on chipping or knife and wire cutting depends predominantly on the film thickness of the glue. The film thickness is very thin in many products and components and thus the available working surface for most end-effectors is insufficient. An alternative is to separate glued joints by overstraining; this is independent from the film thickness. The following forms of fracture can result by overstraining (Figure 7b):

- adhesive breaks, i.e. the break takes place between material surface and glue film,
- cohesive breaks, i.e. the break takes place in the glue film,
- mixed breaks, i.e. adhesion breaks and cohesion breaks take place at the same time, and
- material breaks, i.e. the material breaks first.

For reducing the remanufacturing effort it is advantageous to generate adhesive breaks. In the case of cohesion and mixed breaks the glue remnants have to be removed from both parts before they can be joined again. In the case of part replacement it is advantageous if the glue sticks completely on the part to be replaced. A field of application for this is in glued car headlights, where only the front glass is damaged.

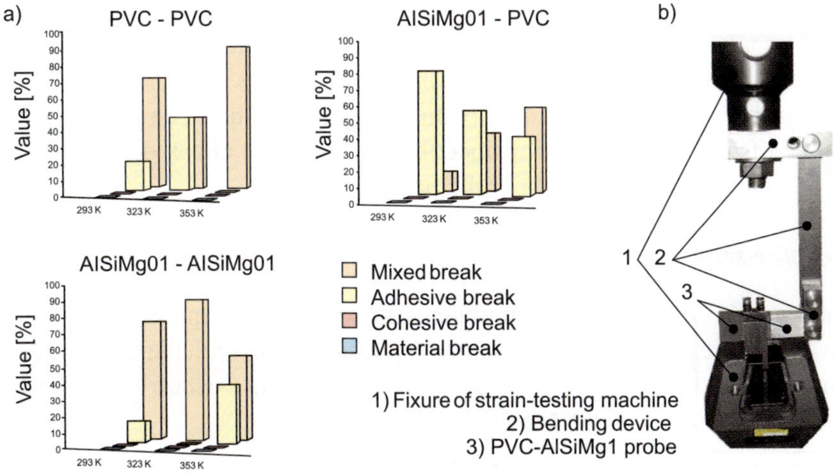

Fig. 8: a) Distribution of breaking forces depending on material and temperature
b) Test set-up

Besides the desired break form, i.e. an adhesive break, the forces and the energy should be as low as possible during the separation process. This can be achieved by working principles such as bending or splitting which strain only a few adhesive and cohesive bonds instead of the complete glued area. For the determination of the suitability of these working principles as well as the influence of heat input on the breaking behaviour among others a bending device was developed and experiments were undertaken. A stove was used for warming the test parts. To determine the required heating time to achieve certain temperatures in the glue film, the temperature of drilled test parts has been measured during a warming-up process with a thermocouple. The test parts have been made of hard PVC and AlSiMg1 and glued with UHU Plus Endfest 300. The strength of the test parts were determined by the strain-testing machine Zwick Z150. Figure 8a and 8b illustrate the test set up and the results of the experiments.

In these and all other experiments undertaken, the reliability to break the adhesion of a joint by overloading and applying heat energy is insufficient. A possibility to increase the proportion of adhesive breaks is to change the surface treatment. However, a lower adhesive strength between part surface and glue increases the danger of an infiltration of the gluing layer with contaminants and thus of product failure.

To what extent and by which chemical substances the adhesive strength of a joint can be weakened to support the separation process depends on the specific joint, i.e. on the material of the parts, the glue, and the conditions during its life cycle.

Although many glued parts can be separated without destruction by overloading, the effort for the adaptation of the components after the dismantling is very high. Tools based on overloading must have end-effectors with a low degree of form independence, since the tool does not attack directly at the glued joint but at the surfaces of the bonded parts.

Flexible Handling

Handling of products and components is sometimes required during and always before and after the loosening operation. This is challenging for automated disassembly processes, because the amount of occurring shapes of products and components is extensive.

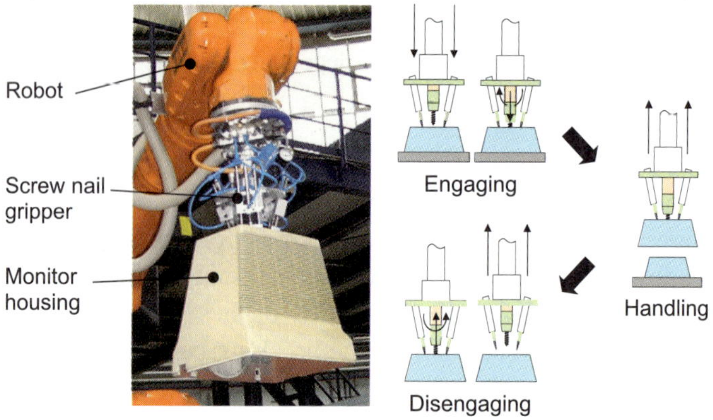

Fig. 9: Flexible handling tool and its working principle

Therefore, flexible handling tools such as a screw nail gripper shown in Figure 9 are vital. The tool generates a new active surface by grooving a thread into the part utilizing the form closure to carry or transmit loads [14]. This working principle destroys the engaged part and can only be employed if the part is rugged enough to carry its self-weight and the weight of the attached parts. Fields of applications are e.g. handling of housings made of plastics.

5.2.2 Disassembly Tool Kit

Resource consumption for required tool functionality and overall tool costs for assembly and disassembly can be reduced by using modular tool kits. Furthermore, the development time for providing new tool functionality, e.g. required for new products to be assembled or disassembled, can be also shortened. Particularly in the field of disassembly, modular tool kits are advantageous, because of low product similarity, the unpredictable product conditions, and thereby the required multiple tool functions [15].

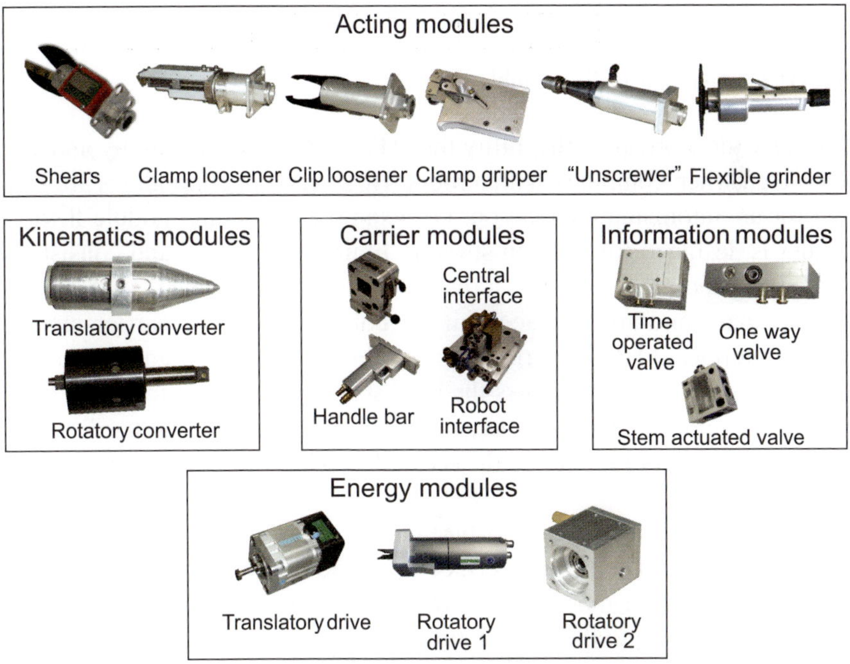

Fig. 10: Module groups of the DTK

A modular tool kit, the Disassembly Tool Kit (DTK), was developed to provide multiple tool functions with high efficiency by little resource consumption. The DTK has a modular architecture and consists of the module groups: acting, carrier, energy, information and kinematics (Figure 10). Some of the developed modules have been prototypically implemented for evaluation purposes.

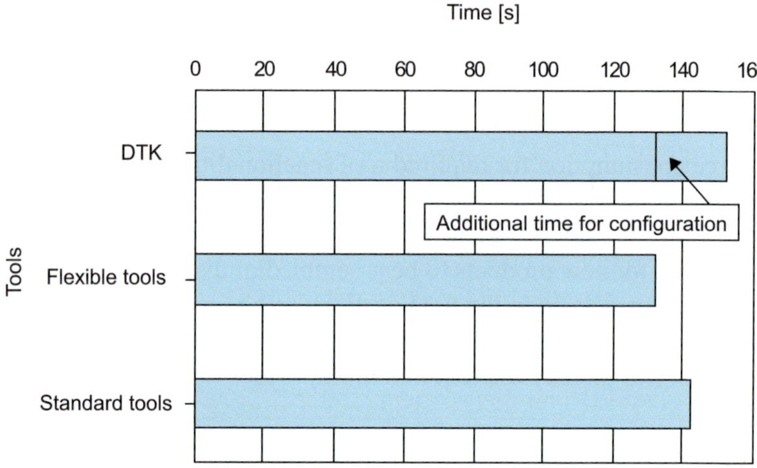

Fig. 11: Comparison of disassembly time

To provide a broad applicability the DTK can be used manually and also automated when combined with a robot. However, not all modules are suitable for an automated application. For example the acting module flexible grinder is designed for tasks in strongly limited workspace and consists of a fist-size grinding unit to which the grinding disc is attached, a 1.5 m long flexible shaft, and an interface unit for coupling with the carrier module central interface. Only the grinding unit, the flexible shaft, and the arm of the tool operator need to access the workspace. The central interface module, the carrier module handle bar, and the energy module rotatory drive can be outside of the product to be disassembled. This will not be possible in a robot application.

The carrier module robot interface can be attached to the tool changing system of a robot. The robot interface locks the central interface, the acting as well as the energy modules and transfers compressed air and current via the central interface to the information modules.

The information modules drive the energy and acting modules with compressed air and supply the Life Cycle Units of these modules with current. Just one push button for activating the air pressure flow is integrated in the handle bar to keep the operation of a configured tool for the shop assistant as simple as possible. Some configured tools require a different drive of one or more acting and energy modules. For example the information module time control can be used, if an acting and an energy module shall be driven in series as is required for the tool configuration partly destructive unscrewer. First the time control module passes the compressed air to the impact module for notching, and after a few seconds to the rotatory drive module for unscrewing. By selection of different information modules, the switching time, the volume flow and the distribution of the compressed air can be varied. In this way the tool partly destructive unscrewer can be configured into the tool chisel. The tool configuration non-destructive unscrewer can also be changed into the tool driller by attaching the acting module driller and if required a different revolution can be achieved by changing the information module. The action and energy modules are equipped with LCUs to ensure the modules' availability and process reliability [16].

The applicability and the performance of DTK were evaluated by a comparative investigation: motor and discharging pump of washing machines were disassembled applying conventional tools, flexible tools, and the DTK. Main constraints for planning and execution of the disassembly were the following [3]:

- Dismantling has to be done manually by one person.
- One machine after another has to be disassembled due to space restriction.
- Other parts than the motor and the discharging pump are allowed to be damaged.
- The risk of injury for the worker has to be minimal, thus e.g. all edges must be smooth after processing.
- The motor and the discharging pump has to be disassembled as fast as possible.
- The set-up times have to be as low as possible.

In the following the results are shown with respect to the utilization rate and the total process time (Figure 11 and Figure 12) .

As the figures above show, the utilization rate of most modules is higher than the rate of flexible and standard tools. However, the process for dismantling the motor and the discharging pump takes slightly longer with DTK, resulting from the additional handling effort to configure modules to a tool. The more different functionalities for loosening operations are required and the longer the deployment time of a tool is, the more beneficial is the application of the DTK.

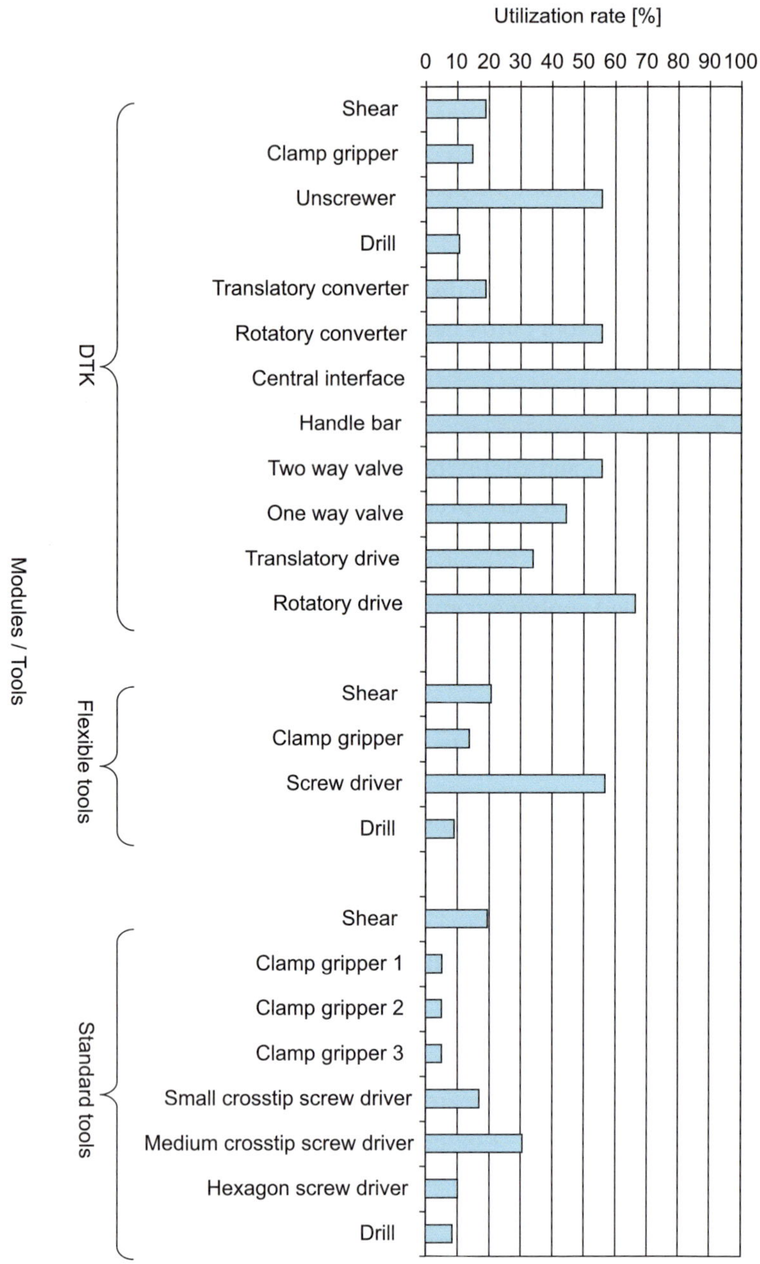

Fig. 12: Utilization rate of modules, flexible tools and standard tools

References

1 Rebafka, U.; Pfeiffer, R.: Schrauben statt schreddern. VDMA Nachrichten 08/2001, S. 44-46.

2 Rebafka, U; Keil, T.; Seliger, G.; Stenzel, A.: Flexible Disassembly tools. In: Proceedings of the 2001 IEEE International Symposium on Electronics and the Environment, Denver, USA, 07.-09.05.2001, pp. 30-35.

3 Rebafka, U.: Beitrag zur Entwicklung modularer Demontagewerkzeuge. Dissertation TU Berlin, 2003.

4 Wagner, M.: Prozessmodell für die Kraftübertragung durch neue Wirkflächen zur Entwicklung geometrietoleranter Demontagewerkzeuge. Dissertation, TU Berlin, 1997.

5 DIN 8583: Manufacturing Processes Joining; Classification, Subdivision, Terms and Definitions. Deutsches Institut für Normung, Beuth Verlag, Berlin, Köln 2003.

6 Braunschweig, A.: Automatisierung der Demontage. Habilitation, TH Ilmenau, 2002.

7 Bauer, C.-O.: Handbuch der Verbindungstechnik. Carl Hanser Verlag, München, Wien 1991.

8 Seliger, G.: Chances and Challenges for Remanufacturing. In: Proceedings of Polish-German Workshop on Lean Remanufacturing. Wroclaw, Poland, 20.-21.04.2006, pp. 3-21.

9 Keil, T.: Informationstechnische Integration hybrider Demontagesysteme. Dissertation, TU Berlin, 2004.

10 DIN 8505: Soldering and Brazing; Classification of Processes, Terms. Deutsches Institut für Normung, Beuth Verlag, Berlin, Köln 1979.

11 Stenzel, A.; Seliger, G.: Flexible Desoldering Station for Automatic Disassembly. In: Proceedings of the International Symposium on Sustainable Manufacturing 1999. Shanghai, China, 1999, pp. 144-150.

12 Seliger, G.; Rebafka, U.; Stenzel, A.; Zuo, B.-R.: Process Model Based Development of Disassembly Tools. Journal of Engineering Manufacture, Vol. 215, Part B, pp. 711-722.

13 Niemeier, J.: Entwicklung und Verifizierung eines Prozessmodells für das Einzelpunktlöten in der Elektronikfertigung. Dissertation TU Berlin, 1998.

14 Stenzel, A.: Beitrag zum flexiblen Greifen in der Demontage. Dissertation, TU Berlin, 2001.

15 Seliger, G.; Consiglio, S.; Odry, D.; Zettl, M.: Development of Intelligent Modular Tools for Disassembly. In: Proceedings of 15th International CIRP Design Seminar 2005, Shanghai, China, 22.-26.05.2005, pp. 182-187 (CD).

16 Buchholz, A.: Zustandsorientierte Instandhaltung von Standardkomponenten mit Life Cycle Units. Dissertation, TU Berlin, 2005.

5.3 Clamping Systems

Eckart Uhlmann, Ingo Früsch, Berlin, Germany

Clamping of workpieces in flexible manufacturing systems of any kind demands a clamping technology, which allows fixing a large variety of products. Flexible clamping systems must be able to deal with different geometric dimensions and shapes of the manufactured workpieces. The more as agile manufacturing and construction processes require an increasing flexibility towards changes of products. In a parallel development the complexity of manufacturing systems is rising.

The time being necessary for the clamping process has to be reduced to a minimum. Especially the set-up-time of a clamping system in case of changing the type of workpiece has to be as short as possible. To reach fully automated manufacturing processes, many improvements are still to be do. Atomization is essential for production plants in highly industrialized countries as capacities have to be used in a more efficient way.

Another request towards clamping systems in modern manufacturing is an optimal functional linkage to the logistic environment in production plants. In the present situation the placement and exchange of workpieces is mostly done by manual work. Automated pick-and-place-systems for exchanging and clamping workpieces are characterized by a high complexity and often cause problems concerning the reliability of a manufacturing system. Additionally the costs of investments for these auxiliary systems reduce the productivity of the manufacturing system. Because of these reasons fully-automated clamping processes are not common in plants with a non-determined variety of products.

In the recent past different concepts of machine tools based on parallel kinematics have been developed. Since in 1994 on the IMTS in Chicago where Ingersoll Inc., Rockford, Illinois and Giddings & Lewis Inc., Font du Lac, Wisconsin showed their new types of machines, a rapid development began. New concepts and machines where designed and produced for different purposes. Now a clamping system is developed, which uses the flexibility of parallel kinematics to clamp and position a large variety of workpieces.

For the disassembly of used goods, clamping devices are necessary, which have to work under different conditions in comparison to clamping devices in the production of new products. Based on the economic purpose to process many different products in one factory, the task is to clamp a wide range of one product type of different producers into one and the same clamping device. Furthermore, the condition of these old products is unknown. It varies within a wide range. The geometrical changes caused by usage, such as deformations, or surface changes such as corrosion, pollution, coatings and lacquering have to be considered. Also, the formation of variants has to be considered resulting from usage through modifications, missing parts, destructions, etc. Characteristics such as component elasticities, components stabilities, and the stabilities of such disassembly products are hard to predict.

The result of thus emerging possibilities of variant development is that the lot size „1“ has to be expected for the products to be processed in the disassembly factory, since each product is significantly different from the others in disassembly. The aim of scientific disassembly leads to the requirement of clamping devices with the highest possible flexibility. The clamping devices have to be flexible with respect to shape and geometries of the objects to be clamped. The aim is to generate methods for developing of clamping devices for prismatic, rotationally symmetrical, and of any other geometries as well as to realize them [1, 2, 3].

Clamping devices in industrial disassembly systems must be flexible regarding the dimension and the shapes of the workpieces.

Bases for the choice, planning and design of the clamping systems for disassembly are aside from efficiency, quality and productivity:

- the capability to clamp a wide range of products,
- the capability to clamp workpieces independent from the current condition,
- the capability to support the handling,
- the capability to operate in manual and automated systems.

Clamping is an auxiliary function in the disassembly process. These disassembly processes can be subdivided into pre-processes, main processes and post processes. Main functions are all separating operations. Auxiliary functions help indirect with the fulfilment of the disassembly processes. These are processing steps like handling, clamping, or measuring, which operate in auxiliary process time. The duration of these auxiliary functions has to be reduced. In particular the set-up time for clamping systems has to be minimized, because of the small lot sizes in the disassembly. Another requirement is the optimal integration into the logistic environment [1, 3, 4].

5.3.1 Dodekapod – Parallel Kinematic with 12 Axes

The kinematic structure of the most developed hexapodic machine tools is designed to move a spindle or a kind of tool placed on the Stewart-Gough-platform by its six legs. The workpiece is fixed below or in front of the moved tool. Hexapod applications, which do not move a tool-system, are mainly simulators or positioning systems. To move a clamped workpiece by a parallel structure in a disassembly process is the purpose of the discribed system. The aim is to combine the kinematic system of a machine tool and the function of its pick-and-place module into one system. This new type of clamping system can be called autonomous clamping system.

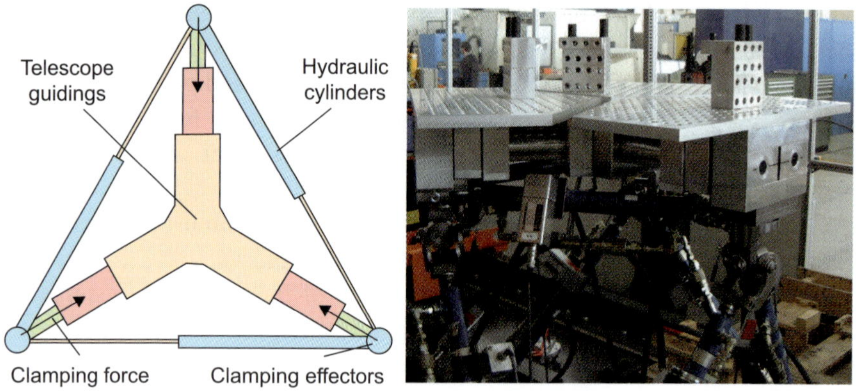

Fig. 1: Structure of a triangular clamping platform and assembled

An autonomous clamping system contains the aspects of autonomous picking and placing of workpieces in a manner that no human decision or aid is intervening the process of clamping. The general clamping process mainly includes the spatial positioning and orientation of the workpiece, fixing of the workpiece and the transduction of the loads from the manufacturing process. In the recent past there were several attempts to design clamping systems which were able to set up automatically for different workpieces. Due to the variety of workpieces, the clamping system and its effectors or jaws must have the ability to reach a wide stroke in different geometric formations. For this purpose a structure which can be described as a hydraulic triangular clamping platform is suitable (Figure 1).

The ability to adapt to different geometries is reached by the independence among the hydraulic cylinders. For this reason the structure embodies three degrees of freedom, allowing to clamp different types of geometries. The three clamping-effectors or jaws located at the points of the triangle can be moved like rotatory clampers as well as centric and parallel clampers. By changing the effectors an adaptation to most surfaces can be achieved. The hydraulic principle allows controlling the clamping force. Additionally in a setup process the jaws can be adjusted and adapted in a fixing grid on the platform (Figure 2).

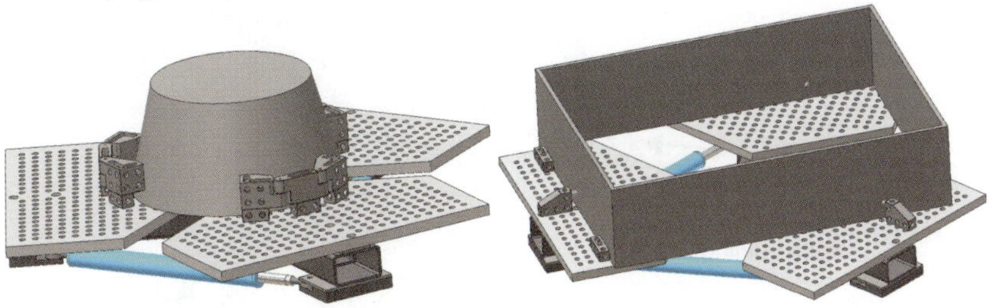

Fig. 2: Clamping platform with different jaws for different workpieces

Apart from the adjustment functionality the clamping platform needs guidings which generate a plane movement and inhibit rotational displacement of the knots where the effectors are located. The double-telescopic design of the guidings corresponds to the construction elements of cylinders and leads to a kinematic range, which gives no restrictions to the cylinder movement. The guidings must be dimensioned in order to sustain the forces of the manufacturing processes and torques as well as the generally much higher loads of the gripping processes. If a workpiece is clamped, all three hydraulic pistons are under pressure. In this case clamping platform and workpiece form a unit, which behaves as a Stewart-Gough-platform [2, 3].

Parallel Kinematics for Clamping Systems

In consistent continuation the rack of the clamping system is built as a framework construction (Figure 3). The structure is inspired by the design of hexapods, but goes beyond them in its kinematical abilities: The Stewart-Gough-platform is further developed to a clamping platform as described above [5]. The base points of the six hexapod legs are not connected to a fundament, but to mobile joint knots which can be shifted through a plane triangle of hydraulic cylinders. The design concept of the new structure leads to a self containing and basement free machine.

Hence a structure results, which is an oktaeder with 12 movable legs. It can also be regarded as a hexapod with two geometric-flexible platforms, called „Dodekapod“. Analog to the name „hexapod“, the name is composed of the Grecian words „dodeka“ = twelve and „pod“ = leg [1].

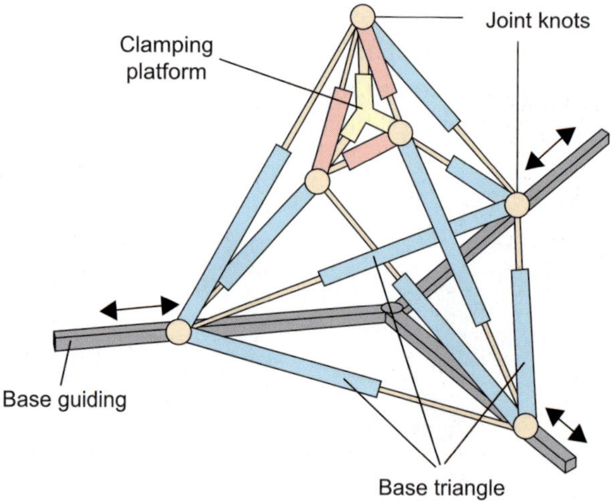

Fig. 3: Geometric structure of a dodekapod

Drives

In discussing the geometric structures of parallel machines and the kinematics including the degrees of freedom the mechanical components these machines are built of are of importance. Amongst the drives of hexapods rigid struts, lengthen-variable driven struts, plates and shells are to mention. Different drive-concepts are to be named, first of them the direct linear motors, which are vehemently in development.

Amongst all drives, hydraulic cylinders show the most favorable proportions in size and weight compared to the height of produced forces. Beyond that, the hydraulic principle is frequently used in clamping technology. For this reason hydraulic cylinders were used in the "Dodekapod" (Figure 4). The disadvantages are that hydraulic cylinder show less rigidity and allow only limited speed due to the compressibility of oil.

For a sufficient stiffness of hexapods their frames need to be rigid enough not to deflect under load. Because of lever-effects the tensions in the structure can be much higher than the external loads. Due to the closed framework of the octahedral dodekapod the inner loads do not reach into the fundaments. Therefore the static requests at the fundament and the basement guidings are not high [1, 3].

Fig. 4: Dodekapod, parallel kinematic with 12 axes

Joints

The design of the joint knots of the dodekapod stucture deserves special attention. The kinematical requests are extremely high. Each joint knot is connecting five components: a guiding element, two spherical rotating cylinders, and two plane swiveling cylinders. Figure 5 shows six of many possible arrangements.

One of the ideas of parallel structures (like hexapods) is to have merely compression and tension loads in driven leg components. These structures then promise to embody the best proportion between their own weight and the useable forces or handled weights. This can only be achieved when the pivots of particular rotating linear drives are placed together as close as possible. The aim is to concentrate the axes in one single certainly fictive point. As a result no torque or bending effects occur external of the joint knots. So the guidings at the fundament are charged only with a minimum of forces. As a second result of the force concentration the tensions inside the joint knots can be reduced to a minimum, which leads to a lighter design.

The geometric restrictions of joints have to be reduced to reach the needed kinematical compatibility between the four connected hydraulic cylinders. But the desired concentration of force-vectors can only be approximately realized without restricting the geometric capability of the whole system. Thereby two cylinders must be able to swivel a plane in an angle of about 45 degrees and the two spherical cylinders must be able to work in a half sphere. Compared to hexapods the enlarged working area of the Dodekapod requires joints which have a higher kinematical capacity.

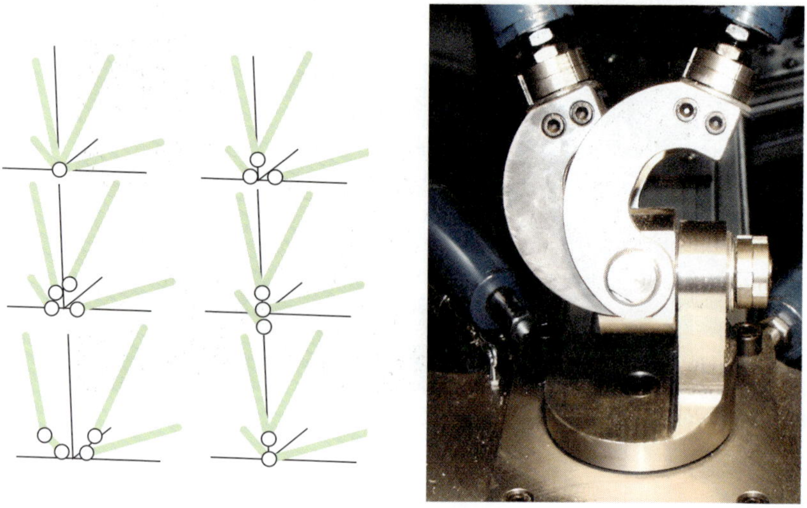

Fig. 5: Joint knot with guiding and four hydraulic cylinders

For practical mathematical description and numerical control of the Dodekapod, each single joint element must offer a spherical rotation around one ideal point. In the design of the joints the cardanic principle was extended for one additional revolute joint, in the baseplate of the knot joint. This enlarges the swiveling range of a connected leg to a cone with an angle of over 200 degrees (Figure 5).

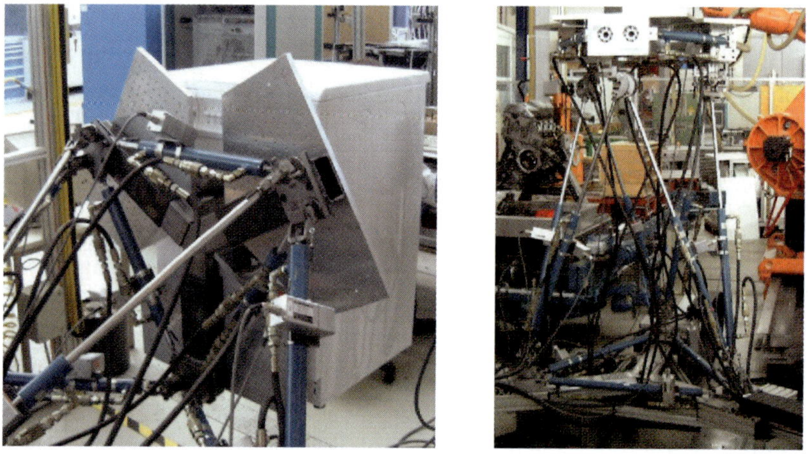

Fig. 6: Clamping platform rotated and at highest position

Thus it is possible to rotate the clamping platform more than 90 degrees around between the base guidance (Figure 6). This circumstance offers the possibility for additional functionality for parallel structures. It shows that the parallel kinematics need new specific joints and design elements due to their specific characteristics. Some special joints are necessary, in order to be able to increase the workspaces [3, 6, 7, 8, 9].

Kinematics

The gripping system includes 12 driven axes. For clamping a workpiece the movement of the three upper clamping cylinders is necessary. In dependence on the clamped workpiece the cylinders find their positions. Their three lengths determine the geometry of the clamping platform, which influences the possible moving range of the dodekapod. These lengths are parameters for the movements and the workspace of the Dodekapod. Therefore the moving abilities of the dodekapod depend on the clamped workpiece. The nine remaining axes can be used for movement. Clamping platform and workpiece are regarded as one solid body from a kinematical point of view.

The system is kinematically triply redundant. Within wide areas of the workspace movements can be produced by altogether four different sequences of cylinder movements. In the core workspace it is possible to move along every path and to reach each point by four different cylinder movements. At the peripheral zone of workspace the motion possibilities are reduced to be less redundant. At the edges of the motion possibilities each movement of the clamping platform can only be done with one sequence of cylinder movements. Apart from the enlargement of the workspace the additional possibilities of movement can be used to optimize other quantities concerning the system performance. The system can be moved in such a way, that the rigidity of the structure in each handling position is sized to a maximum. Movements into a desired position can be controlled to shortest processing time. The structure also permits operation with a reduced number of axes, e.g. hexapod kinematic is a special case. The control of a Dodekapod necessarily demands higher performance from the control units compared to conventional serial and orthogonal machine tools [3, 9].

Control System

The movements of the Dodekapod are controlled by a PC-based control-system, which has an open and modular structure. The position control is realized by twelve specific positioning controller cards, one for each axis.

The communication between the Dodekapod and the plc of the disassembly system is realized with profibus interface protocol.

If the workpiece is clamped, the lengths of the clamping platform axes are used for the parameterization of the movements. Because of this depending on the geometry of the clamped workpiece different movement-programs must be generated. Hence the programming system is unitized. In particular the program includes function-modules for different clamping strategies, different disassembly processes and different handling operations.

This modular structure of the movement programs has a further advantage: the fast adjustment at disassembly process changes. The programs are generated depending on the presetting commands from the plc [3, 9, 10, 11].

Workspace Analyses

One important feature of the Dodekapod structure is the possibility to vary the geometry of the base platform. For that purpose additionally three driven axes are arranged. The variation of the base platform allows on the one hand a shifting and on the other hand an enlargement of the workspace. Thus a big drawback of hexapod structures, the proportionally small workspace, is minimized.

The kinematical structure of the Dodekapod is represented by a mathematical model. This model consists of the geometrical design and the algorithms for the coordinate transformations.

The constant orientation workspace (cow) describes all three space coordinates (x, y, z) which can be reached with a determined orientation of the clamping platform. The comparison of the workspaces of a Hexapod and the Dodekapod can be made on the basis of different criteria (Figure 7). The extent describes the difference between the maximal and minimal positions the centre of the clamping platform along the Cartesian coordinate axis can be reached. For the x- and y-axis no difference exists. Concerning the z-axis a huge advantage can be determined. The extent of the Dodekapod is approximately 60 % larger than the extent of a Hexapod.

Another method to evaluate the workspaces is to compare the volumes. The volume of the constant orientation workspace of the Dodekapod is more than 30 % larger than a Hexapod.

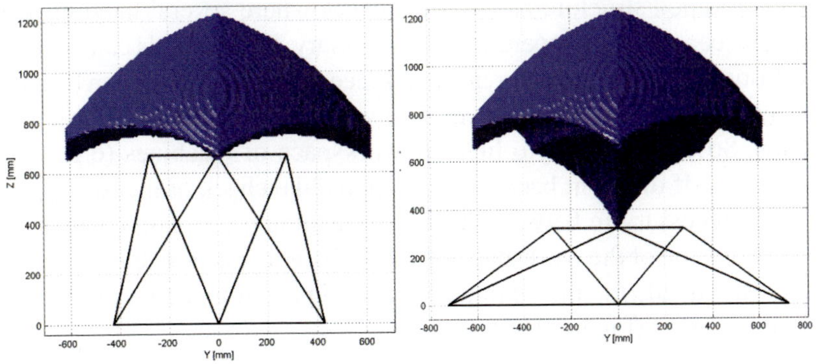

Fig. 7: Workspace (cow) comparison of a Hexapod and the Dodekapod

The Dodekapod has an additional important advantage. The clamping platform can be swivelled more than 90 degrees relating to the x-y-plane. Figure 8 shows such a conFigureuration of the axes to swivel the clamping platform [3, 9, 12].

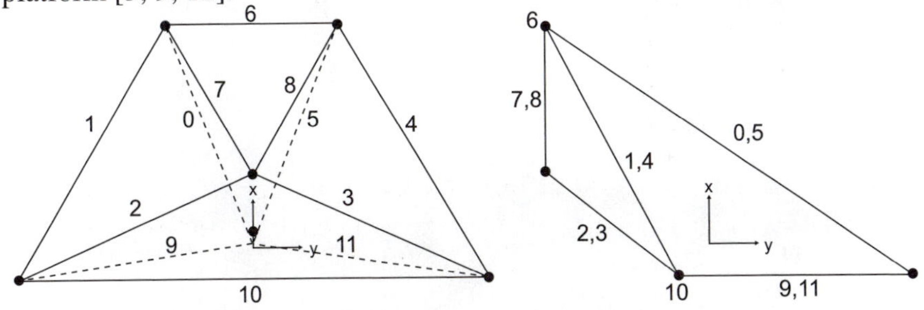

Fig. 8: Axes configureuration to swivel the clamping platform more than 90 degrees

5.3.2 Deployment of Parallel Structured Clamping Systems

Supporting processes in manufacturing like the clamping process and the movement of workpieces during a machining process to produce a geometric surface or shape have different requests. For the machining process the attention generally has to be laid on the static, thermal, and dynamic rigidities of the machine tools. For the handling of components or sub-assemblies usually three translatorical axes suffice. The possible workspaces of machine tools and handling equipment differ grossly. Therefore, the tendency is to generate a system which is either pliant and restricted in workspace or complicated and costly.

A Dodekapod structure has its advantages where several handling processes between phases of operation are necessary. Dodekapods can be used in production systems with different workspaces. The workpiece can be moved back and forward quickly between two workspaces of manufacturing subsystems. So it is possible to link the workspace of machines (e.g. robots) and humans. If different handling operations like check or measuring processes are necessary in between different machining processes, which must be made outside of the workspace of a tool-system, it is possible to move the workpiece outside the machining area and to pose it in a comfortable position for a worker (Figure 9).

Fig. 9: Simulation of disassembly with human and robot interaction

Handling equipments, which are characterized by more than three degrees of freedom, are able to adjust workpieces in machine tools. Thus also Dodekapods are suitable for such processes. The fields of deployment for the Dodekapod clamping system are wherever the requirements of machining function and handling function are similar. I.e. parallel structures like the Dodekapods can cope with manufacturing situations demanding small requests of accuracy of the machining as well as high requests at hand-ling functions such as adjusting and clamping processes.

The Dodekapod was designed as an flexible clamping system in a disassembly system. It serves for the handling of used products like washing machines during the disassembly process. This process is characterised by a mixture between manual and automated activities by robotic systems. For safety reasons the working areas of disassembly workers and robots must be separated, whereas the washing machines stay clamped during the transport between the areas. The dodekapod is able to pick up washing machines from a palette and can move it to the different working areas of disassembly (Figure 10).

Fig. 10: New clamping system with 12 axes based on a parallel structure

The dexterity of the dodekapod is immense. In this special case, it offers a working volume of nearly 5 cubic meters at a footprint of 5 square meters. The Dodekapod is based on hydraulic cylinders therefore stiffness, speed and accuracy are limited by the compressibility of the oil. For demands of higher dynamic stiffness Dodekapods with other drive concepts can be designed [8, 9, 12, 13, 14].

Configuring Parallel Structures for Disassembly

A crucial advantage of parallel structured kinematics is the large portion of similar components. In particular, the Dodekapod is designed as a low cost handling system for the disassembly, because the accuracy requirements in disassembly systems are often less important. This opens up the possibility to configure and reconfigure the handling system depending on the requirements of the disassembly process.

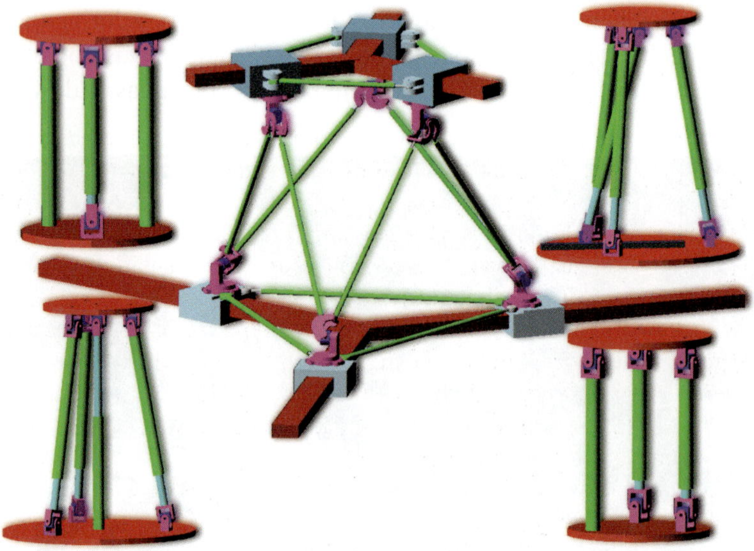

Fig 11: Multi body simulations of parallel structures with different degrees of freedom

If the disassembly process, e.g. the unscrewing of a combustion engine, require only a movement in one plane or a swivel around one axis, the Dodekapod is overpowered. Other parallel structures, which consists of the same components, like hydraulic drives, joints and guidances can configured and controlled by the same control system (Figure 11).

References

1 Seibt, M.: Disassambly Factories II – Processes and Tools. In: Proceedings of 4th World Congress R'99 – Recovery, Recycling, Reintegration. Genf, Schweiz, 05.02.1999.

2 Uhlmann, E.; Seibt, M.: Flexible Spanntechnik. In: Proceedings of the Kolloquium zur Kreislaufwirtschaft und Demontage. TU Berlin, 2000, S. 21-26.

3 Uhlmann, E.; Früsch, I.: Handhaben mit Parallelkinematik. In: Proceedings of the 1. Berliner Runde: Prozess, Struktur und Simulation rund um die Werkzeugmaschine. Berlin, 16.-17.02.2006.

4 Spur, G.; Uhlmann, E.; Seibt, M.: A New Type of Gripping System with Parallel Kinematics. Production Engineering Vol. V/2, 1998.

5 Merlet, J.-P.: Les robots. Edition Hermes, Paris 1990.

6 Uhlmann, E.; Früsch, I.: Dodekapo – Parallelkinematik zum Spannen und Bewegen. In: Proceedings of the 3. Dresdner WZM-Fachseminar „Chancen für Parallelkinematiken einfacher Bauart". Dresden, 2001.

7 Uhlmann, E.; Härtwig, J.-P.; Friedrich, T.; Früsch, I.: RP – Modelle in der Demontagesimulation. In: Futur – Mitteilungen aus dem Produktionstechnischen Zentrum (PTZ) Berlin. IPK Berlin und IWF der TU Berlin, 2003.

8 Uhlmann, E.; Friedrich, T., Früsch, I.: Possibilities for Improvement of the Automated Disassembly. In: Proceedings of the Global Conference on Sustainable Product Development and Life Cycle Engineering. Berlin, 2004, pp 171-174.

9 Uhlmann, E.; Friedrich, T.; Früsch, I.: Strategies for Increasing the Efficiency of Disassembly Systems. In: Proceedings of the 11th International CIRP Life Cycle Engineering Seminar: Product Life Cycle – Quality Management Issues. Belgrade, Serbia 2004, pp. 77-81.

10 Uhlmann, E.; Spur, G.; Seibt, M.: Steuerung eines parallelen Spann- und Handhabungssystems mit zwölf Achsen. In: Proceedings of the 2. Chemnitzer Parallelkinematik-Seminar. Wiss. Skripten, Chemnitz, 12./13.04.2000.

11 Uhlmann, E.; Spur, G.; Seibt, M.: Anforderungen an Steuerungen von automatisierten Spannmitteln in der Demontage. VDI-Bericht, VDI Verlag, Düsseldorf 2000.

12 Uhlmann, E.; Schäper, E.; Früsch, I.: Simulation komplexer parallelkinematischer Maschinen. In: Futur – Mitteilungen aus dem Produktionstechnischen Zentrum (PTZ) Berlin. IPK Berlin und IWF der TU Berlin, 2/2005.

13 Uhlmann, E.; Härtwig, J.-P.; Seibt, M.: Pilot Disassembly System for Home Appliances Using a New Twelve Degrees of Freedom Parallel Manipulator. In: Proceedings of the 33. CIRP International Seminar on Manufactural Systems. Stockholm, 2001.

14 Uhlmann, E.; Härtwig, J.-P.; Friedrich, T.; Früsch, I.: Flexible Handling in a Disassembly System. In: Proceedings of the Colloquium on e-ecological Manufacturing, Berlin, March 27.-28.2003.

5.4 Cleaning Technologies

Eckart Uhlmann, Adil El Mernissi, Robert Hollan, Berlin, Germany

According to the German norm DIN 8592 E cleaning is defined as the removal of unwanted substances from the surface of workpieces up to a required, agreed, or possible degree. Amongst others there are mechanical, chemical and thermal cleaning processes. In the mechanical cleaning process the cleaning effect is realised by a relative movement of tool and workpiece. Examples are brushing and blasting. There are two possibilities of chemical cleaning. On the one hand the contamination can be solved with solvents and on the other hand the cleaning effect can be carried out by chemical reaction. The third class of cleaning processes is the thermal cleaning. Thereby the cleaning effect is based on a defined, time depending change of temperature.

5.4.1 Challenges

Industrial cleaning technologies, especially for the cleaning of hard surfaces, are very important in manufacturing, maintenance, repair and recycling. At the moment, mainly chemical and mechanical cleaning processes are used. These processes are characterised by a lack of flexibility regarding contamination and basic material, by an abrasive and corrosive influence on the component to be cleaned and by insufficient performance against the background of rising quality standards. The complexity and amount of the surface treatment equipment and also the time and energy consumption is very high due to necessary follow-up procedures as washing and drying.

Moreover, chemicals, solvents and sound emissions lead to health risks for workers and environment. The contaminated sewage from surface treatment processes has to be treated and collected sufficiently. Otherwise this would lead to water and soil pollution.

New environment friendly cleaning processes become more and more important. Research is beginning to develop a number of new dry-cleaning processes, like for example laser, plasma, and environment friendly blasting processes. The European industry has recognised the urgent need for action and cleaning technologies are currently in a state of change. It can be assumed that the substitution of conventional cleaning technologies will continue in the next years and that the application of new technologies will increase steadily.

The aim of cleaning in disassembly is to reduce the exposure of operators to harmful substances and to supply cleaned components to further manufacturing steps in remanufacturing and/or product recycling [1]. The requirements for cleaning processes are very complex due to a large variety of shapes, materials and impurities. Conventional methods can often cause damage to the component to be cleaned and can fail to remove the impurities completely.

Alongside technically demanding and economically successful disassembling processes, an efficient cleaning technology is required to manage this amount and diversity of products. Furthermore this technology has to deal with unknown impurities generated during the utilization of the product.

For a modern cleaning technology in the field of disassembly, the whole cleaning process must be investigated in terms of the required purity degree for the subsequent process stage. Requirements placed on novel cleaning technologies aim at environment friendly, economical, fast, flexible and reliable production. A new strategy has to be developed including the ideas represented in Figure 1.

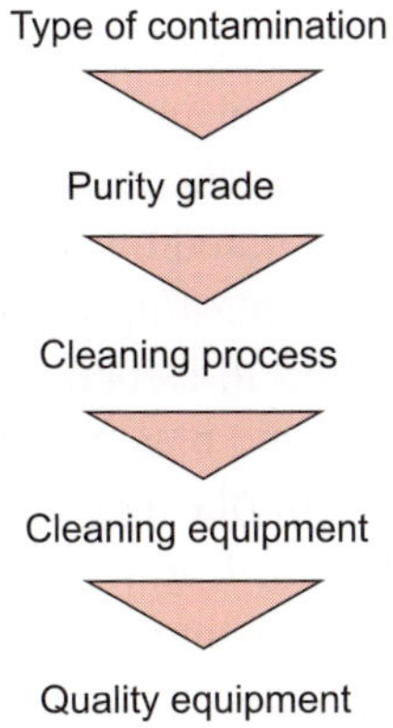

Fig. 1: Process steps for cleaning within disassembly

Cleaning surfaces usually includes the removal of reaction, sorption and contamination layers. Hereby, the bonds, i.e. the adhesion forces between the impurity and the base material, must be eliminated. The adhesion to the surface depends on different physical-chemical interactions, causing different bond forces [2]. The bond strength is determined by the bonding energy. It is based on the fact that energy is required for its splitting. Its size depends on the existing bond type. Hereby, several bond types may act simultaneously. Oxide layers and old paint layers might represent an exception, often forming a compact coat [3].

The existing impurity degree and the necessary purity degree must be determined first to provide the basis for the selection of the adequate cleaning process and equipment. For disassembling-specific cleaning flexibility, variability and a modular structure are of utmost importance. Finally the cleaned component has to pass a quality inspection regarding the achieved purity degree, the surface roughness, the sub-surface influence, the residual stress, the deformation or other quality criterias.

For cleaning in disassembly new environment friendly cleaning processes for flexible application in disassembly were investigated. In this context several blasting techniques and mechanical cleaning processes for defined applications were qualified.

5.4.2 Blasting

Blasting techniques such as compressed air blasting and laser cleaning prove to be more than alternative. Their advantages are low working forces, omni directional processing and arbitrary starting points. In addition, they are very flexible in terms of material, contamination and coating.

Compressed air blasting is used in industry for preparation of metallic surfaces and for cleaning in order to the reuse and further use of products. Different blasting agents are accelerated by compressed air and lead trough a nozzle onto the workpiece. The blasting agents are classified according to hardness, size, shape, and material. Depending on the hardness ratio between the blasted good's material and blasting agents, different changes occur on the blasted surface. Thereto, blasting media of varying hardness are added according to the application, leading to material removal from the workpiece. The influence of the blasting medium on the machining result can be determined by the cleaning performance, the surface roughness, and the topography. Based on investigations on the removal of lacquer coatings on different materials, a range of blasting media were compared in technological and economical terms. Blasting media were tested, such as round and edged cast steel, round glass beads, corundum, and duroplastic, as well as the volatile blasting medium dry ice in its typical pellet form.

It can be clearly shown in Figure 2 that the cleaning performance depends on the type of blasting medium and the material. The highest cleaning performance of 1.92 m^2/h was achieved with glass beads with DC01 A as workpiece material. The lowest cleaning performance was achieved with round cast steel as blasting medium and DC01 A as workpiece material.

The cleaning performance was in the range of performances achieved with dry ice blasting.

After blasting the workpiece materials with compressed air blasting, an impact on the surface roughness could be detected. The comparison of material surface roughness after blasting with durable blasting mediums and dry ice pellets is represented in Figure 3.

The comparison of the blasting media is carried out on the basis of the values detected with standard parameter setting. The biggest surface roughness was achieved in the case of blasting with corundum followed with edged cast steel and round cast steel. In the case of blasting with glass beads and, in particular, with duroplastic the surface roughness is smaller. In the case of blasting with dry ice pellets, the surface roughness is even smaller. These results correspond to those of unblasted steel samples.

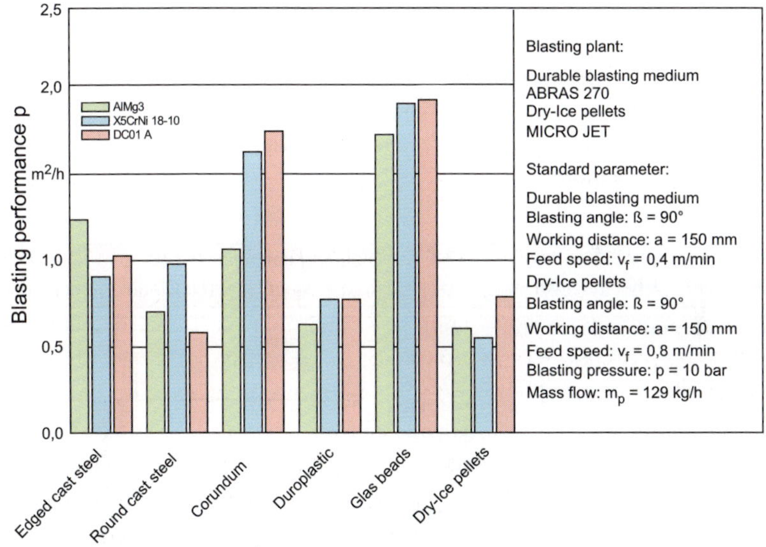

Fig. 2: Maximum cleaning performances after blasting of AlMg3, X5CrNi 18-10 and DC01 A with the tested blasting media

In order to carry out an economic assessment, a cost calculation system was developed for the comparison of the cleaning costs of compressed air blasting with the tested blasting media. The manufacturing costs per hour are the sum of the machine rate per hour, the labor costs per hour and the tool costs per hour.

Figure 4 shows the cost calculation system for cleaning. Point of departure was a cleaning performance found in technological investigations, and the one-way usage of the blasting medium. The disposal costs of the blasting medium residues were not considered in this calculation.

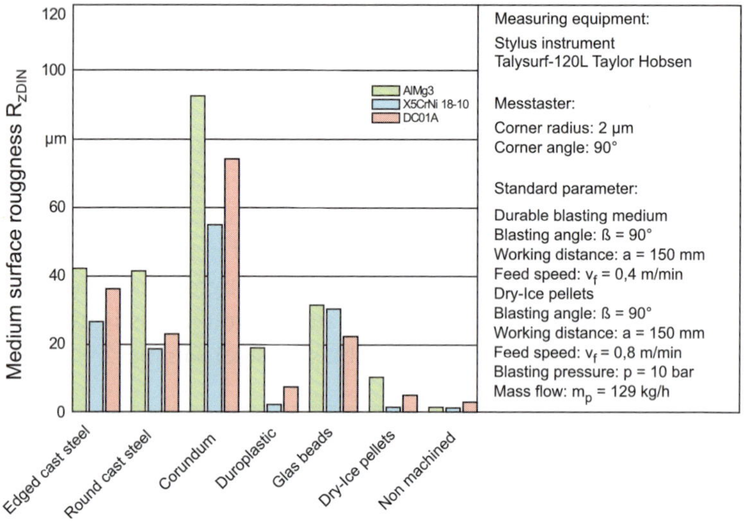

Fig. 3: Comparison of medium surface roughness after blasting with exanimate blasting medium

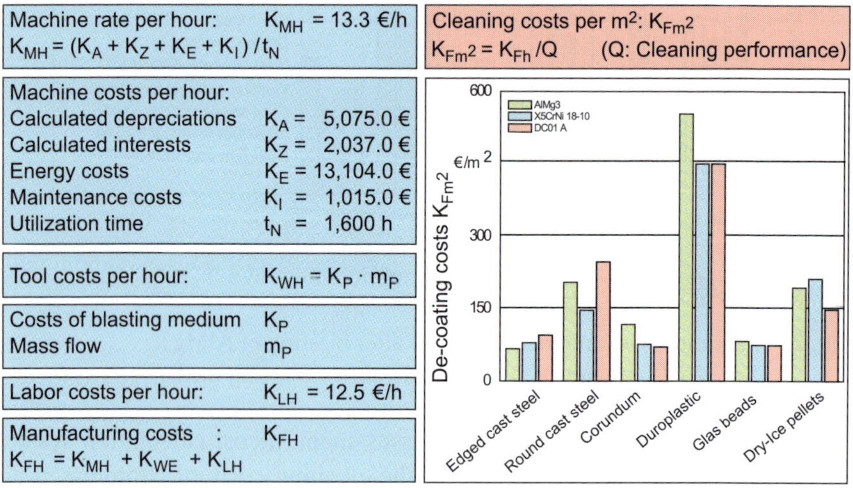

Fig. 4: Cost calculation system for cleaning and comparison of the cleaning costs of the tested blasting media and materials

Furthermore, Figure 4 shows the costs for de-coating per square meter. The calculation is based on the maximum cleaning performance in technological investigations. It can be clearly seen that the cleaning costs of duroplastic as blasting medium are the highest with 500 €/m². This is a result of the high tool costs, making up 90 % of the manufacturing costs. Moreover, the cleaning performance of duroplastic is very poor.

Blasting with edged cast steel, corundum, and glass beads accumulate the lowest costs. In the case of corundum however, tool costs and material consumption are higher than in the case of edged cast steel. This is due to the higher cleaning performance of corundum. The cleaning costs of round cast steel and dry ice pellets are between 150 €/m² and 250 €/m². Considering the disposal costs, the cleaning costs of dry ice pellets are similar to those of edged cast steel, glass beads, and corundum.

Blasting with Carbon Dioxide CO_2

In the field of cleaning technology several blasting media are used, witch due to the new legal, ecological and economical frameworks were limited. In this context new environment friendly, flexible and economic rentable blasting media are to be developed and qualified for cleaning. In the recent capital the relevant of dry ice blasting such as in the term of the quality of the cleaned surface as the economic term could be shown, that dry ice blasting comparable or better as the investigated blasting media. In the following blasting with solid carbon dioxide and laser cleaning, as one other blasting process, with active mechanisms, process variants and applications in the disassembly were described.

Carbon dioxide is odourless, colourless, inflammable, electrically non-conducting, and chemically inert. However, it supersedes oxygen and affects heart and breathing rate. Since it is heavier than air, sufficient air supply must be ensured when used. The Earth's atmosphere contains 0.03 % carbon dioxide, altogether approximately 735 billion tons. The carbon dioxide used for blasting is a waste product of chemical processes.

In contrast to water, CO_2 has no liquid phase at ambient pressure. Depending on the temperature, it is either gaseous or solid. Thus, solid carbon dioxide does not melt, but sublimes. It is therefore called dry ice.

The active mechanisms of blasting with carbon dioxide are based on three effects (Figure 5):

- The thermal effect leads to a regional cooling of the part where the pellets strike the surface. As a result elasticity is lost, and the adhering coating embrittles and shrinks. Cracks are formed. Due to the different thermal expansion coefficients of coating and substrate, the bond dissolves when the adhesive energy is exceeded. The coating partially chips off.

- The mechanical effect due to kinetic energy of pellets and air stream contributes to the removal of the coating.
- The sublimation effect is based on to the sudden increase in volume by 600 fold when the pellets strike the surface of the part. Gas flows underneath the adhering coating. The removal of material is hence based on a combined thermo-mechanical effect.

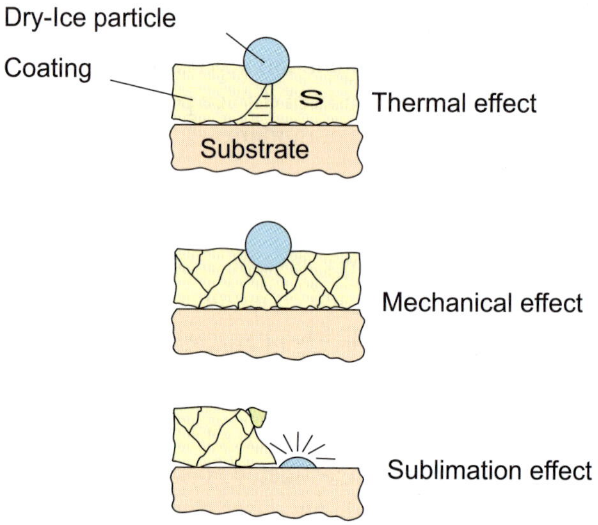

Fig. 5: Active mechanisms of blasting with carbon dioxide

The properties of CO_2 in combinations with the active mechanisms during blasting provide specific advantages. The blasting agent sublimes when impinging on the surface to be cleaned. The gaseous carbon dioxide can be released to the atmosphere as a natural component. Thus, there are no secondary contaminations in the process and only the removed homogenous impurity has to be disposed. This ensures short cleaning times without disassembling and drying.

The low hardness of solid carbon dioxide allows the blasting of mechanically sensitive and heterogeneous materials with only small abrasive and corrosive influence on the substrate. The process is still flexible in terms of impurity thickness and material. Further properties of CO_2, such as low electrical conductivity and bacteriostatic effect, lead to further advantages in electrical engineering and the food industry.

As any other blasting process, blasting with CO_2 reaches only visible parts; there is no access to undercuts. Also electrostatic charging may occur as a result of the blasting agent rubbing against the tubes during transport.

Thus, it cannot be applied in explosion-proof areas. Specific disadvantages of blasting with solid CO_2 are the high noise exposure, the unpredictability of the cleaning result due to combined active mechanisms and primarily the high operating costs.

The process variants of blasting with solid carbon dioxide are dry ice blasting and CO_2-snow blasting. For the production of dry ice pellets liquid carbon dioxide with approximately 20 °C and 57.3 bar is expanded to atmospheric pressure. Due to the Joule-Thomson-effect during the expansion it cools down and CO_2-snow is generated. With help of a hydraulic stamp the CO_2-snow particles are pressed through a mould and cylindrical dry ice pellets are formed. Pellet parameters that effect the cleaning process are density, hardness and shape.

For CO_2-snow blasting liquid carbon dioxide is used and solid CO_2-particles are generated within the blasting process. There are two principles, the expansion at ambient pressure and at blasting pressure. In CO_2-snow blasting expanding at ambient pressure (Figure 6a) liquid CO_2 is expanded directly through specially designed two-component blasting nozzle. Solid CO_2-snow particles are generated and afterwards accelerated by gaseous CO_2 and a compressed air coat jet. The two phase supersonic jet is blasted on the surface to be cleaned.

If the expansion takes place in the blasting pressure (Figure 6b) the liquid CO_2 is expanded into the pressurized agglomeration chamber. The generated CO_2-snow can thus agglomerate to larger particles, which are subsequently accelerated by compressed air and the nozzle.

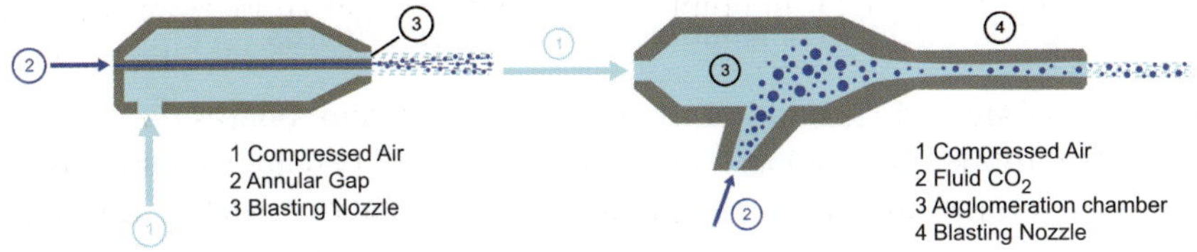

Fig. 6: a) CO_2-snow blasting (expansion at ambient pressure)
b) CO_2-snow blasting (expansion at blasting pressure)

Removal of Paint from Metal Components

One important application of dry ice blasting is the removal of paint from metal components (Figure 7). For this purpose, the parameters for processing different materials were optimized. The aim was to find the optimum settings for the removal of different types of paint with different coat thicknesses. In this context the influence of the setting parameters blasting pressure, feed speed, working distance, blasting angle, mass flow of dry ice pellets on the removal rate, surface topography and surface roughness are determined. For paint removal with dry ice blasting a round and a flat nozzle were tested [4].

Fig. 7: Paint removal with dry ice blasting

De-coating and Cleaning of Mounted Printed Circuit Boards

The de-coating and cleaning of mounted printed circuit boards can be done as well by dry ice blasting as by CO_2-snow blasting (Figure 8). Objective is the reuse of components and printed circuit boards. The proccss is applied for the removal of conformal coatings within repair and product recycling and before the de-soldering process of damaged or expensive components.

With regard to dry ice blasting, the blasting pressure will be reduced to 5 bar and the dry ice mass flow limited to 80 kg/h. The technological investigations showed that de-coating and cleaning of the usually commercial epoxy, polyurethane, and silicone coatings from mounted printed circuit boards is possible. Thereby a damage of the printed circuit boards was not detected [5].

The main part of the technological investigations of de-coating and cleaning of mounted printed circuit boards by CO_2-snow blasting deals with the process parameter optimization for different coating systems. For this, the most frequently used coating systems were used. The coatings were applied and hardened according to the processing regulations. The examined coatings systems were PU resin systems, epoxy resin systems, acrylate resin systems, alkyd resin systems and silicone rubber systems. Eight of eleven examined coating systems were applied in the dipping process and three in the casting process. All coating systems applied in the dipping process can be removed with CO_2-snow blasting. Thereby the damage of printed circuit boards can be excluded by the choice of suitable parameters. The materials applied in the casting process cannot be removed due to their high coating thickness and their high elasticity. The coating layer thickness decreases the thermal effect. Besides, the cast materials have a high elasticity so that the abrasive effect on the material is negligible [6].

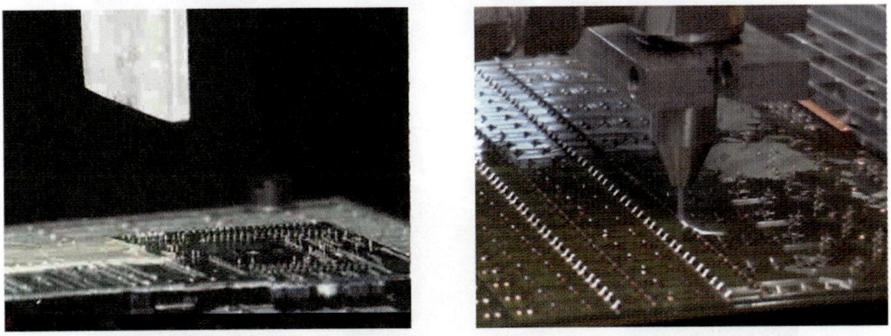

Fig. 8: Cleaning and de-coating of mounted printed circuit boards by dry ice blasting and CO_2-snow blasting

Cleaning of Exchange Engine Components by Dry Ice Blasting

The seal between the components of automotive engines is provided with silicone. In repair and in the maintenance two problems occur caused by the silicone. The disassembly of the components is hampered by the bonding effect of the silicone and the seal remainders are hard to remove. Till now there is no satisfying technology to solve this problem. At present, the silicone seals are scraped off manually. However, this is very time-consuming and expensive. There is no known chemical solvent or cleaning agent for silicone. Due to its elastic characteristics it is not possible to remove silicone completely by blasting with sand, glass or steel. Moreover, these procedures lead to damages of the components. In addition, the geometry of the components is very complex.

The experimental investigations on the removal of silicone seals were executed at components made of an aluminum magnesium alloy (Figure 9). For parameter optimization the mass flow of dry ice, the blasting pressure, the working distance, the blasting angle as well as the feed speed were varied. To detect possible material impairments through the dry ice blasting the surfaces and pre-surfaces of the materials were analyzed after the processing. This was performed with the help of a surface measuring laser and a scanning electron microscope. No significant modifications of the surface roughness and the material structure could be proven [5].

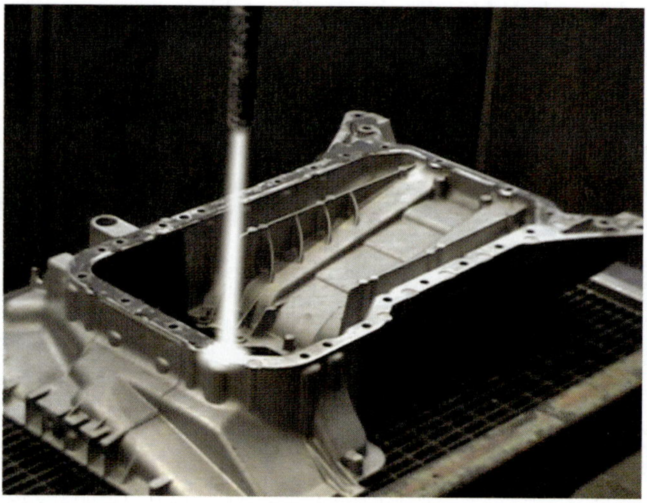

Fig. 9: Removal of silicone seals by dry ice blasting

It was shown that the removal of the silicone seals with dry ice blasting is technically feasible, and cost savings are expected compared to present technologies. The renewed assembly of the engines after the processing is possible.

Fig. 10: Cleaning of different components of exchange engines and old vehicles by dry ice blasting

Moreover, different components of exchange engine and old vehicles were cleaned by dry ice blasting. In this context cylinder heads, pistons and outlets with strongly adherent contaminations were investigated (Figure 10). These components were contaminated with hardly adhesive, carbon-containing impurities. Thereby in the line with technological investigations the parameter mass flow on dry ice pellets, blasting pressure, feed speed, blasting angle and working distance for different components and contaminations were varied. As result a complete removal of the contaminations was realized.

Laser Cleaning

In recent years, an increasing interest in using laser technology for surface cleaning was observed. Compared to the conventional methods of surface cleaning laser offers significant advantages such as:

- contact-free and hence force-free cleaning,
- cleaning of sensitive surfaces,
- low thermal and mechanical influence of substrate,
- high precision,
- no blasting medium and no chemicals used,
- high grade of automation and control, and
- selective cleaning.

The process of laser cleaning is based on local and contact-free influence of pulsed or continuous laser beam. The process relies on the fact that damage threshold of the contamination is lower than of the underlying substrate material. The energy density of the laser pulse is adjusted so that it exceeds the damage threshold of the contamination layer, but is well below the damage threshold of the substrate material. In this way the surface structure of the underlying material is not damaged. Once the surface is clean, further use of the laser will have no harmful effect. The process is therefore self-limiting making laser cleaning very suitable for the cleaning of sensitive substrates.

Different active mechanisms appear depending on composition and thickness of the contamination, and basic material properties. For the cleaning process the following mechanisms are important:

- ablation by evaporation or decomposition of the contamination layer, and
- ablation by thermally induced stresses or by a laser beam-induced shock wave.

The laser light is rapidly absorbed by the top few microns of the contamination instantaneously vaporizing this thin layer and forming highly compressed plasma (unstable high-pressure ionized gas). The creation of this plasma generates a shock wave which propagates into the bulk contamination breaking it up and ejecting the contamination as fine particles. The ejected particles are then captured by an exhaust extraction system.

Often several mechanisms operate simultaneously. The respective portion depends on the material properties and the setting parameters, in particular of the wavelength, power density, and time of interaction. The absorption of the contamination and the basic material is of special importance for the cleaning process. If the contamination absorbs the laser beam well and the underlying substrate material exhibits a small absorption factor, the process of removal is stopped automatically with the impact of the laser beam on the reflecting substrate [7].

Laser De-coating of Compact Disks

The audio-CD was developed by Philips and Sony in 1982. Today, every compact disk has the same construction and measurements. The design and dimension of a CD-DA (Compact Disk Digital Audio) have been specified in IEC 908 specification. All subsequent types of CDs are based on this specification. Due to its handiness, quality and flexibility, CDs have been rapidly disseminated as a cheap mass store of information. CDs pressed in industrial manufacture consist of a high-purity plastic sheet from polycarbonate with a diameter of 120 mm and a thickness of 1.2 mm. The sheet is coated with a 40 nm to 50 nm aluminium coating and a protective varnish of a thickness of 10 μm. Due to the high lot sizes, a considerable amount of plastic waste, varnish residues and old metal pile up every year. CDs are disposed of in large numbers and after very short use, especially those with a limited validity such as product catalogues, phone books, corporate information and software. Defective pieces and excess production add to this, as well as CDs that have to be destroyed for data protection reasons. Due to the recycling economy and waste legislation established in 1996 and the new EU waste law for information technology, industry is urgently seeking solutions to integrate CDs into a product and natural resource cycle or an economically efficient recycling process [8].

The further use of the material would be another possibility. This includes shearing up of the coated CDs and processing the polluted plastics granulate to a variety of plastics products. This granulate does not meet the quality requirements for a material recycling in CD production. There is however a number of processes in plastics industry that make use of the high-purity polycarbonate, in such a way that the de-coated polycarbonate bodies could be integrated into sensible resource cycle.

In the course of process optimization, a technological data base was developed. A de-coating time of 5 s per CD was reached with the installation prototype [9]. Given manufacturing costs of 29.8 €/h, the costs for de-coating are estimated at 0.041 € per CD (Figure 11). Since the removal in relation to the surface may further increase with future laser systems, the economic efficiency of the process is clearly proven.

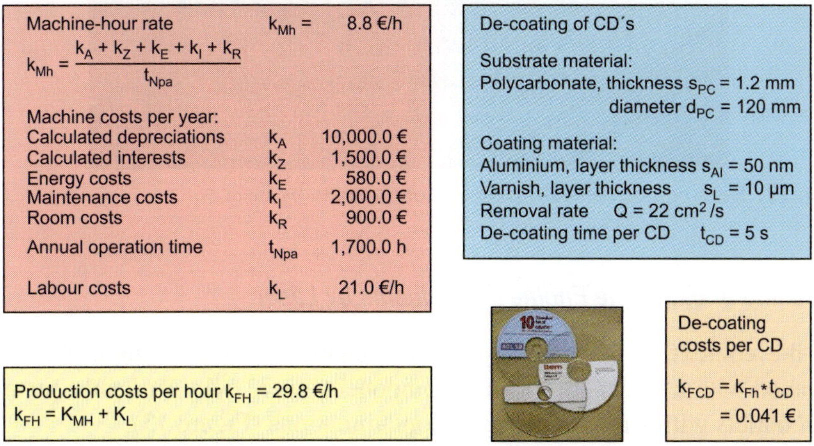

Fig. 11: Assessment of economic efficiency for laser cleaning of CDs

Removal of Paint from Metal Components by Laser

The experimental investigations on the removal of paint from metal components with laser (Figure 12) covered the determination of the relevant setting parameters with regard to the cleaning result and the thermal and mechanical influence on the surface.

For the realization of experimental investigations an electro-optically pulsed Q-switched Nd:YAG solid body laser with an infrared laser radiation at a wave length of 1,064 nm was used. The distance of points, pulse frequency, feed speed, beam forming and the beam performance were defined as relevant setting parameters. The distance of points determines the space between machined areas by the pulsed laser system. For the evaluation of the mechanical and thermal influence the roughness was measured and the processed surfaces were observed with an optical microscope. As a result, the modulation of feed speed and the beam performance shows a significant influence on the roughness.

Fig. 12: Removal of paint from metal components by laser

Cleaning of Exchange Engine Components by Laser

In the context of a feasibility study exchange engine components were cleaned with a laser. The cleaned components were cylinder heads, pistons and outlets with strongly adherent contaminations (Figure 13).

Fig. 13: Cleaning of piston from exchange engine by laser

These components were contaminated with carbon-containing impurities. The cleaning of the examined exchange engine components by laser was successful. The contamination was completely removed. A damage of the basic material was prevented by an exact adjustment of the relevant parameters.

5.4.3 Brushing

The principle of brushing is the mechanical stripping or grinding off of mainly solid impurities from the component surface. The brush hardness must be adjusted to the surface to be cleaned. The more the impurities adhere to the surface the more rigid and harder bristles have to be used. It must be checked, whether damaging the component surface by abrasion can be tolerated. Bristles of metal wire, natural and plastic fibres are used for brushing. In order to increase the abrasive effect, bristles of plastic fibres are impregnated with coarse sand. The relative movement between bristles and the component surface is usually achieved by rotating the brush. The brush can be guided over the component manually or automatically.

Brushing represents a sub-group of mechanical cleaning. According to DIN 8592:2003-09, brushing is "the removal of impurities by means of brushes, or similar devices".

In industrial cleaning, brushes are often used in the case of profiles, belts, plane components, or cast parts. Hereby, mainly hard brushes are used which remove impurities through a strong abrasive effect, but also small amounts of the base material. In some cases very soft brushes are used for a gentle surface cleaning.

The cleaning effect on difficult-to-access areas such as boreholes, blind holes, grooves and corners, strongly depends on how much the brush can adjust to the workpiece and on the pressure between workpiece and brush.

5.4.4 Flexible Cleaning Centre

The application fields and the range of products are just as manifold as the cleaning processes. To fulfil this spectrum, high flexible instruments must be developed. In addition to information systems to assist the choice of the adequate cleaning process, the integration of several flexible cleaning processes was investigated. The aim is the design and planning of a high flexible cleaning centre, which includes selected cleaning processes. Thereby the combination and interaction of different cleaning processes is possible. In this context the blasting techniques blasting with solid and durable blasting media, dry ice blasting, CO_2-snow blasting, laser, brushing and section cleaning processes were combined in one system.

The term centre is derived from machining centre. A machining centre describes a machine tool system in the field of forming, cutting, and joining. Alongside automated tool changers, machining centres are characterized by their own tool storage. The integration of different machining operations in one single machine allows extensive machining while the workpiece is clamped only once. Hereby, machining takes place via numerical controls. Additionally, the machine may comprise measuring and control functions. Centres are preferably used in small and medium batch manufacturing, since they can be quickly adapted to varying manufacturing tasks due to the high grade of automation. These systems are very flexible and different workpieces can be manufactured consecutively with a minimum machine idle time.

The following properties can be derived from the above-mentioned features:

- automated tool-changer,
- tool storage,
- different machining operations, and
- numerical control.

The cleaning centre to be designed must thus integrate different cleaning processes, which can be numerically controlled. Tools required for the cleaning processes such as nozzles and brushes are stored in storage with an automated changing function. The cleaning centre was designed according to the VDI 2221 German standard of systematic design.

The actual functional structure of the flexible cleaning centre, which represents the starting point for the subsequent solution finding, is generated by joining the sub-functions (Figure 14). The common feature of all subfunctions is that they bring together the workpiece and energy, the cleaning process taking place in the contact area. The energy is electrical, pneumatic and mechanical energy.

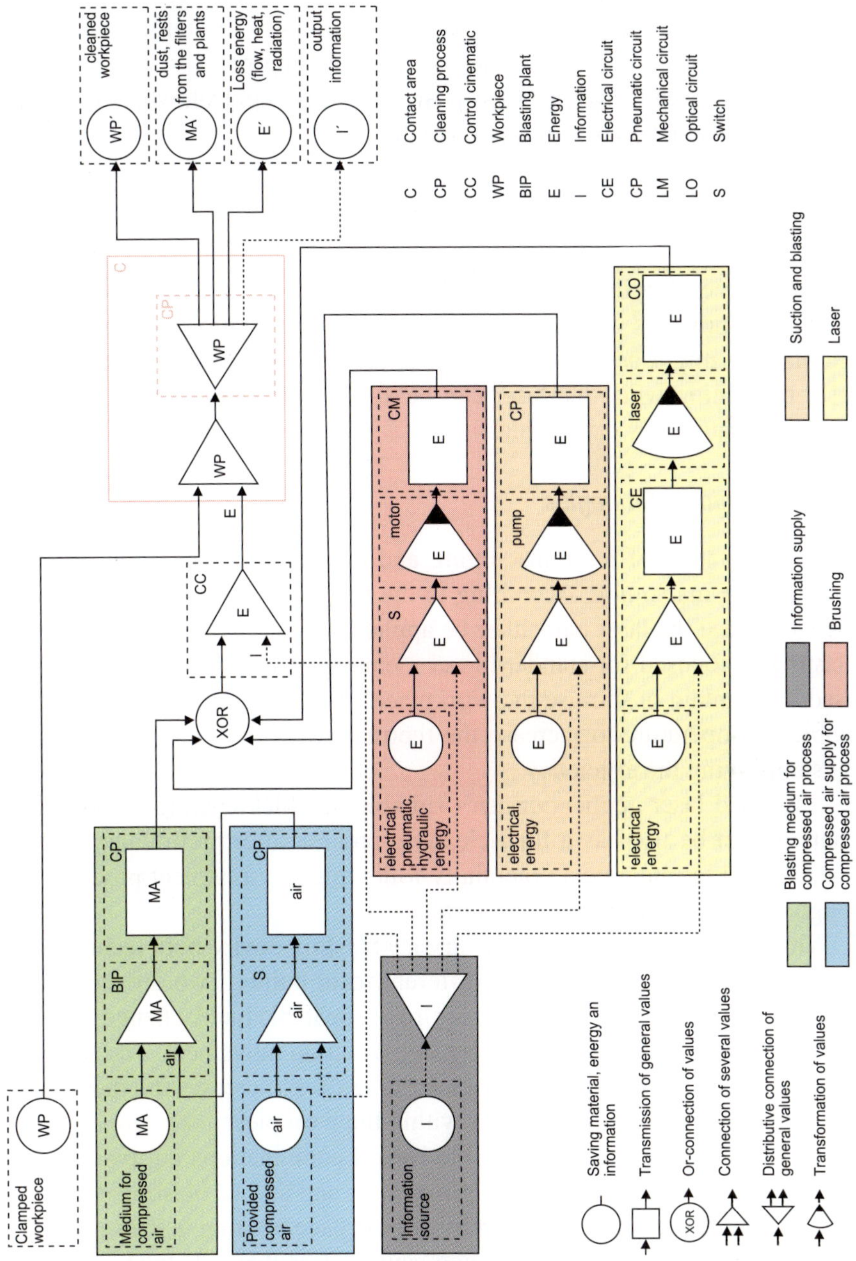

Fig. 14: Complete functional structure of the flexible cleaning centre

Since several cleaning procedures are linked with the centre, a device is needed to handle the single processes. According to the information it receives, it clamps a tool and carries out the cleaning process. The source of information for the whole process is an integrative control, which drives all sub-processes and the handling kinematics.

5.4.5 Laser Assisted Dry Ice Blasting

The objective of hybrid dry ice blasting-laser processing is to increase the material removal as well as the area-related cleaning ratio. Furthermore dry ice blasting might be used as coolant to reduce the temperature caused by laser machining while the laser can be applied as energy source avoiding low temperatures by dry ice blasting.

Hybrid Machining Strategies

The combination of laser machining and dry ice blasting offers different strategies of machining. According to the relative position of the laser and the workpiece the laser can either be applied unfocused to assist the dry ice blasting or focused for standard laser cleaning. An unfocused laser can be used to maintain or to raise the workpiece's surface temperature above the ambient temperature to increase the thermal effect during dry ice blasting thus improving the efficiency.

A focused laser application isn't suitable for thick coatings or contaminants. It can be used as a final cleaning process after preliminary dry ice blasting and to integrate a potentially following pre-treatment process, e.g. roughening or structuring of the surface.

With regard to the laser and the dry ice blasting device both technologies can be applied in the same or in different focal points. Two differing focal points avoid any interference but require a repeatable quick change between the two stand-alone-technologies by an oscillating movement. The same focal point for both technologies is easier to realize, increases the thermal effect and reduces the thermal stress within the workpiece.

As a result there are four possible hybrid combinations but two reasonable hybrid machining strategies. While laser and dry ice blasting do not affect each other measurably, a laser assisted dry ice blasting in the same focal point makes sense saving energy and causing less thermal stress compared with two different focal points.

The application of a preliminary cleaning by dry ice blasting followed up by a precision laser cleaning process in two different focal points offers the possibility to focus the laser machining upon difficult-to-remove residues of the preliminary cleaning process.

Hybrid Cleaning Device

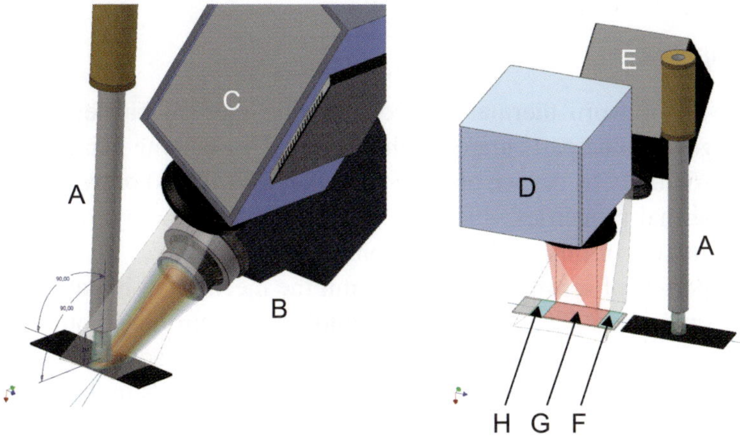

Fig. 15: Hybrid cleaning device concepts for laser assisted dry ice blasting (left) and precision laser cleaning after preliminary dry ice blasting (right)

The concept of laser assisted dry ice blasting shown in Figure 15 consists of the dry ice blasting nozzle (A), a unfocused laser device (B) and a thermography camera system (C). To measure the surface temperature is important to control the thermal effect as well as the thermal impact on the workpiece's material. The concept of a precision laser cleaning after preliminary dry ice blasting shown in Figure 15 consists of the dry ice blasting nozzle (A), a focused laser system with a 2D-scanner (D) and a CCD-camera system (E). Due to different focal points of dry ice and laser, residues of the preliminary cleaning can be detected (F) and focused by precision cleaning (G) before a finally quality check (H).

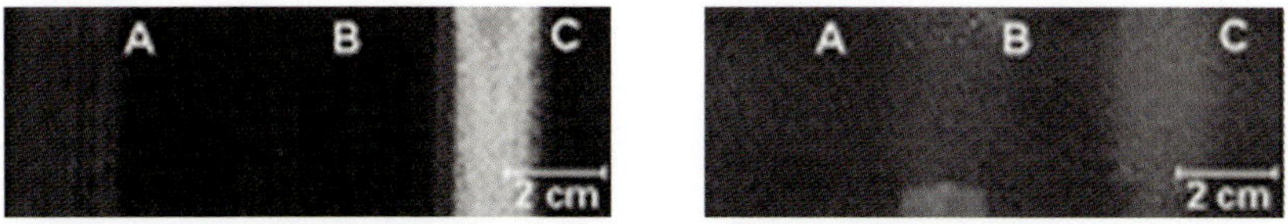

Fig. 16: Cleaning results of a PUR-2 components varnish coated specimen (left) and a rusted specimen (right)

Figure 16 shows exemplarily the de-coating and cleaning results of laser assisted dry ice blasting. The defocused laser (A) and the dry ice blasting (B) were applied to compare the results with the hybrid technology (C). Only the feed speed was adapted to the kind of specimen. The results show a strong synergetic effect of the hybrid technology. Merely the removal of the last percentage of rust is inefficient. In this regard a specialized process has to be added, e.g. precision laser cleaning [10].

Types of Application

The removal of worn thermal sprayed ceramic thermal barrier coatings (TBC) of gas turbine parts are of high interest. This coating can hardly be removed efficiently by dry ice blasting (low pellet hardness) or by laser processing (thermal properties, thickness of the coating). The hybrid concept offers a reduction of time and dry ice consumption.

The properties of sensitive materials limit the mechanical impact as well as the thermal stress caused by any technology. Controlling both effects they can be balanced so that sensitive materials may be machined, too. The cleaning of plastic based complex three dimensional moulds in the automotive industry might exemplify for sensitive materials [11].

References

1 Bold, J.: Trennende Schutzeinrichtungen für Werkzeugmaschinen zur Hochg eschwindigkeitsbearbeitung. Dissertation, TU Berlin, 2004.

2 Kollek, H.: Reinigen und Vorbehandeln. Curt R. Vincentz Verlag, Hannover 1996, S. 23, 29-47.

3 Müller, K.-P.: Praktische Oberflächentechnik. Friedr. Vieweg & Sohn, Braunschweig 1999, S. 35, 43-45.

4 Uhlmann, E.; Elbing, F.: Reinigen und Entschichten mit CO_2-Trockeneisstrahlen. In: tagungsband zum Schwerpunktseminar Reinigen / Abtragen, Institut für Werkstoffkunde, Universität Hannover, Hannover, Germany 1998.

5 Uhlmann, E.; Elbing, F.; El Mernissi, A.: Applications for Eco-Efficient Surface Machining with Dry Ice Blasting. In Proceedings of the 1st International Conference on Design and Manufacture for Sustainable Development, Liverpool, England 2002, pp. 195-208.

6 Uhlmann, E.; El Mernissi, A.: Arbeitsschutz und Sicherheitstechnik in der Demontage. In: Arbeits- und Ergebnisbericht, Teilprojekt A4, Sonderforschungsbereich 281 – Demontagefabriken zur Rückgewinnung von Ressourcen in Produkt- und Materialkreisläufen. TU Berlin, 2003, S. 129-160.

7 Liebelt, S.: Analyse und Simulation des Laserstrahlschneidens von Faserverb undkunststoffen. Dissertation, TU Berlin, 1998.

8 Uhlmann, E.; Axmann, B.; Elbing, F.: Cleaning and De-coating in Disassembly. In Proceedings of the 4th World Congress R `99 – Recovery, Recycling, Reintegration, Geneva, Switzerland 1999, p. III.51-III.56.

9 Elbing, F.; Spur, G.; Uhlmann, E.; Krieg, M.: Industrial Cleaning Technologies for Hard Surfaces: Dry Ice Blasting and Laser. In Proceedings of the 40th International Detergency Conference IDC `01, Strasbourg, France 2001, pp. 156-165.

10 Uhlmann, E.; Hollan, R.; El Mernissi, A.: Hybrid Dry Ice Blasting-Laser Processing: Nd-YAG-Laser assisted Dry Ice Blasting for De-Coating. In: Proceedings of the 8th International Conference on MIT'2005, Piran 2005, pp. 163-166.

11 Uhlmann, E.; Hollan, R.; Veit, R.; El Mernissi, A.: A Laser Assisted Dry Ice Blasting Approach for Surface Cleaning. In: Proceedings of the 13th CIRP International Conference on Life Cycle Engineering, Leuven 2006, pp. 471-475
.

5.5 Hybrid Disassembly System

Günther Seliger, Eckart Uhlmann, Thomas Friedrich, Robert Harms, Christian Sönnichsen, Berlin, Germany

The tools and processes that were developed to conduct destructive and non-destructive disassembly processes were to be integrated in pilot systems. The central interest was to find out the how the high flexibility requirements of disassembly could be met in a system that combines automated with manual workstations. The economical feasible degree of automation was to be elaborated. The system should comprehensively represent disassembly processes and the interwoven dependencies.

5.5.1 Requirements

Disassembly is characterized by its high variability. Different product types and variants, uncertain product condition and product volumes as well as volatile volumes are major challenges for a disassembly system.

Tasks like the recognition of uncertain product types and conditions, the handling of various shapes and sizes, the destructive and non-destructive separation of parts joined by different technologies or special operations like cleaning can only be handled with the realization of novel tools as well as processes and control strategies.

In spite the technological abilities of modern production systems the human cognitive ability may be necessary in order to fulfill all the flexibility requirements regarding disassembly economically.

The handling and disassembly of non-rigid components as well as the disassembly of disassembly friendly joining elements, for instance, can only be economically realized by manual processes. Hence, a hybrid system concept, which is a combination of manual, partly automated and automated processes, is required.

Based on that, disassembly tools, processes and equipment had to be developed and configured in a pilot disassembly system. Goal was to prove the technological and economical feasibility of automation supported disassembly for a high amount of different products ranging from electronic goods, household appliances to automotives.

In order to realize a system that is extensible and fulfils the requirement of comprehensive representation of disassembly processes and can easily be reconstructed, following requirements should be met.

A modular structure is required. Therefore the disassembly system should been divided into the following modules:

- automated destructive disassembly station,
- automated non-destructive disassembly station,
- manual work stations,
- tool changing system,
- clamping and handling system,
- conveyor system,
- sensor system,
- control system.

The division of automated destructive and non-destructive disassembly stations is favourable because of different requirements for the exhaust and safety equipment. Within the modular concept, automated and manual work stations are linked via a flexible conveyor system. Therefore it has to allow the transport of products to a manual disassembly station, alternatively to each automated work station. The reason for this requirement lies in the process reliability of automated disassembly processes. Therefore, a manual follow-up procedure has to be made possible. An uncertainty is the non-determined process time within automated disassembly stations. The disassembly of a component can need more time than expected. Therefore bypass and buffer facilities need to be included.

Based on those requirements one disassembly system for white goods and mechanical aggregates and one for electronic consumer goods was realized. They will be presented in the following sections.

5.5.2 Layout Variants

The conceptualization of a pilot disassembly system presents the generalized choice of specific solution out of a set of alternatives. Important elements of choice were represented in a morphological box. Based on the different modes of operation e.g. ring- or star-shaped 30 layout variants could be developed (Figure 1). In addition, the arrangement of robots, tool change system, working areas and transportation systems were compared during the development of the layout variants.

The main goal of the realization of the pilot disassembly was the development of operation resources like handling devices or transportation system, the development and testing of novel tools and operations for the manual and automated disassembly as well as the verifying of researched knowledge. Thus, the pilot disassembly system should have the features to disassemble different kind of products, goods with varying conditions and differing lot sizes.

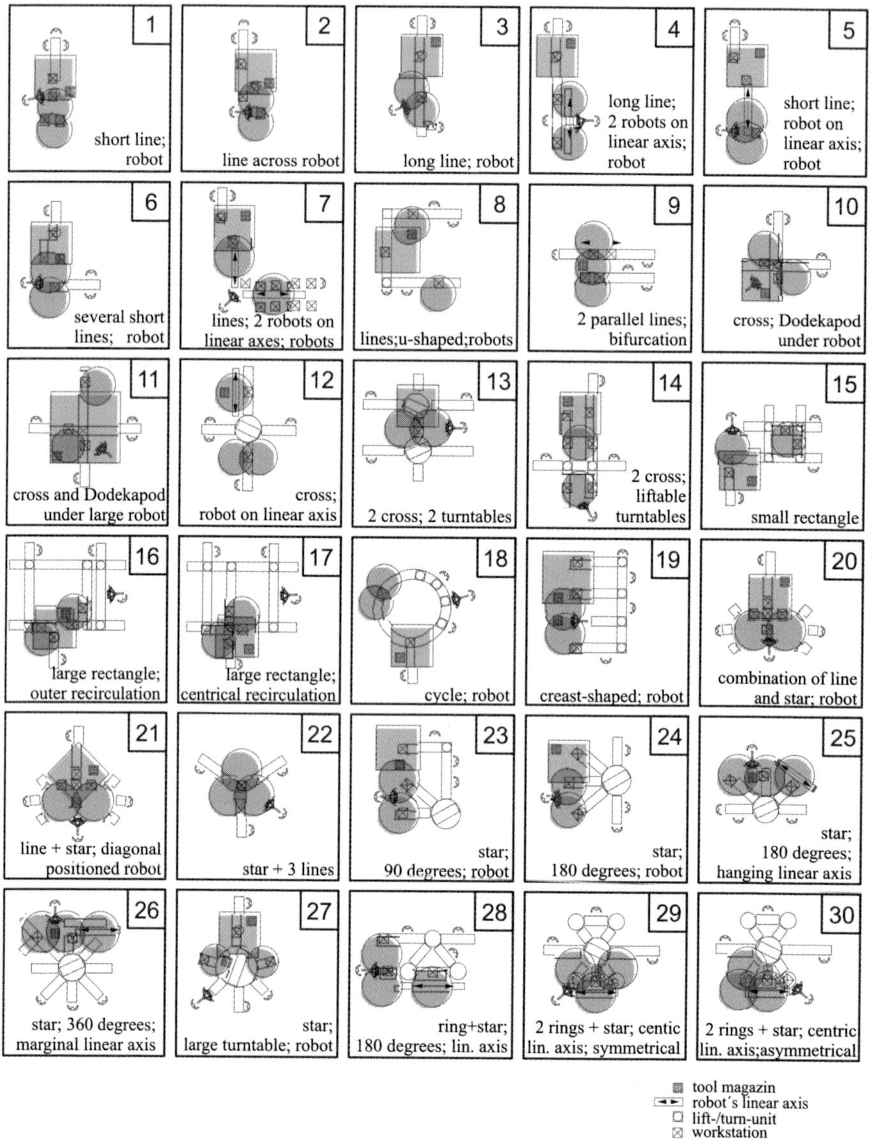

Fig. 1: Layout variants of a pilot disassembly system

Furthermore, the pilot disassembly system should not be product-specific but oriented at a typical structure of a product group. The layout variants were evaluated using a use-value analysis. The favoured flexibility was achieved by the use of hybrid mode of operation, the use of exchangeable work piece carriers as well as the integration of different separation, handling and control techniques.

Finally the layout version 29 was the solution with highest value according the set criteria (Figure 2) [3, 7]. That layout can be used for the pure manual and automated disassembly in a cell or line mode of operation as well as for the hybrid disassembly in a cell, line or network mode of operation. The final layout is composed of a central turntable which is connected with a star-shaped conveyer system. The automated area (bottom) and the manual area (top) shape a triangular-shaped cycle so that two external chains can be realized. Because of the central turntable, it is possible to change quickly between the automated and manual work areas which support hybrid disassembly. The turntable should be able to turn simultaneously two work piece carrier, so a bottle-neck is avoided. The chosen layout version is very compact. A costly conveyance system that serves for reefed exclusively can be omitted.

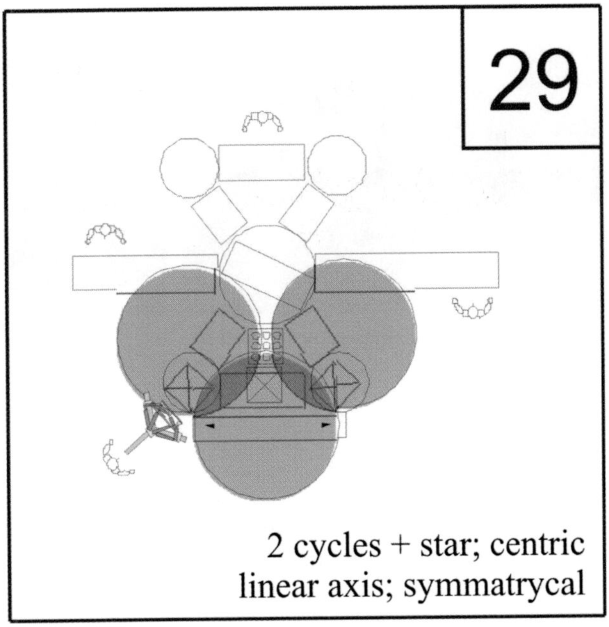

Fig. 2: Layout solution of the disassembly system

A central tool change system is possible due to the appropriate arrangement of the robots. The Dodekapod is arranged in a way that it can cooperate with two robots or work as a manipulator for manual work station. Two small turntables at the automated area provide additional degrees of freedom for the automated decentralized disassembly on both work stations. Destructive and non-destructive disassembly tools are used for the automated disassembly.

Based on the rough planning the detailed planning was conducted. System components for handling transport and tools were specified and chosen. Subsequently the system components were adapted to the predetermined basic conditions of the installation location.

A graphical-dynamical simulation software has been used to evaluate and allocate the working spaces of the robots and to detect possible collisions.

Figure 3 shows a simulation of the chosen layout consisting of a star-shaped structure with two bypasses at the front and the back of the Figure. The pallets carrying the disassembly goods can either move to the front working area, which is the area for automated disassembly, or to the back area for manual work. One advantage of this layout is, that it allows a linear throughput as well as direct transport from an automated process to the manual working area.

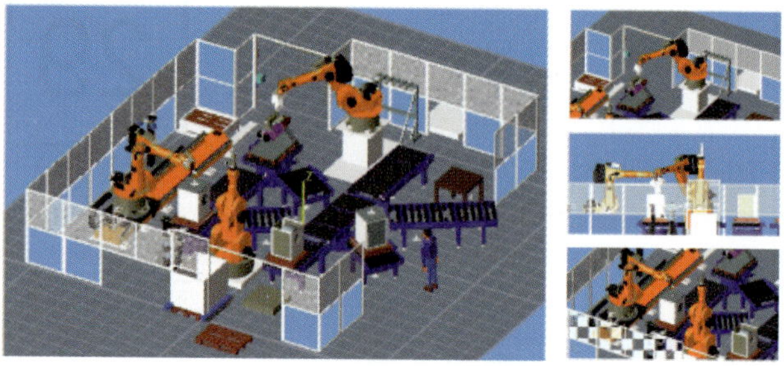

Fig. 3: Simulation of the pilot disassembly system

5.5.3 Equipment and Facilities

The system was equipped with two manually and automated work stations. Within the modular concept, automated and manual work stations are linked via a flexible conveyor system (Figure 4). The tubes, the wires on the rear panel and the cover plate of a washing machine for instance are disassembled on a manually work station. The rest of the washing machine is disassembled on the automated work stations. Furthermore, a novel handling system was realized. It represents an interface between the manual and the automated workstations.

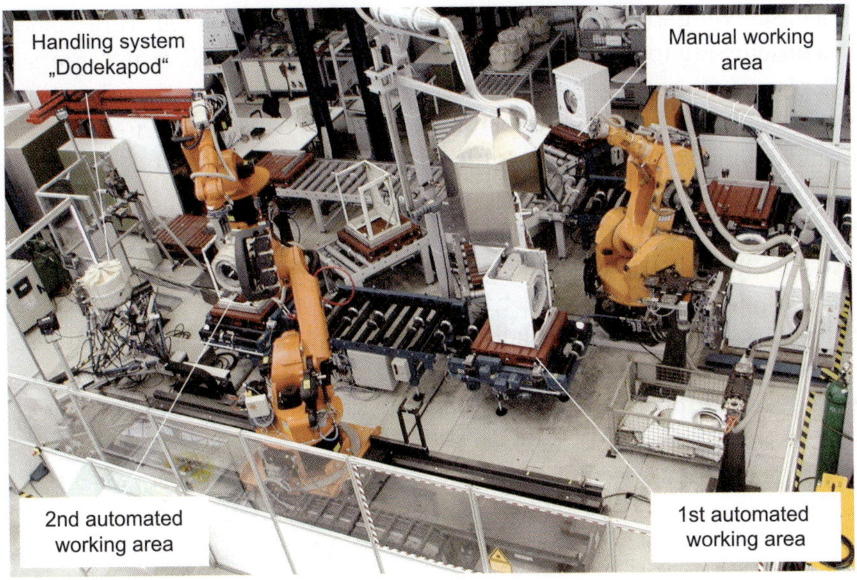

Fig. 4: Pilot disassembly system

For the transport system, roller conveyors were selected from a modular system. Polymer pallets were used, providing an economic transport solution. Three articulated robots have been arranged symmetrically. Two robots were chosen because more than 80 kg have to be removed in addition to the weight of tools components. The central robot can be positioned along a linear axis measuring 4 m length. This enables the robot to operate at different work stations. The robot's tasks are supporting operations, e.g. the removal of the washing machine's housing panels. All robots are fitted with a uniform tool changing system. The system provides standardized connections for power, air pressure, and data transfer.

In addition to the mentioned three robots, the developed gripping manipulator called "Dodekapod" has been integrated into the pilot system. The manipulator allows the movement of the disassembly good into an ergonomically favourable position for manual disassembly as well as placing the object into the working space of robots.

Beside the necessity of a flexible layout of disassembly systems there is a need for flexible tools. As a new approach the pilot disassembly system includes a variety of non-destructive and destructive disassembly tools as well as grippers for handling parts or components. The system is equipped with the following robot-operated tools:

- a unscrewing tool for non-destructive disassembly,
- a unscrewing tool for partly-destructive disassembly,
- a plasma arc cutting tool for cutting metal surfaces,
- an abrasive cutting tool for cutting solid metal parts,
- a hydraulic blade cutter for cutting hard material and non-conducting materials,
- a vacuum suction pad for the handling of flat surfaced parts,
- a two finger gripper for the handling of parts with parallel or cylindrical surfaces,
- a contour clamp for the handling of parts with complex shapes and
- a screw nail gripper for flexible handling of thin plastic or metal
- a specific gripper with different kinds of gripping functionalities parts as well as
- the modular Disassembly Tool Kit (DTK).

5.5.4 Unscrewing Process

Unscrewing is the main process in the disassembly of many products. In automotive engines for instance most of the connections are realized by bolts. In the usage phase, which can last more than 15 years, parts and fasteners of the used products can be soiled, corroded or damaged. The corrosion of threads may result in extremely high breakaway torques, breaking bolts or tools during the unscrewing. A lost or damaged bolt may have been exchanged with another bolt of the same thread but different length of engagement or different spanner flats. For example a M6 bolt with the same thread comes with a 10 mm hexagon, an 5 mm Allen, a Torx E10 or a cross recessed head. Hence, knowledge gained from automated screwing and assembly processes cannot simply be copied for unscrewing used products.

A realized non-destructive unscrewing tool was developed (Figure 5). It consists of a robot-handled component with all sensors and actuators needed for the unscrewing, the controller which processes all data from an information system, the camera the sensors. It also provides a user interface.

As a basis for the robot-handled component, an industrial screwing drive with integrated torque and incremental revolution sensors is used. The signals are connected to analogue outputs. With digital signals, 16 programs can be called. In a program the revolution angles, speeds and minimum or maximum torques up to 320 Nm can be defined. Parallel to the axis of the drive a calibrated Charged Coupled Device (CCD) camera system with integrated illumination has been implemented. Because it is impossible to cover the sun and workshop light, six halogen lamps have been arranged around the camera to ensure a stronger illumination than all disturbing external influences. As described above, the engagement of tool and bolt head is a delicate process depending on the concentricity. External tolerances from the work piece carrier or the product can be adjusted by the image processing, but play in the drive and tool exchange, that means tolerances in the unscrewing system itself, can not be calibrated and not be recognized. Purchasable quicklocks for the change of bits do have slackness due to tolerances in the drives of the nuts (bit holder). To allow a precise and quick tool exchange a clamping jaw, based on the '7/16" Hex' standard, has been developed. Three jaws, sliding on beveled guidings are actuated by a pneumatic cylinder, coaxial to the axis of rotation. Due to the beveling, the hexagon in the center opens or closes until the hexagon nut drive is clamped without play (Figure 6).

Fig. 5: Robot handled unscrewing system

When the products enter the disassembly cell, it has to be identified either by a human or an electronic identification, e.g. a life cycle unit. Dependant on the product type and variant, the disassembly plan will be provided an information system. The disassembly plan contains all relevant data needed for the disassembly processes. It consists of the sequence and removal directions in which the parts of the product have to be disassembled. Bolts are characterized by their thread diameter, length of engagement, the bolt head, meaning the shape and size of their spanner flats and their location in the product. The location of the bolt, in product coordinates, is transferred from to the handling robot via OPC (Figure 7).

Fig. 6: Beveled jaws for hexagon bit drives

The robot transforms the product coordinates to robot coordinates and moves the camera system over the expected location of the bolt. The captured picture is transferred to the unscrewing controller. Dependant on the expected type of bolt head, a template will be loaded and the image analyzed for this template. The fine location, transformed from camera to robot coordinates is transferred via Profibus, back to the robot. Figure 8 (left) shows a rusty bold head, hardly to detect and Figure 8 (right) a detected metallic clean one. As shown above, dependant on degree of staining, the image analysis is not always capable to identify and locate the bolt head. For this case, a human interface as shown in Figure 9 has been implemented on the unscrewing controller. It displays the image on a screen and allows an operator to locate the bolt head with a mouse click at the image.

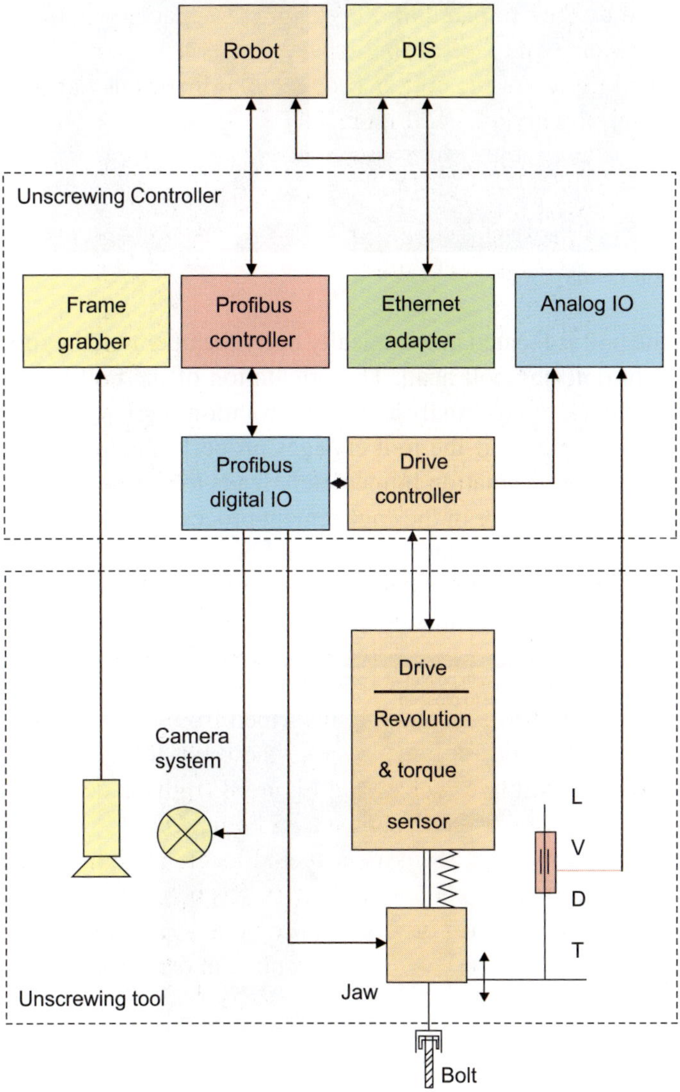

Fig. 7: Structure of the unscrewing system

If a bolt is missing, the operator can proceed to the next disassembly process, in case of a bolt with unexpected spanner flats, he is able to choose the compatible unscrewing nut tool from a menu.

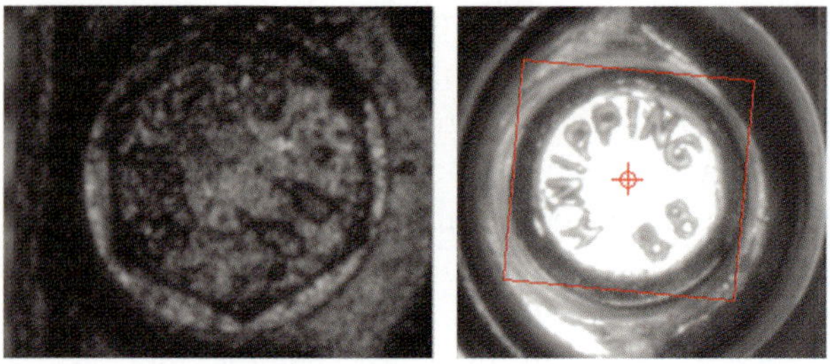

Fig. 8: Bolt heads

When the bolt is located automatically or by the operator, the robot moves the tool right onto the bolt head. The orientation of the bolt head has been determined by the image analysis but the rotation angle of the tool is not known, so its uncertain if the tool engages properly. In 30 % of the cases, starting the loosening rotation immediately leads to a slipping between tool and bolt head and a failure of the engagement process and damage of the bolt head. With an oscillating rotation of the tool between ±60°, programmed in unscrewing drive, the tool slips onto the bolt head. If the torque, during the oscillation, reaches approximately 10 Nm, the engagement was successful.

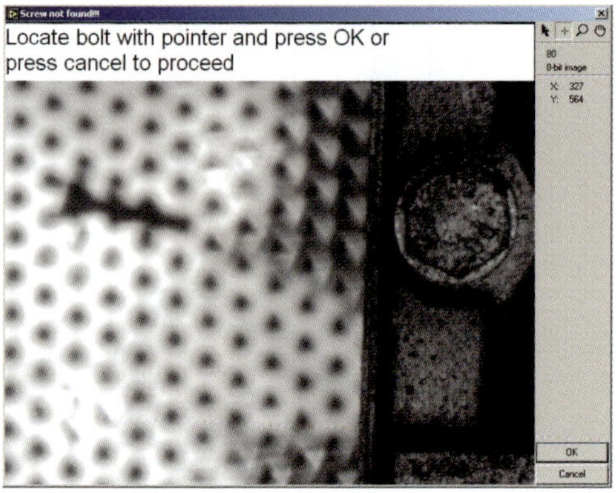

Fig. 9: User interface

While loosening, the bolt head travels a linear distance equal to its length of engagement. For acquiring this data, a LVDT (Linear Variable Displacement Transducer) has been implemented. When the gradient of its signal becomes zero, the revolution can be stopped exactly and the bolt drawn out of the hole. A broken bolt or a bolt with a worn out thread does not move out during rotation and is indicated with the slope of the signal. Analyzing both the torque, delivered by the unscrewing drive and the travelling distance, more data about the process can be gained. If the initial breakaway torque, typical for each thread diameter, is not reached, the bolt has not been fastened properly. A much too high breakaway torque with no travelling distance is a sign for a broken tool or broken bolt. After the bolt has been loosened properly, it has to be drawn out of the hole. In approximately 20 % of the cases the nut-integrated magnet is not strong enough because the thread is stuck in the hole. With rotating the bolt while drawing out, when using a permanent magnet to pull the bolt out of the hole, a gripper is used to remove the bolt from the nut. The robot moves the unscrewing tool to the gripper that the bolt shaft is between the clamping jaws. The gripper closes, the robot moves up and the bolt remains in the gripper. The state of the closed gripper is detected by a proximity switch. An activated switch means a fully closed gripper and is an indicator that the bolt has not been gripped and is still in the hole. If a the bolt head remains in the nut, due to a crack, the next engagement processes will fail and an operator has to remove the bolt head manually.

5.5.5 Adaptation to Different Products

After the system was developed, it was specifically adapted to the disassembly requirements of the product spectrum of household appliances and automotive aggregates. Those product groups were chosen as they represent a large portion on recyclable and reusable products. As a particular example, washing machines and combustion engines were chosen. They provide a number of typical characteristics for the mentioned product groups. Whereas washing machines are predominantly viable for material recycling and thus destructive disassembly is sufficient, the combustion engine is eligible for reuse and remanufacturing thus non-destructively disassembled [3].

The experiences of the disassembly of conventional washing machines were used in order to develop a prototype of a disassembly friendly washing machine [1].

Washing Machine

For better understanding Figure 10 shows the components of a washing machine. Market studies have shown that motors, pumps and valve units can be reused as spare parts [3]. The capacitors should be disassembled when they contain Post Chlorinated Biphenyl (PCB). If the counter weights made of concrete are disassembled, smaller shredders are necessary. Additionally a separation of the concrete from the shredder fractions is not applicable. For all other materials, a good separation helps to increase the earnings.

At the system entry, the products are fed in and manually prepared. Here, all external parts like the power cable or a cover panel are removed and flexible parts on the inside like cables which could lead to complications during automated disassembly are disassembled manually. These tasks are hardly to automate; therefore no robot is integrated at this point. At the first automated work station, a robot cuts the side panels out of the case using a plasma cutting tool.

The aim is to quickly gain access to internal components, e.g. the oscillation system, by destructive cutting. This method was selected because the case is not intended for reuse. Cutting and welding of side panels is an easily and by that economically automatable process,which encourages the application of robots at this point. Due to safety requirements, only a mechanic or automated device comes into consideration for holding and handling the disassembled parts. In the next work procedure, the entire oscillation system is automatically separated from the remaining frame, performed by the two robots.

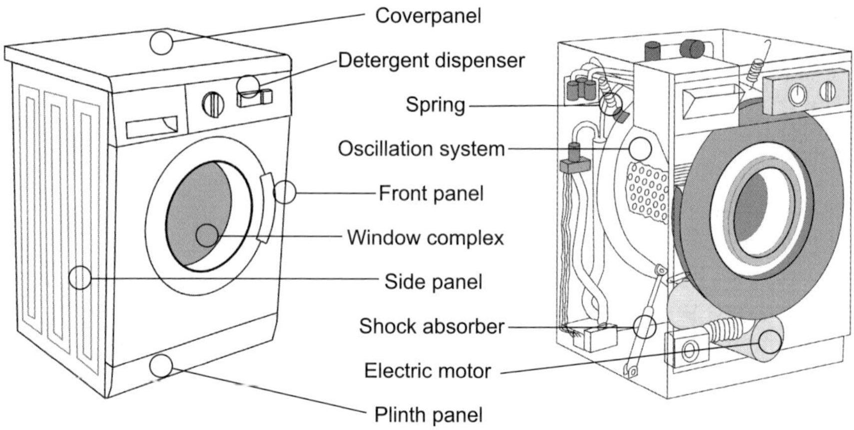

Fig. 10: Components of a washing machine

Here, the operating robot uses a hydraulic shear for cutting the springs and shock absorbers. A hydraulic tool was selected because high cutting forces are reached with solid hardened material which is often used for springs and shock absorbers in washing machines.

New variants are fitted with plastic shock absorbers which makes cutting by conductive beam tools impossible. Again, a destructive method was chosen because the sales profits don't justify a non-destructive method, e.g. by automated unscrewing.

The supporting robot is equipped with a specially designed screwing grabber, a flexible handling tool, which creates its force-effective surface independently from the surface structure. It is very applicable for thin metal or plastic parts of unknown geometry. These materials are often used for lye containers in washing machines. The containers are then moved to a recycling station. Once the oscillation system is removed, it is passed on to an innovative manipulator, the Dodekapod. The manipulator can position and hold the disassembly object for easy manual disassembly. The removed oscillation system is disassembled in a combination of manual procedures and an automated unscrewing and removing of parts. All components of the oscillation system except the motor will be reutilized.

The loosening of the screws is realized by a robot-controlled unscrewing tool. This tool can be applied independently from the geometrical structure of a screw head, because it creates the force-effective surface itself. The unscrewing procedure is supported by an image processing system. The disassembled parts are to be removed by another robot. The washing machine's frame remains on the pallet and is conveyed to a manual work place. There, parts for which a demand as spare parts exists, for example the pump, will be separated for reuse. The reaming parts transported out of the system for a further reutilization.

Combustion Engine

Many of the components of a combustion engine, cylinde heads or oil pumps, are to be reused and may not be destructed during the disassembly. The industrial disassembly is mainly conducted manually. Automation of the disassembly operations has not been implemented, yet [4].

Based on a test-disassembly the disassembly sequences for the manual and automated disassembly could be generated. In the same time all relevant data were documented, like a CAD-Model of the engine, the dimension of components or the position of screws. The attachment parts of engine are manually disassembled, like the spark plug, the injection system or the induction pipe. The rest of the engine without attached parts is automatically disassembled, like: the cylinder head cover, bearing cover or the cylinder head. The automated disassembly is realized by the unscrewing of the bolts with the aid of a proprietary developed unscrewing tool as well as by the handling of the different parts with a flexible gripper.

The realized unscrewing tool consists of a robot-handled component with all sensors and actuators needed for the unscrewing, the controller which processes all data from the Disassembly Information System. As a basis for the robot-handled component, an industrial screwing drive with integrated torque and incremental revolution sensors is used. The realized gripper was developed to conduct different kinds of gripping functionalities for the disassembly of an automotive engine. With the use of only one gripper, all components, which are automatically disassembled, can be handled.

Case Study Mobile Phone

Proving that the developed concepts and processes are economically and technologically applicable for products with different characteristics than those of washing machines or automotive engines a hybrid robot cell for disassembling of mobile phones was developed and tested. The objective of the disassembly was the future adaptation and reuse of the product. The technological developments had to focus on non-destructive processes.

The assembly structure of a product is a key factor for the assessment of potential disassembly processes. It determines the options of adaptation and therefore the involved disassembly processes. Based on an analysis a principle process chain has been modelled as displayed in Figure 11. Here, the most relevant processes for cellular phone remanufacturing are included.

Obsolete cellular phones are typically acquired and stored. Following, accessories such as chargers and batteries are removed from the phones. They are registered, sorted, tested and cleaned separately before being deposited in the accessories warehouse. The cellular phones are registered according to brand and model by detecting the International Mobile Equipment Identity (IMEI) number that enables a one-to-one identification of the devices. The IMEI number is always provided as a bar code and can be detected using commercial scanners. Next, the phones are sorted according to brands and types. While the majority of phones are tested according to functional and cosmetic criteria, phones that are not suitable for reuse due to reasons of age or obvious substantial deteriorations are directly assigned to external material recycling processes. Following, the phones are disassembled and the separated components are cleaned, usually using detergents such as alcohol. If necessary, spare parts that are either disassembled from old phones or externally obtained are fed before the phones enter the reassembly process.

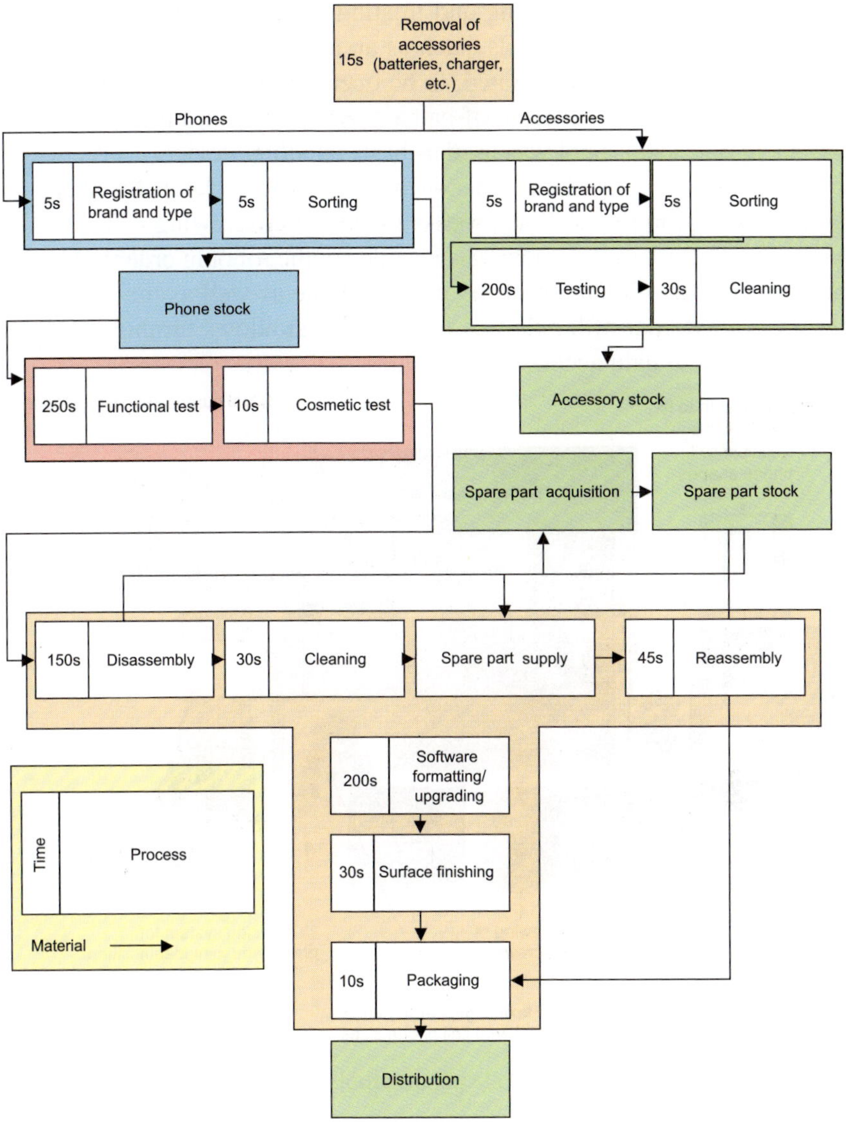

Fig. 11: Cellular phone remanufacturing processes chain

With respect to process optimizations, concentrating on disassembly, as one of the time consuming operations, seems reasonable. To reduce this cost factor, a hybrid disassembly system for cellular phones was developed (Figure 12). Here, innovative tools like flexible clamping or gripping devices have been developed and implemented.

These contribute to handling and treating cellular phones, independent of size and shape. Further tools, such as an unscrewing device and a multifunctional gripper are integrated, to be operated by a robot, after manual operations such as the removal of batteries. As shown in Figure 12 a four-axis adept one scara robot was used as basis handling device.

It controlled the handling tools such as a flexible clamp for shape independent gripping as well as process tools like unscrewing or leveraging systems. A CCD-Vision system was installed at the robot in order to identify the type and the orientation of the mobile phone as well as the location of screws. An external fifth axis was integrated that allows turning the disassembly object to gain access from all sides.

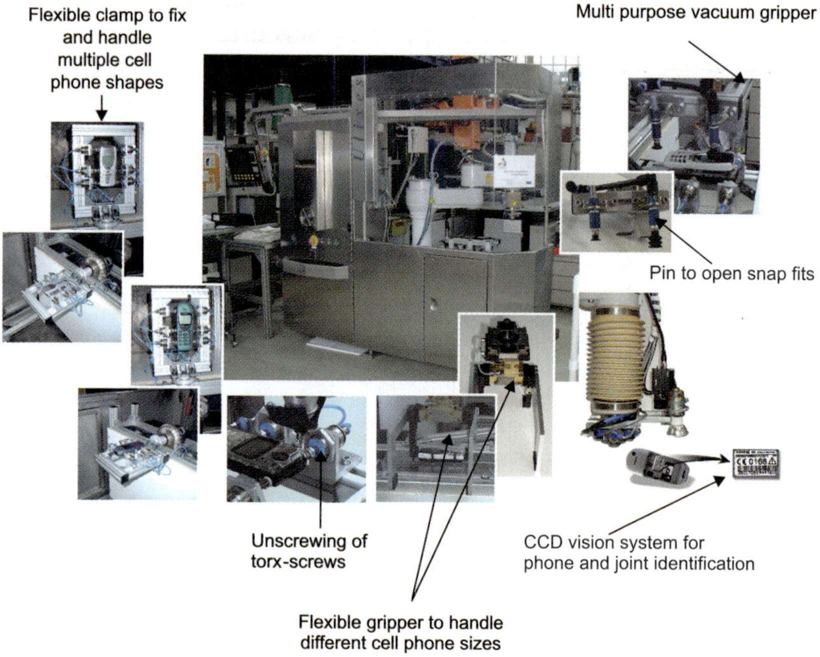

Fig. 12: Disassembly cell for mobile phones, tools

5.5.6 Information and Control

Disassembly processes can be categorized in recognition, handling, separation and special operations. In contrast to assembly processes, those processes have to cope with a lot of uncertainties owing to a wide product variety, uncertain product amounts and product conditions.

In manual operated systems human workers are able to adapt quickly to differing conditions, e.g. a changed joining element will lead to the use of a new tool. Commonly used control systems, which are intended to fulfill the requirements of assembly systems, are not capable of coping with that kind of increased demand for adaptability.

In order cope with product variations, the developed disassembly control system consists of modularized control programs using parameters for the control modules. In order to react on unknown and various product conditions, the system's condition flexibility was increased using a hybrid disassembly system concept.

Human intelligence is key factor for securing the disassembly processes. Hence, its integration was to be taken into account while designing a control system. If a requested disassembly process cannot be accomplished by an automated process, then the process is finished in a manual workstation.

This causes a new scheduling for the entire disassembly system. Product quantities are separated into different production lots. In order to handle various lot sizes, the control system generates a new control program for each lot. This is possible down to a lot size of 1.

The control structure of a disassembly system was developed adapting existing structures, which are used for conventional production system control. The hierarchical and decentralized approach is commonly used. Furthermore, standards were applied to the networking infrastructure as well as the hardware which is used to establish network connections. For the seamless integration of all levels of automation, a gateway was integrated between the network structures used. In order to allow multi vendor software integration, a commonly used interface were applied, e.g. OLE for Process Control (OPC).

A new approach in the field of control techniques was the development of a flexible control system able to generate the control code online. In order to develop a flexible disassembly control system, the technical feasibility is considered in conjunction with the generation of disassembly control sequences. Existing control algorithms for manufacturing systems are adapted to the needs of disassembly. Beyond that, a modularized and parameterized control approach was applied to the cell control of a prototypical disassembly system. Existing approaches for modular control structures, the description of a new approach as well as an interaction between cell and robot controllers were needed [6].

An important contribution for an efficiently working hybrid disassembly system is the integration of the information technology throughout the whole system. The purpose of this integration is to provide all areas of the disassembly system with relevant data, so the disassembly can be executed autonomously, reliably and safely.

The developed Disassembly Information System (DIS) is based on a database management system. It is characterized by a central organization of the data and a separation of data administration and user administration, thus enabling a multi-user environment. Also, it enables for application oriented views on the data base. In order to realize the real time requirements of the automation technology, a distributed database system has been used [7].

The product, process and equipment databases were implemented in a conventional object-relational database system. It serves for the administration of disassembly object data, disassembly process data and disassembly equipment data. It is responsible for the acquisition and storage of the product, process and equipment data provided by the manufacturer or the disassembly operator. The database system is equipped with standard interfaces to communicate with applications on the planning level, e.g. production planning and control systems. A direct connection to the production data acquisition system was also implemented [7].

The production data acquisition was realized in a real time system with data base character and according interfaces to the equipment technology. The separated administration of the production data is necessary, because the automation technology has different requirements regarding data volume, number of data packets and reaction time. In order to enable the production data acquisition for an integration in the information system, it consists of both a database part for interaction with the object-relational database system and a processing part for the acquisition of the data on execution level [7].

The communication management system serves as communication platform for the used automation equipment of the disassembly system. It administrates all data connections of the automation devices with the information system and between themselves. OPC is a standard that is available for almost all automation devises and easily usable with personal computer based software applications [7].

The web server enables for a platform independent information representation. It can provide methods for the local analysis of the disassembly data [7].

Modular Robot Programming

Because the automated disassembly is an important part of the hybrid disassembly, methods for the programming of industrial robots have to be developed for users, which may not be specially qualified. A modular robot programming of industrial robots is a method for easy programming without advanced knowledge. In addition the programming is supported by a human machine interface for the programming and the easy manipulation of the programs. As a result of the high variability of the use goods, the programs have to be adapted on the conditions of the products.

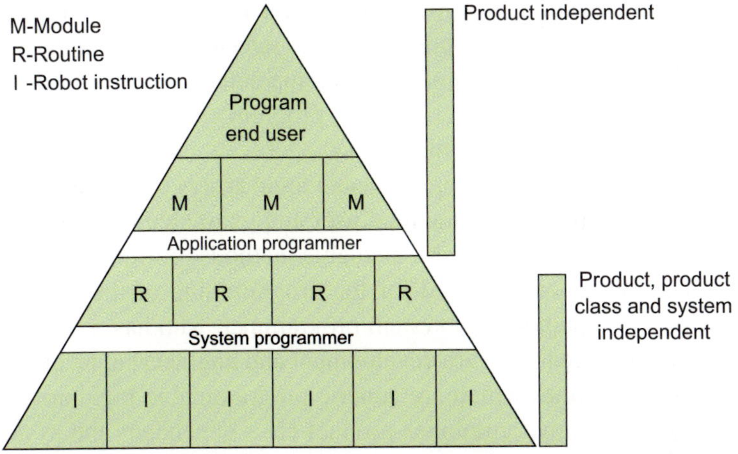

Fig. 13: Programmable logic arrays of a robot program

Furthermore, new methods for the programming of industrial robots were developed, because the cost for programming of small lot sizes is a key factor of the commercial examination. Beyond it, programming methods have to be developed for an end user, who was not explicitly qualified for the programming of industrial robots. A commercial use of industrial robot in small and medium-sized enterprises should be enabled. The efficiency of the programming and the flexibility of the use can be increased with the aid of modular programming and the parameterization of the programmable logic arrays. A quicker adaptation of production process changes can be achieved by the generation of robot programs composed of modules, which are a result of a recombination of programmable logic arrays. In addition parameterized programmable logic arrays are provided, which can use for a product class. The generation of a program is realized by the use of two different programmable logic arrays, like the routine and the modules (Figure 13).

The use of robot instruction is the first level of the robot programming, when an industrial robot is textually programmed. With the aid of robot instruction the robot can be controlled with an advanced programming language. The second level of a robot program is the function module. This module is product and system independent and describes function units, e.g. the activation of a tool or sensor request. A function module can be realized by the combination of several routines.

The function module is linked with the database and consequently system and product dependent. A repeating operation is presented by a function module, e.g. a tool change or grip process for a disassembly step. The fourth level for the realization of a robot program is the combination of several function modules to a robot program for a selected disassembly process. This robot program is product and system dependent, but it is product class independent. That means with one robot program can be disassembled different product types of one product class.

The programming of the programmable logic arrays by different key personnel enable a generation by an end user and guarantee the operability of the combined robot programs. All available methods of robot programming (online and offline) can be used for the programming of the programmable logic arrays. The task of a system programmer, who has a consolidated knowledge in the area software development and engineering technology, is to develop the routines which are generic programmable logic arrays. The application programmer combines product class dependent and system independent modules which base on the developed routines. Furthermore, the application programmer develops modules which are linked with a product and system database. These modules represent separate and interdependent programmable logic arrays. The end user, e.g. a skilled work man, can combine a robot program as a result of his technology knowledge and without wide knowledge of robot programming.

The generic programmable logic arrays are adapted to a product class and to a disassembly system as a result of the linkage. Thereby the variables of the programmable logic arrays are linked with the corresponding data of a database, which guarantee the operability of the arrays [8]. Furthermore, modules are generated, which are developed for a product class and respect special intrasystem features. An adaptation of the programmable logic arrays and the programs are enabled by the adaptation of the database, in which the end user doesn't need an advanced programming knowledge.

5.5.7 Conclusion

The integration of the developed tools and processes in hybrid disassembly systems has proven the technological feasibility of automated disassembly. The required flexibility and adaptability which is a consequence of the high variety in product types, conditions and volumes, however is only economically realizable in a hybrid disassembly system that combines human intelligence and accuracy and speed of automated systems. Hybrid systems require an adapted structure that provides the interfaces between manual and operated workstations.

The control and information system of a hybrid disassembly system has to be fully integrated in order to provide required information at the right time and the right place. If these requirements are met the disassembly can be conducted economically and safe and serve as a basis for responsible and sustainable development.

References

1 Basdere, B.; Friedrich, T; Härtwig, J.-P.: Fallbeispiel: Entwicklung und Realisierung eines Prototyps einer demontagegerechten Waschmaschine. Umweltgerechte Produktentwicklung: ein Leitfaden für Entwicklung und Konstruktion. Beuth Verlag, Berlin 2004.

2 Basdere, B.: Beitrag zur Steigerung der Nutzenproduktivität von Ressourcen durch anpassen von Mobiltelefonen. Dissertation, TU Berlin, 2004.

3 Härtwig, J.-P.: Verfahren und Systeme zur Demontage komplexer Gebrauchsgüter. Dissertation, TU Berlin, 2005.

4 Seliger, G.; Uhlmann, E.; Friedrich, T.; Harms, R.: Pilot Disassembly System for Automotive Engines. In: Proceedings of the 39th CIRP International Seminar on Manufacturing Systems (ISMS 2006). Ljubljana, Slovenia, 2006, pp. 63-66.

5 Uhlmann, E.; Seliger, G.; Härtwig, J.-P.; Keil, T.: Pilot Disassembly System. Production Engineering - Annals of the German Academic Society for Production Engineering, Karlsruhe, Ausgabe I/1, 2001, S. 83-88.

6 Kim, H.-J.: Beitrag zur dynamischen Prozessplanung und Generierung von Steuerungssequenzen für flexible Demontagesysteme. Dissertation, TU Berlin, 2004.

7 Keil, T: Informationstechnische Integration hybrider Demontagesysteme. Dissertation, TU Berlin, 2004.

8 Seliger, G.; Uhlmann, E.; Friedrich, T.; Harms, R.: Realization of an adaptive modular control for a disassembly system. In: Proceedings of the 6th IEEE International Symposium on Assembly and Task Planning (ISATP 2005). Montreal (Canada), 2005, compact disc.

6 Planning for Remanufacturing and Recycling

The previous chapters have shown new approaches and innovative processes and tools developed for increasing the efficiency of disassembly. The manifold of potential technological realizations considerably expands the solution space available for planners. The adequate balance in selecting along a manifold of different criteria gains importance. Changing product types, changing numbers of components on demand and supply side as well as different product conditions after respective usage phases have to be considered. Software tools can support planners by identifying feasible solutions and evaluating more alternatives in shorter planning time.

Facility planning applies integer linear programming approaches for the dimensioning and structuring of disassembly systems. The solution alternatives are evaluated by discrete event simulation. For given facility layouts and demand and supply conditions, program planning is performed by integer linear programming on the example of flat screen monitors. Mathematical tools prove to be a valuable aid for complex planning and scheduling in disassembly.

6.1 Facility Planning

Günther Seliger, Sebastian Kernbaum, Berlin, Germany

The planning of mainly manual disassembly systems has been a field of intensive research for more than a decade [1, 2, 3, 4, 5]. The European Waste Electrical and Electronic Equipment directive and the End of Life Vehicle directive have triggered an increasing demand for cost efficient high capacity disassembly systems. The previous chapter has shown that the automation of selected disassembly processes offers the potential of increased performance for common manual disassembly systems. Highly flexible disassembly equipment have been developed, and commonly used manufacturing equipment have been applied in automated disassembly. Hybrid disassembly results in more complex planning processes. It has to focus strongly on the planning of capacities and their utilization as well as on cost for investment and operation. Due to the high proportion of the fixed costs in automated systems and the high risk of idle capacity cost caused by low utilization or low availability, the balancing of the system's loads gains importance. Tools are required in order to efficiently select good out of manifold potential solutions for hybrid disassembly systems.

The objective of the presented approach is to support the experienced planner in the design of a disassembly system by means of tools for modelling and evaluation. The planner will be enabled to rapidly design disassembly sequences by using a dialogue based interface to match modelled disassembly products (product model) and disassembly processes (process model) with equipment and tools (resource model). The resulting qualitative and quantitative description of the disassembly process model represents the disassembly sequence: an integrated data set of product, resource and process attributes which is an essential basis for the systematic modelling of any disassembly system.

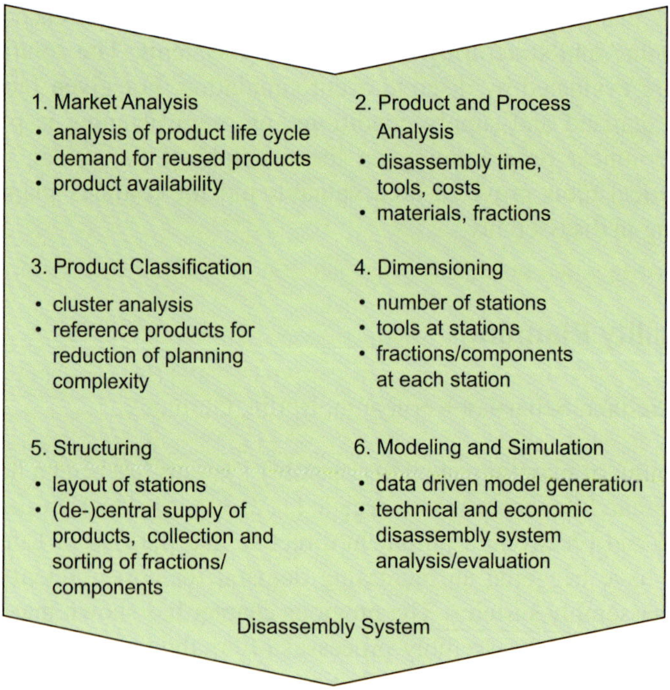

Fig. 1: Six-step approach for the planning of disassembly system

Such sequences need to be adequately represented in a database structure to support the decisions necessary for designing disassembly systems in a later step. Applicable software tools need to be used in order to administrate product, resource and process models, to describe the disassembly sequence based on these models and to find similarities in between those sequences. In a later step, these grouped or clustered sequences can be used for the design and subsequent evaluation of alternative disassembly systems, thus sequences need to contain information on performance and cost. An adequate software tool is discrete event simulation, which allows the evaluation and visualisation of processes and structures as well as supports the balancing of loads in the system. Figure 1 depicts the basic elements of the approach addressed below.

On the basis of the requirement analysis for successful disassembly and under the consideration of existing approaches for disassembly system design, a general six-step approach for the planning of disassembly systems has been derived. The first step requires a detailed market analysis regarding the supply and potential reuse market in order to determine the economic framework for disassembly. A methodology for the evaluation of recycling and reuse suitability of a product design is presented in [6, 7].

6.1.1 Product and Process Analysis

In the second step, a detailed product and process analysis is required in order to derive planning relevant data. The methods applied are described in the following and an exemplary application on mobile phones is presented in [8].

For the rough planning of the disassembly process, a product model for the disassembly object and a resource model for the equipment are applied. The product model contains a selected disassembly plan consisting of disassembly steps derived from the product condition model presented in the product development chapter. Each step contains the following information: disassembly operation, fraction, precedence, type and amount of the joint, and joining elements. The description of the disassembly steps from the disassembly plan give a rough impression of the process to be planned yet does not provide an explicit solution.

The resource model distinguishes between functions for separation and manipulation (Figure 2) and considers developments and experiences described in the chapter processes and tools for disassembly. Separation is described in two levels using functions according to DIN 8580 and VDI 2860 [9, 10]. Manipulation is also described in two levels, differentiating between manual, mechanical and automated manipulation on the first level, and fundamental types of manipulation, i.e. cartesian, cylindrical, polar and joined-arm for automated manipulation on the second level.

The resources for the execution of the operations are being differentiated into effectors and kinematics. Effectors fulfill separation and handling processes, kinematics move, position and orientate effectors. Effectors and kinematics are modeled as Modular Equipment Classes (MEC). The required information in order to link the product and the resource model to derive a process model, i.e. the disassembly sequence, is provided by the following MEC attributes: performance, weight (respectively resilience towards weight), and cost.

Empirical data on equipment properties for manual, mechanical and automated separating and handling have been documented during case studies. These equipment and process data are adequate for the rough planning of disassembly systems. Attributes for MEC and disassembly objects have been derived based on the experience and are used to describe the product and resource models.

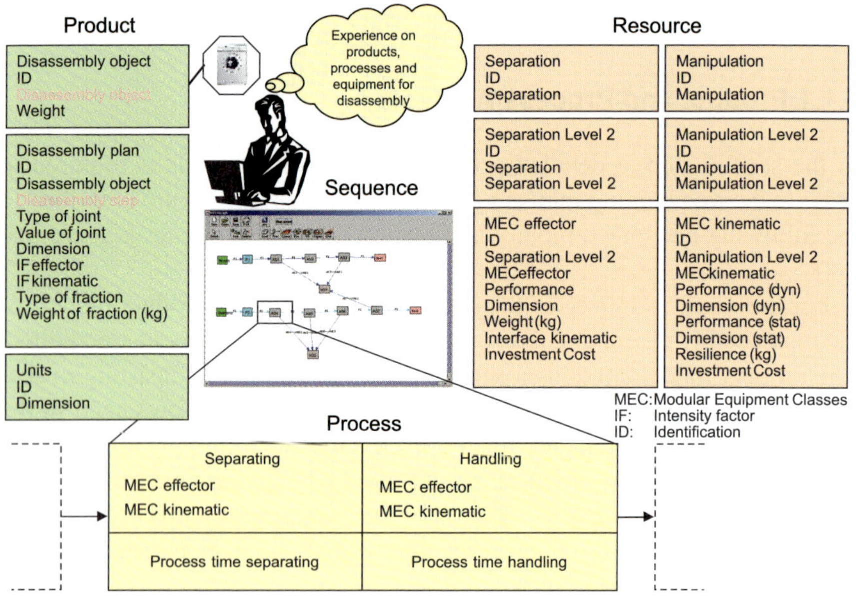

Fig. 2: Product, resource and process model for sequence planning

Sequence Planning

The objective of the planning of disassembly sequences is to match Modular Equipment Classes (MECs) with disassembly steps and to derive a rough estimate of the process time. A disassembly sequence planning tool developed in Java™ supports the planner of a disassembly system in the phase of sequence planning. Figure 2 presents the data structure used by the tool. Functions for separation as well as for manipulation and consequently corresponding MECs are selected by the planner. Each disassembly step is specified by its separating process and an optional handling process. The latter is required if a fraction needs to be handled during or after the separating process. Processing the parameter values from MEC attributes and disassembly objects allows a rough estimation of the processing time for each disassembly step.

The description of disassembly sequences for several disassembly objects and the calculation of corresponding process times provides the basis for the later design of flexible work stations. Process times for separating and handling result from the linkage of product and resource model and a supplementary evaluation by the planner. The product model of the disassembly object contains qualitative and quantitative information for the separating process. The resource model provides information on the performance of MECs. Any separating or handling process requires the combination of an effector and a kinematic.

Separating operations are differentiated into dynamic and quasi-static operations considering the motion of the effector by the kinematics for total performance or not.

S = Disassembly sequence	t = Time
E = MEC effector	T = Separating
K = MEC kinematic	H = Handling
l = Performance	i = Intensity factor effector
s = Performance (stat)	j = Intensity factor kinematic
d = Performance (dyn)	$E_u \in$ {piece/s; mm/s}
w = Value	$K_s \in$ { positioning/s}
u = Dimension	$K_d \in$ {mm/s}

It is assumed that quasi-static separating processes e.g. unscrewing or cutting require only low kinematic support so that the effector determines the time for separating. The manipulation of the position and orientation of the effector using the selected kinematic is addresseded by the performance parameter K_s. Depending on the number of operations S_w in each disassembly step, the process time for the quasi-static operation is calculated with Equation 1.

$$t_T = S_w \left[\frac{S_i}{E_l} + \frac{S_j}{K_s} \right] \tag{1}$$

Dynamic separating processes such as abrasive and plasma cutting require kinematic support throughout the process. The lower performance of effector and kinematic defines the separating performance.

The process time for the dynamic separating operation is calculated with Equation 2 based on the length S_w of the joint.

$$t_T = \frac{S_w}{Min\left\{ \frac{E_l}{S_i} ; \frac{K_d}{S_j} \right\}} + \frac{1}{K_s} \tag{2}$$

The handling of a fraction can be necessary during or after the separating process. Primarily, its consideration is important in order to assign Modular Equipment Classes (MEC) to disassembly steps. The performance of the effector E_l and the performance of the kinematic K_s are used to calculate a rough estimate for the process time (Equation 3).

$$t_H = \left[\frac{1}{E_l} + \frac{1}{K_s} \right] \tag{3}$$

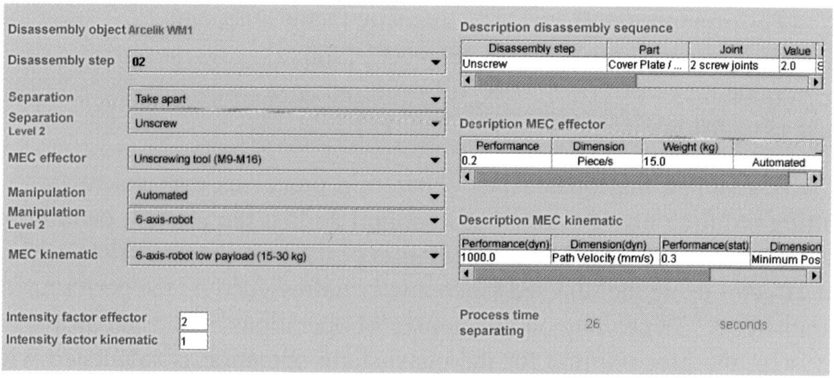

Fig. 3: Dialogue for sequence planning

Figure 3 depicts a dialogue in the developed tool that enables the planner to rapidly design disassembly sequences through a dialogue based interface to the modelled disassembly products, processes, equipment and tools, taking into consideration rules for matching disassembly objects with equipment. Data is stored in a MS-Access™ database.

The appropriate selection and combination of MEC for each disassembly step is performed by the experienced planner and results in a disassembly sequence that offers an adequate estimation of the overall disassembly process time. By adjusting the intensity factor i for effectors and j for kinematics the planner is able to apply his process knowledge to improve the process time estimation. By planning a handling process, the planner can decide whether the calculated process time should be added to the separating process.

6.1.2 Product Classification

The planned disassembly sequences give the basis for the further rough planning of disassembly systems. In the third step of the planning approach, the main focus lies on the systematic consideration of the product diversity [1]. Test disassemblies of different products such as television sets and washing machines show that disassembly sequences not only differ between manufacturers but also between product variants from the same manufacturer. Grouping disassembly sequences under the consideration of product attributes such as type and weight of fraction, and process attributes such as required MEC and process time is a promising approach.

The objective of cluster analysis is to sort objects into groups, or clusters, so that the degree of association is strong between members of the same cluster and weak between members of different clusters. Figure 4 depicts the modus operandi. Hierarchical cluster analysis is a widely applied tool in disassembly planning. Product and production oriented methods have been developed [3, 5, 11]. Hentschel [3], suggests a product oriented fuzzy-based method. Von Werder [5] and von Westernhagen [11] consider product and process attributes. Their approaches differ in the selection of attributes and distance functions. Von Werder and von Westernhagen group objects for manual disassembly using the attributes fractions, joining elements, separating operations, separating equipment, process times and disassembly depth. A manual disassembly system is planned for a homogeneous group of objects. Disassembly steps and specified equipment can be subsumed for one workstation. Automated processes are not considered.

The challenge in cluster analysis is to scale disassembly relevant attributes and select adequate distance functions and cluster algorithms. Binary scales such as fraction or no fraction and metric scales such as weight of fraction are easy to process. Attributes that cannot be described using binary or metric scales require indirect scales, e.g. to describe the precedence of disassembly steps.

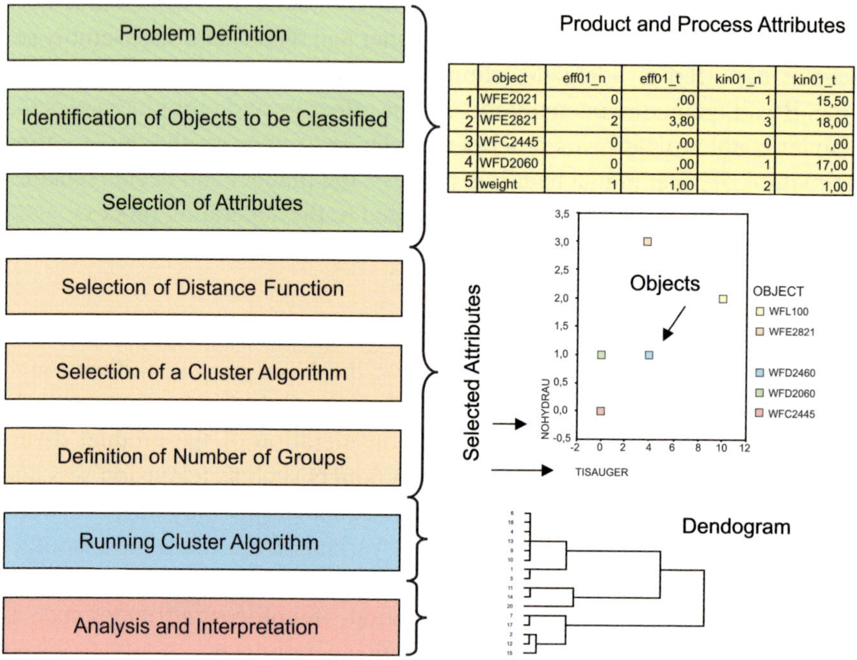

Fig. 4: Cluster analysis

Disassembly sequences consider manual as well as automated processes. Assigning MECs for each step enables a more detailed determination of distances between process and product specific disassembly sequences. Different disassembly objects are comparable based on the resemblance of their sequences. The direct linkage between disassembly step and the selected MECs enables clustering based on process and product specific attributes. These attributes are effectors and kinematics for separation and handling operations as well as process times and fractions.

A tool developed in Visual Basic for Application (VBA) serves as a connection between the database and the standard software tool SPSS™ that is used for the grouping of sequences. The tool allows the selection of sequences (objects) that are described by previously defined attributes. Only metric attributes are used. Each individual MEC, e.g. grinding effector, unscrewing effector, 6-axis-robot, is represented by two attributes: the number of MECs and the accumulated process time for each MECs regarding the selected sequence. Each fraction class, e.g. PP, Steel, PP+Steel, PVC, PVC+Cu, is represented by the attribute number of fractions.

Using standard software like SPSS™, which offers a large variety of implemented distance functions and cluster algorithms, reduces the development effort. The software tool SPSS™ also has the possibility to transpose the spreadsheet and to weight attributes in order to set preferences. Each disassembly sequence is represented by a row, whereas the last row is reserved for a factor to weight the attributes in the columns. For every MEC selected from the database, two columns are required: number (n) and process time (t). Fraction classes are represented by one column for each class.

Depending on how the attributes are scaled very different groups can result from the application of an algorithm. Cluster methods presented in [5, 11] apply a maximum of two distance functions and only one cluster algorithm. Experience in other fields of application show that there is not only one perfect function or algorithm. The presented tool can be applied to group sequences using different distance functions and cluster algorithms in a very easy to use and fast manner. The same data can be used to assign disassembly sequence clusters, i.e. disassembly objects, to existing disassembly systems. Also, the specification of disassembly sequences by means of MEC and fraction classed comprises of decisive information required for the design of disassembly systems.

6.1.3 Dimensioning

The general approach for the dimensioning a disassembly system can be described as follows: For a given yearly disassembly volume, with a set of disassembly objects, which are described by disassembly sequences, with attributes like resources and processing time, the minimum number of disassembly stations and assigned resources needs to be determined.

A variety of factors adds to the complexity of the described problem. Precedence constraints need to be considered, as some disassembly tasks require completion in a particular order, which is defined by the disassembly sequences. In addition, some disassembly tasks cannot be carried out at the same disassembly station due to operational safety restrictions, e.g. manual tasks cannot be assigned together with automated tasks, and technological restrictions, e.g. plasma and water jet cutting cannot be assigned to the same station.

In the following, a mathematical model addressing the above described dimensioning problem is introduced. For the sake of clarity the formerly named disassembly sequences are considered to be products that require disassembly processes. The disassembly system is configured for a given set of products (P) considering disassembly processes (D). The resources Modular Equipment Classes (MECs) and stations (S) are considered as the systems cost drivers.

The objective in the planning of the system is to minimize the system cost while considering the constraints such as cycle time, precedence of processes of every product, operational safety and technological restriction.

The problem is formulated as an Integer Linear Programming (ILP) model. The objective function is given in Equation 4, MC represents the cost for a MEC (M) that is assigned to a station (S) using the binary decision variable SM∈{0 = not assigned, 1 = assigned} and SC represents the cost for a station that is opened using the binary decision variable SO∈{0 = not opened, 1 = opened}.

$$MIN\left(\sum_{s=1}\sum_{m=1} MC_m \times SM_{s,m} + \sum_{s=1} SC_s \times SO_s\right) \tag{4}$$

In order to be used in a ILP solver, the needed data from the database have to be prepared. The total time for separating and handling is considered to be the Disassembly Process Time (PDT). A MEC within this model represents both, the effector and the kinematic required for separating, yet not the equipment required for handling.

First, the Cycle Time (CT) for the system is derived for a given production performance (products/day) and available working time, e.g. 360 products/day and 900 minutes/day working time for each resource, results in a cycle of 150 seconds. It is assumed that all PDTs are shorter than the underlying CT. If CT is shorter than the longest PDT, the CT needs to be extended and the production performance (products/day) needs to be estimated again. The stations within the disassembly system are arranged in a strictly sequential order, therefore parallel stations are not considered.

The constraints given in Equation 5 avoids, that Disassembly processes (D) of a Product (P) assigned to a Station (S) exceed the previously derived Cycle Time (CT) by using the binary decision variable PDMS with PDMS∈{0 = not assigned, 1 = assigned} and PDT for the required disassembly process time.

$$\sum_{d=1}\sum_{m=1} PDMS_{p,d,m,s} \times PDT_{p,d} \leq \mathrm{CT} \qquad (5)$$
$$\forall s, \forall p$$

Equation 6 guarantees, that every Modular Equipment Class (MEC) that is required for a Disassembly process (D) and Product (P) is assigned to a station. In PDM each MEC (M) is represented by an integer.

$$\sum_{s=1} PDMS_{p,d,m,s} = 1 \qquad (6)$$
$$\forall p, \forall d, \forall b \big| b = PDM_{p,d}$$

Equation 7 guarantees that only one MEC (M) is assigned for each Product (P) and Disassembly process (D) at each Station (S).

$$\sum_{m=1}\sum_{s=1} PDMS_{p,d,m,s} \leq 1 \qquad (7)$$
$$\forall p, \forall d$$

Equations 8 and 9 guarantee, that any two MEC (M) that are assigned to one Station (S) are not conflicting with each other, using the binary variable MM with MM∈{0 = conflict, 1 = no conflict} and the binary variable SM with SM∈{0 = not assigned, 1 = assigned}. The variable X represents a large positive number.

$$\sum SM_{s,m1} + SM_{s,m2} + X \times MM_{m1,m2} \geq 1 \qquad (8)$$
$$\forall s, \forall m1, \forall m2, X >> 1$$

$$\sum SM_{s,m1} + SM_{s,m2} - X \times MM_{m1,m2} \leq 1 \qquad (9)$$
$$\forall s, \forall m1, \forall m2, X >> 1$$

Equation 10 guarantees that a Station (S) will be opened if a MEC (M) is assigned to it, using the binary decision variable SO with SO∈{0 = not opened, 1 = opened} and PDMS.

$$SO_s \geq PDMS_{p,d,m,s} \qquad (10)$$
$$\forall s, \forall m, \forall p, \forall d$$

Equations 11 and 12 guarantee, that if no MEC (M) is assigned to a Station (S) in PDMS, the station will not be opened.

$$SO_s + X \times \sum_{p=1} \sum_{d=1} \sum_{m=1} PDMS_{p,d,m,s} \geq 0 \quad \forall s, X >> 1 \tag{11}$$

$$SO_s - X \times \sum_{p=1} \sum_{d=1} \sum_{m=1} PDMS_{p,d,m,s} \leq 0 \quad \forall s, X >> 1 \tag{12}$$

Equation 13 guarantees that if a MEC (M) is assigned to a Station (S) in PDMS, this MEC will be assigned to a station in SM. SM is required in the objective function to calculate the cost for the assigned MEC.

$$SM_{s,m} \geq PDMS_{p,d,m,s} \quad \forall s, \forall m, \forall p, \forall d \tag{13}$$

Equations 14 and 15 guarantee that if no MEC (M) is assigned to a Station (S) in PDMS, no MEC will be assigned to a station in SM.

$$SM_{s,m} + X \times \sum_{p=1} \sum_{d=1} PDMS_{p,d,m,s} \geq 0 \quad \forall s, \forall m, X >> 1 \tag{14}$$

$$SM_{s,m} - X \times \sum_{p=1} \sum_{d=1} PDMS_{p,d,m,s} \leq 0 \quad \forall s, \forall m, X >> 1 \tag{15}$$

Equation 16 guarantees that no more than a pre-defined number (MN) with MN∈{Integer} of MEC (M) are assigned to one Station (S).

$$\sum_{m=1} SM_{s,m} \leq MN \quad \forall s \tag{16}$$

Equation 17 guarantees that precedence relations are considered.

$$\sum_{m=1} \sum_{s=1} S \times PDMS_{p,d+1,m,s} - S \times PDMS_{p,d,m,s} \geq 0 \quad \forall p, \forall d | d < \max(D) \tag{17}$$

The described ILP-Model was implemented using the solver Lingo™. The data required for applying the ILP-Model are derived from the disassembly sequences [12]. For selected products, i.e. disassembly sequences, the data regarding number and precedence of processes, process time and assigned equipment is converted into matrices within a MS-Excel™ spreadsheet. Lingo™ imports data and exports decision variables from and to MS-

Excel™ using the Object Linking and Embedding (OLE) interface. The decision variable PDMS – describing which products, processes and equipment are assigned to which station – is generated for main and side disassembly lines. Results of the dimensioning phase are:

- number of required disassembly stations and
- assigned disassembly resources (MEC).

6.1.4 Structuring

In the phase of structuring, the fifth phase of the system planning, an efficient layout with minimal lead time and maximal utilisation needs to be derived for the designed disassembly system. The structuring phase is comprised of spatial structuring, i.e. layout design, and of temporal structuring, i.e. the determination of the systems process flow [13].

The generation of layout alternatives is one of the most critical steps in facility planning, since the selected layout establishes physical relationships between activities [14]. As a final layout is selected from the alternative layouts, it is important that a large number of different high quality alternative designs are considered. The number of possible layouts for any real-life problem can be quite large. Furthermore, the specification of constraints and the establishment of an accurate objective function are tasks that can be accomplished only fuzzily at best. For this reason Askin and Standridge come to the following conclusion [15]: "The design process does not ask for the selection of the best design from all possible feasible designs. Rather, it asks for a selection to be made from among several reasonable designs, in order to satisfy not to optimise".

Therefore, reference layouts were considered for the design of hybrid disassembly systems, which are presented in [16]. These reference layouts re-present feasible design concepts for an efficient disassembly. The objective of the structuring phase is to determine the best location for each disassembly stations determined in the dimensioning phase for each reference layout that is considered. Every reference layout has a number of potential locations for assignment. To be able to assign the required number of disassembly stations, that may vary according to the planning problem, different alternatives are considered for each reference layout. Figure 5 depicts an example of two alternatives for the same reference layout. The number of possible locations for disassembly stations can vary from one to nine.

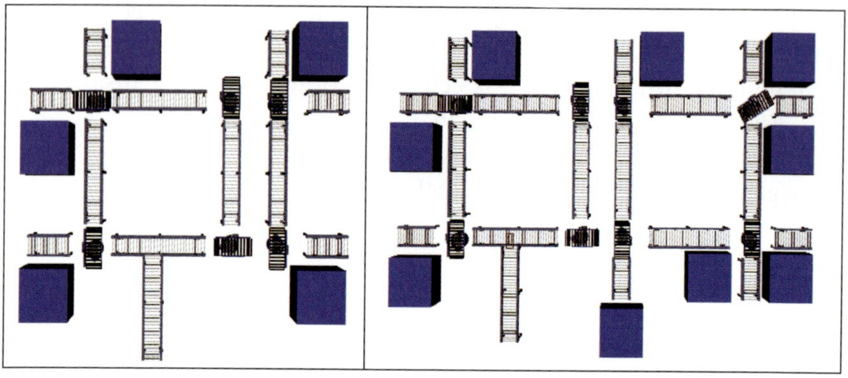

Alternative with up to five locations for disassembly stations

Alternative with up to nine locations for disassembly stations

Fig. 5: Two alternatives for the same reference layout

To find the best location for every considered reference layout an algorithm is formulated as an extended asymmetric quadratic assignment problem. The algorithm considers a distance-based objective and is formulated as well as the dimensioning model as an ILP-Model. The objective is to minimize the sum of flow time distances of every disassembly object for an assumed yearly disassembly performance.

Let m denote the number of disassembly stations to be placed, f_{ij} denote the flow (number of disassembly objects moved per year) from disassembly station i to disassembly station j, and f_{ji} denote the flow from disassembly station j to disassembly station i (asymmetric material flow assumption). The objective function mathematically expressed can be written as Equation 18 where d_{kl} is the distance from disassembly station i to disassembly station j if assigned in the reference layout to location k and l. The distance is hereby measured rectilinearly between the centroids. Y_{ikjl} is a binary decision variable that ensures that a transport between two disassembly stations can only be realized from different locations.

$$MIN\left(\sum_{i=1}\sum_{k=1}\sum_{j=1}\sum_{l=1} y_{ikjl} \times (f_{ij} \times d_{k,l} + f_{ji} \times d_{l,k})\right) \quad (18)$$

Equations 19 ensure that every disassembly station will be assigned to a location in the layout. Where x_{ik} represents the binary decision variable (x_{ik} = 1, if disassembly station i was assigned to location k; x_{ik} =0, otherwise)

$$\sum_{i=1} x_{i,k} = 1 \quad \forall k \tag{19}$$

Equations 20 ensure that exactly one disassembly station will be assigned to each location.

$$\sum_{k=1} x_{i,k} = 1 \quad \forall i \tag{20}$$

The constraints given in Equations 21 guarantee the right assignment of the decision variable y_{ikjl}. It forces the decision variable y_{ikjl} to take the value one, if disassembly station i is assigned to location k and disassembly station j is assigned location l.

$$y_{i,k,j,l} \geq x_{i,k} + x_{j,l} - 1 \quad \forall i, \forall k, \forall j, \forall l \tag{21}$$

The ILP-Model was solved with the solver Lingo™. The advantage of using Lingo™ is the possibility of solution space reduction with the appropriate constraint formulation. The implemented ILP-Model has a good performance, i.e. for up to 20 disassembly stations an optimal solution is found within less than 15 minutes of computing time.

6.1.5 Modelling and Simulation

The structuring phase provides an optimal location for every disassembly station within the considered reference layouts [17]. In the final design step the reference layout offering the best performance, highest utilisation and lowest lead time is determined. An analysis of the dynamic behaviour is required. Discrete-event simulation has proved to be a very efficient method for the analysis of production systems [18]. Results from the dimensioning and structuring phase can be used to model references of disassembly systems. These disassembly systems are evaluated by simulation to determine lead time, capacity utilisation, required buffer sizes, and performance for a representative system load.

Due to the relatively high effort required to build a simulation model a pre-processor for the automated model configuration for the simulation package AutoMod™ has been developed. The topology of the material flow system (e.g. conveyor system) and the locations for possible disassembly stations are generated automatically [19].

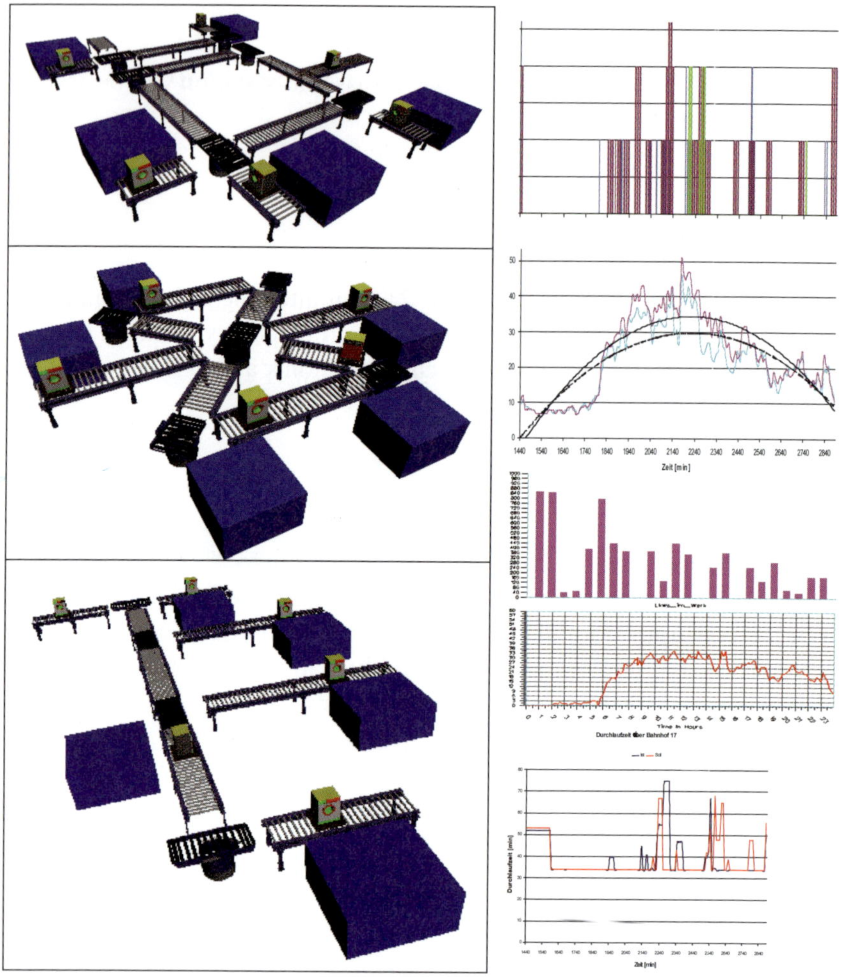

Fig. 6: Simulation based analysis of reference layouts

The results generated by the two applied ILP-Models are exported via the Lingo™ OLE interface to MS-Excel and completed with supplementary data such as type and quantity of products assigned to the disassembly system [12]. The pre-processor reads the data and transforms them to a format which can be read by the simulator. The location of every disassembly station, task sequences, disassembly times, and the routing of every disassembly object are configured for the modelled reference layouts. These configured reference models represent a complete model of a disassembly system and can be evaluated by starting the simulation and interpreting the simulation results (Figure 6).

A performance analysis can be carried out with low effort for every considered reference layout. Using the described approach a 30 day simulation run takes no longer than 10 minutes and identifies the best reference layout under the consideration of economical and technical constraints.

6.1.6 Conclusion

An integrated approach for dimensioning, structuring and simulation based evaluation of disassembly systems was introduced. An implemented ILP-Model supports the planner in the efficient design of alternative disassembly systems. By the use of previously modelled reference layouts and the possibility of automaticaly generating simulation models, the planner is supported in handling the manifold design information. The described models and software module were integrated into the InSane (Integrated Simulation-based Disassembly Plant Design) software tool and presented in [20].

References

1 Seliger, G.; Perlewitz, H.: Disassembly Factories for the Recovery of Resources in Product and Material Cycles. In: Proceedings of Japanese Society of Precision Engineering. Sapporo, Japan, September 1998.

2 Wiendahl, H.-P.; Perlewitz, H.; Bürkner, S.; Seliger, G.: A General Approach to Disassembly Planning and Control. International Journal of Production Planning and Control (PPC) 10 (1999) 8, pp. 718-726.

3 Hentschel, C.: Beitrag zur Organisation von Demontagesystemen. Dissertation, TU Berlin, 1996

4 Seliger, G.; Hentschel, C.; Wagner, M.: Disassembly Factories for Recovery of Resources in Product and Material Cycles, In: Krause, F.-L.; Jansen, H.: Life Cycle Modelling for Innovative Products and Processes. Chapman & Hall, London 1996, pp. 56-67.

5 v. Werder, H.K.: Planung der Demontage elektrischer und elektronischer Altgeräte. In: Fortschr.-Berichte, VDI Reihe 16, Nr. 88. VDI Verlag, Düsseldorf 1996.

6 Kriwet, A.: Bewertungsmethodik für die recyclinggestützte Produktgestaltung. Dissertation, TU Berlin, 1994.

7 Zussman, E.; Kriwet, A.; Seliger, G.: Disassembly-oriented Assessment Methodology to Support Design for Recycling. Annals of the CIRP, 43-1, 1994, pp. 9-14.

8 Seliger, G.; Basdere, B.; Ciupek, M.; Franke, C.: Contribution to Efficient Cellular Phone Remanufacturing, In: Proceedings of CIRP Seminar on Life Cycle Engineering – Engineering for Sustainable Development. Copenhagen, Denmark, 2003, pp. 162-176.

9 DIN 8580: Fertigungsverfahren – Einteilung. Beuth Verlag, Berlin 2003.

10 VDI (Hrsg.): VDI 2860 – Montage- und Handhabungstechnik; Handhabungsfunktionen, Handhabungseinrichtungen; Begriffe, Definitionen, Symbole. Beuth-Verlag, Berlin, Mai 1990.

11 Hessselbach, J.; v. Westernhagen, K.: Systematic Planning of Disassembly with Grouping and Simulation, In: Proceedings of the 2nd International Working Seminar on Re-Use. Eindhoven, Netherlands, 1999.

12 Franke, C.; Ciupek, M.; Seliger, G.: Computer Aided Rough Disassembly Sequence Design (CARDIS), In: Proceeding of the International Precision Assembly Seminar, IPAS 2003. Bad Hofgastein, Austria, March, 2003.

13 Schmigalla, H.: Fabrikplanung: Begriffe und Zusammenhänge. Carl Hanser Verlag, München, Wien 1995.

14 Tompkins, J. A.; White, J.; Bozer, Y. A.; Frazelle, E. H.: Facilities Planning. John Wiley& Sons, Inc., New York 1996.

15 Askin, R.G.; Standridge, C.R.: Modeling and Analysis of Manufacturing Systems. John Wiley& Sons, Inc., New York 1993.

16 Seliger, G.; Uhlmann, E.; Keil, T.; Härtwig, J.-P: Realisierung eines Pilot-Demontagesystems. ZWF Sonderbeilage Demontage, July (2000), S. 20-24.

17 Härtwig, J.-P.: Verfahren und Systeme zur Demontage komplexer technischer Gebrauchsgüter. Dissertation, TU Berlin, 2005.

18 Pedgen, D.: Simulation Environment for the New Millennium. In: Proceedings of the 2001 Winter Simulation Conference. Washington, December 2001, pp. 541-547.

19 Ciupek, M.; Franke, C.; Keil, T.; Seliger, G.: Automatische Modellgenerierung für die simulationsgestützte Leistungsbewertung von Demontagesystemen. In: Proceedings of „Simulation and Visualization“. Magdeburg 2002.

20 Ciupek, M.: Beitrag zur simulationsgestützten Planung von Demontagefabriken für Elektro- und Elektronikaltgeräte. Dissertation, TU Berlin, 2004.

6.2 Program Planning

Günther Seliger, Sebastian Kernbaum, Berlin, Germany

Based on the facilities planned by the approach presented in the previous chapter, the need for developing a suitable master production schedule arises as a next step in a general planning approach. A master production schedule determines the time and amount of products to be processed in a facility. In order to derive a feasible solution out of many possible master production schedules, a level-oriented approach for disassembly control was developed [1]. This concept allows the reduction of the planning complexity and the increase of reactivity by separation of disassembly planning in program and order, also to satisfy short and midterm market demands. Later, a cost minimizing approach for the operations of remanufacturing facilities based on organisational, product, process and equipment frame conditions was realized [2].

6.2.1 Optimisation Model

For the determination of an economically feasible remanufacturing program under consideration of organisational, product, process and equipment constraints, an optimization model is formulated. In the objective function, the revenues of recovered products, components and materials minus the costs for procurement, transportation, storage, collection, testing, disassembly, refurbishment and reassembly as well as for material recycling and disposal are optimized (Equation 1). The objective function is the criterion for the determination of optimal values for the decision variables in the overall multi-period planning horizon.

$$\text{Max} \sum_t \begin{pmatrix} (\text{product revenues})_t + (\text{component revenues})_t + (\text{material revenues})_t - \\ (\text{product procurement costs})_t - (\text{component procurement costs})_t - \\ (\text{transportation costs})_t - (\text{storage costs})_t - (\text{collection costs})_t - \\ (\text{testing costs})_t - (\text{disassembly costs})_t - \\ (\text{component remanufacturing costs})_t - (\text{reassembly costs})_t - \\ (\text{recycling costs})_t - (\text{disposal costs})_t \end{pmatrix} \quad (1)$$

By means of material flow conversation constraints, a continuous assignment of material amounts from source to sink is ensured. Additionally, the compliance with capacity and demand constraints has to be guaranteed.

In the following, the organisational, product, process and equipment related frame conditions are described and the above mentioned optimization model is presented, implemented and applied.

6.2.2 Flat Screen Monitor Remanufacturing

Demand for flat screen monitors today is growing strongly. Used monitors are hardly available due to low market penetration. In 2005 alone, about 137 Mio. desktop CRT (Cathod Ray Tube) and LCD (Liquid Crystal Display) monitors are expected to be sold (Figure 1). With 97 Mio. sold LCD monitors their market share will reach about 71 % [3].

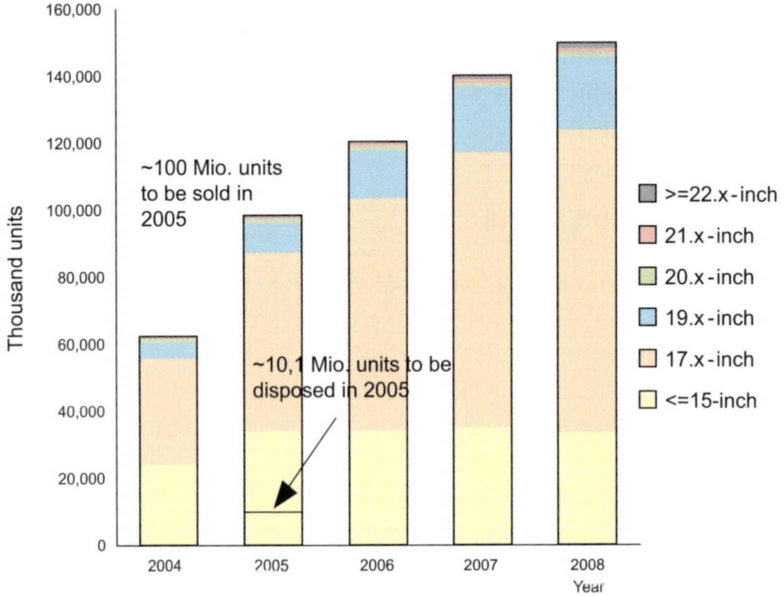

Fig. 1: Worldwide desktop monitor sales (own calculation and [3])

Calculations result in a total of 10,1 Mio. LCD desktop monitors to be disposed worldwide in 2005 [4]. Prices for used and non-functional 15"-18" monitors with control failures from private users – obtained from the online market place eBay™ – usually range from € 40 to approximately € 110, as long as the LCD panel glass is not broken. Defective monitors from leasing returns collected by IT-remarketing companies from institutional users are frequently exported to Eastern Europe, where high demand meets cheap labor. An adequate End of Life (EoL) treatment of these monitors is at least very questionable. From an ecological point of view, the shift from CRT to LCD technology can be considered a step towards a sustainable development in the information and telecommunication sector, if monitors are treated properly at their EoL. A Life Cycle Assessment of desktop computer displays conducted by the University of Tennessee in 2001 compares the environmental and health impacts of the LCD and CRT technologies [5]. As an example, LCD manufacturing is water and energy intensive. The manufacturing life stage, i.e. mainly the glass manufacturing process, accounts for over 75 % of the renewable resources usage. Due to the 60 – 70 % lower weight, when compared to CRT displays, the natural resource consumption of LCD is significantly lower. Energy consumption of LCD during the use is also 60 – 70 % lower than that of CRT displays.

In Article 7 (2) of the European Directive on Waste Electric and Electronic Equipment (WEEE), the reuse and recycling targets for LCDs (group 3) are defined to the 65 % of the product's weight [6]. The Annex II of the WEEE obliges disassembly of all LCDs with a surface greater than 100 square centimeters and for all those with back-lighted, gas discharge lamps, prior to material recycling. More recycling relevant facts for flat screen monitors are summarized in [4].

With a predicted 90 % penetration of the monitor market with flat screen technology by 2010 [7], product returns will be sufficient enough for the development of remanufacturing activities at large scale. The main challenge, as in most remanufacturing activities, is the handling of uncertainty regarding time, quality, quantity, and place of product returns that affects collection, testing, disassembly, reassembly, and the warehousing of spare parts likewise.

Collection

Today, IT equipment returns from leasing contracts are frequently collected by IT remarketing companies. Due to the low market penetration of flat screen monitors, collection from private households is still uncommon. Future collection plans using today's municipal collection hubs are unlikely to guarantee an undamaged take back of flat screen monitors. To obtain access to product returns from private households new take back models, e.g. charity recycling [8], are required. Due to the size and weight of monitors, as compared to mobile phones, collection using postal service is rather unlikely due to the high costs involved. Figure 2 summarises the relevant actors and their relations in remanufacturing and recycling networks that are considered in the developed optimization model.

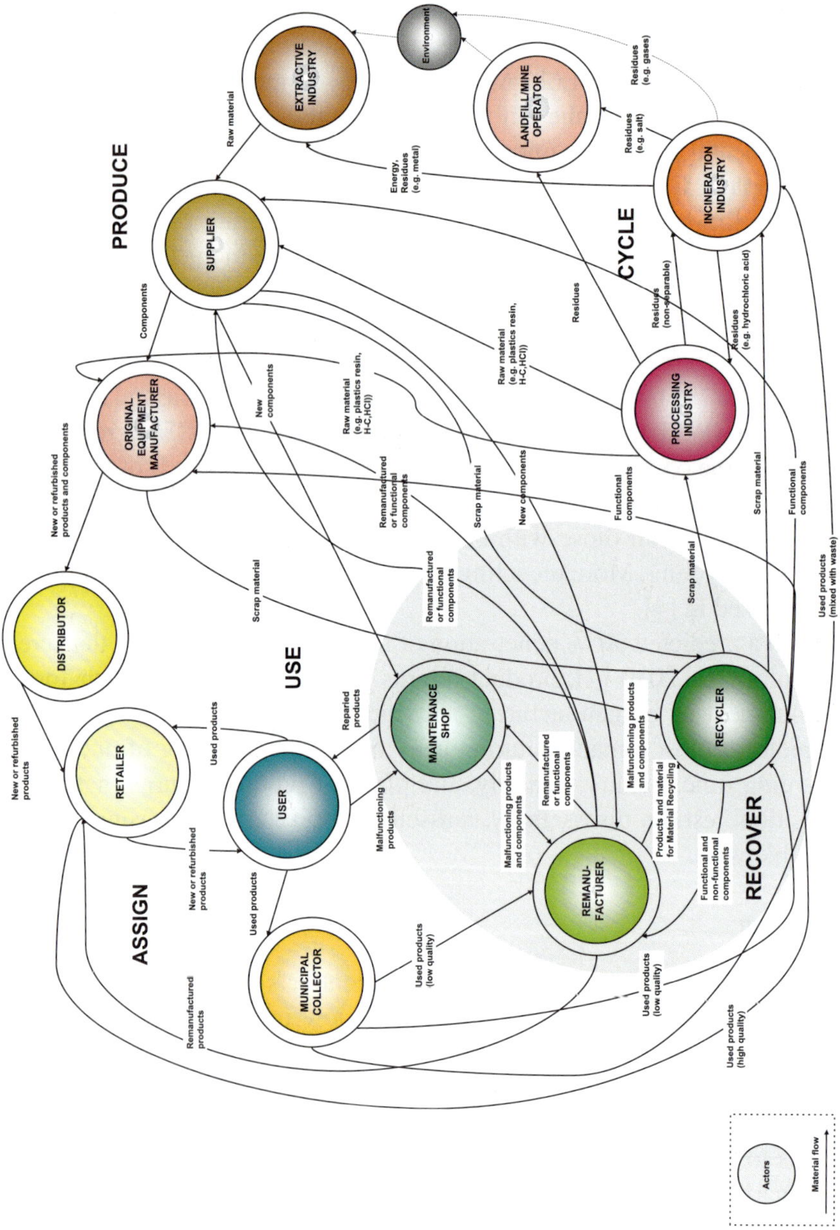

Fig. 2: Complexity in remanufacturing and recycling networks [9]

Disassembly

A disassembly analysis to module level was conducted for 17 different flat screen monitors (Figure 3). Non-destructive disassembly of beyond warranty and non-functional, e.g. broken LCD modules, is usually economically unfeasible with the LCD module being the monitor's primary valuable component. Advanced repair processes, e.g. disassembly of Tape Adhesive Bonding (TAB) for the recovery of driver tabs or the replacement of polarizers are predominantly applied during warranty or for highly integrated LCD modules, e.g., notebooks or industrial control panels. Since it is mandatory by law [6], the disassembly of LCDs from LCD modules is usually carried out destructively.

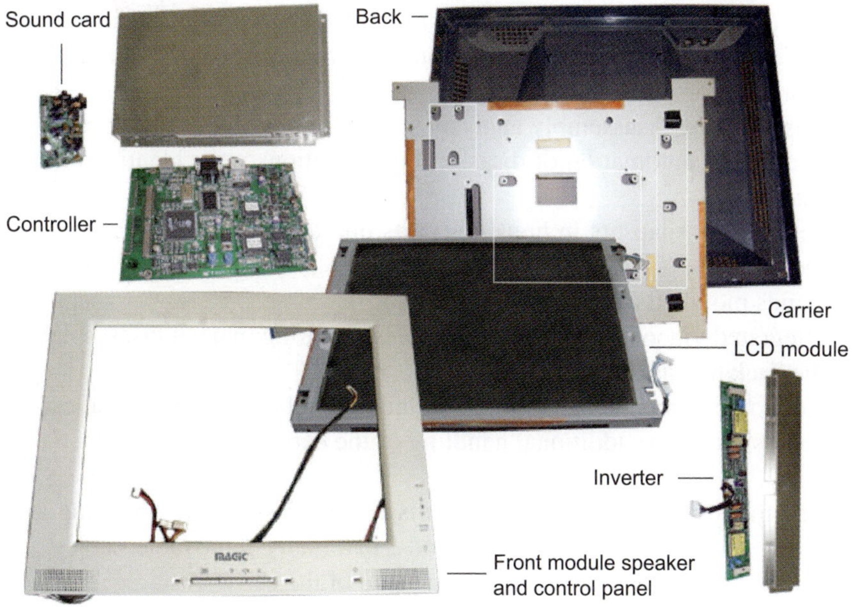

Fig. 3: Flat screen monitor modules without stand

All relevant components, i.e., frame, housing, LCD, foils, Cold Cathode Fluorescence Lamps (CCFLs), light guide and PCBs can be separated within approximately one minute.

Non-destructive disassembly of the monitor to module level is the prerequisite for the majority of successing remanufacturing processes. The most common repair option for the LCD module is the replacement of the CCFLs. Therefore, in the following, CCFLs are defined as a module of the flat screen monitor. Other common repair options for LCD monitors include the replacement of faulty electronic components as well as mechanical components such as plugs.

For a systematic evaluation of the monitor's relevant disassembly properties, the analysis was documented by means of a morphological box. To obtain comparable results regarding the required disassembly effort, a Method Time Measurement (MTM) analysis was conducted for all 17 monitor models. The disassembly times, depending on monitor design, range from 3,6 to 8,7 minutes and the quantity of screws used from 33 to 108 units. Disassembly properties and MTM results were summarized in [10, 11].

Particularly important for cost efficient disassembly is the provision of a suitable disassembly plan under consideration of the acquired test results for each product. A certain order of disassembly steps is usually predefined by the products design. Automation of selected disassembly, i.e. mainly unscrewing processes, can be an option once higher quantities of used flat screen monitors are available.

Based on the disassembly analysis of 17 monitor models basic restraints regarding the automation of disassembly operations were identified as follows:

- very strong snap fits in housings parts due to the strong release torques required,
- stands mounted to the monitor's inside,
- integrated video or power supply cables requiring manual disassembly of the back housing, and
- electronic components mounted on the top and the bottom of the carrier, necessitating the additional handling of the carrier.

The monitors selected for hybrid disassembly offer weak housing snap fits, externally mounted monitor stands, externally removable cables, and electronic components mounted to one side of the carrier.

A prototypical hybrid disassembly system with one automated and one manual workplace connected by three square conveyor blocks was realized. A 4-axis Selective Compliant Articulated/Assembly Robot Arm (Scara) is used for performing unscrewing and handling operations. A manual workplace, originally used for manual testing of mobile phones, was adapted for manual disassembly and reassembly operations and integrated in the system. The system can be extended to one more automated and five more manual workplaces to accommodate further disassembly, as well as testing and reassembly operations.

At the manual workplace, the monitor is positioned on the work piece carrier. Large monitor stands are disassembled manually while small ones remain mounted to the back housing as long as they do not interfere with the accessibility of joining elements. The monitor is transferred to the automated disassembly workplace. A suction gripper is used to pick up the monitor and position it on the pneumatic clamping device. External screws, the monitor back housing and internal components are removed. Metal covers, printed wiring boards and cable connectors are removed respectively by using a two-finger gripper. In case a process fails, the monitor is forwarded to the manual workplace.

Testing

Testing determines, whether a testing object is fulfilling one or more agreed, required or expected terms, especially whether defined error margins and tolerances are met. Electric or electronic devices are basically tested when assembled – alternatively, testing can be applied to components after disassembly.

Testing methods for assembled LCD monitors were systematized in six categories, whose order is derived from the expected time and capital costs for testing. The testing procedure starts with the process that has the lowest effort. If an error occurs during the realization of the testing procedure in one of the testing processes, the next testing process cannot be initiated and options for destructive and non-destructive disassembly arise.

Detailed inspection instructions and inspection records as well as flow charts for visualization were developed for the considered testing methods. These are presented in [12].

6.2.3 Model Implementation and Application

The described optimization model including all constraints – considering organisational, product, process and equipment frame conditions was realized as an optimization model. The developed constraints are illustrated in [2].

The optimization model was implemented by using a solver and a database. The software Lingo™ from the company LINDO Systems was used as a solver, which allows an Open DataBase Connectivity (ODBC) connection to a Microsoft (MS) Access-Database. For enabling platform independent usage, a web-based realisation of the planning tool was selected. It allows browser-based modification of database contents, the initialization of the solver and the analysis of the result. Figure 4 describes the elements and relations of the web-based planning tool. Requests and responses are executed by an Internet Information Services Web Server. A PHP (Hypertext

Preprocessor) is used for the handling of user requests. Database queries and modifications are executed using a MySQL-database driver. Modified data are transferred to a MS Access-Database for the execution by the solver Lingo. MS-Excel functionality was implemented in the web-based planning tool for the graphical evaluation of optimization results. The platform serves as support for the planners by deciding for a suitable master production schedule, by analysing the considered process costs and by balancing the planned capacities.

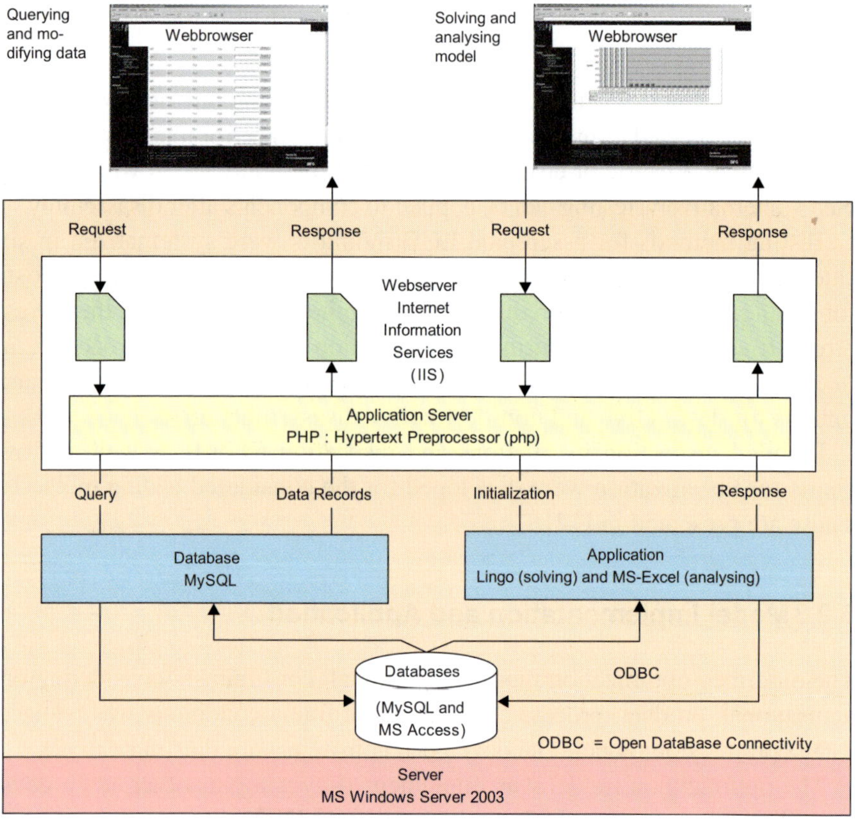

Fig. 4: Web-based planning tool

For the verification and validation of the implemented planning tool, the planning tool is applied on the above mentioned example of LCD Flat Screen Monitors. By means of the approach presented in the previous chapter, a facility for the remanufacturing of LCD Flat Screen Monitors was planned in the first step. Further information on the development of the facility is given in [2]. Figure 5 provides an overview of the system that was used for the development of a master production schedule.

The structuring of workplaces and logistic equipment was developed as a decoupled production line. Transportation and buffering of products and components are realized by dollies. The layout of workplaces is of the U-shape, which allows storage of products and components with a low transportation effort. A comparison of layouts suitable for the disassembly and remanufacturing of electronic equipment is shown in [13].

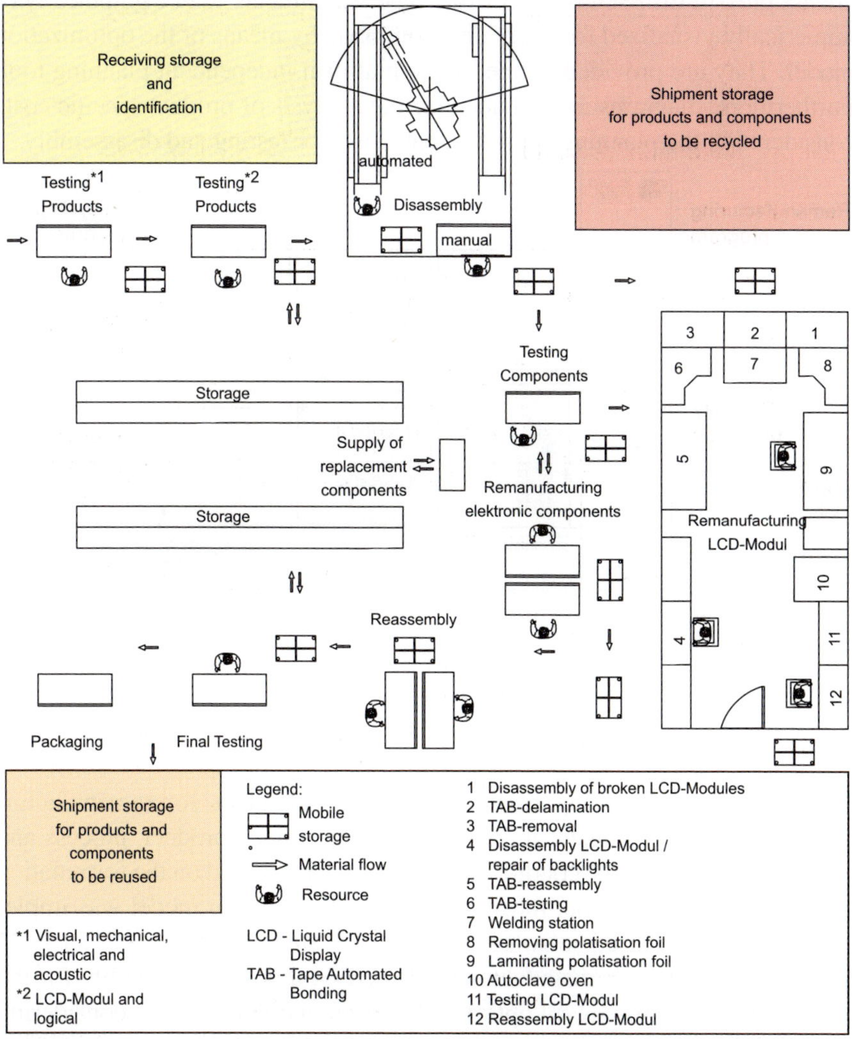

Fig. 5: The layout of the remanufacturing system

After inputing the data into the web-based planning tool, the solver calculates a solution and transfers the results into the MS-Access database. In the next step, the results are reworked by MS-Excel into user-friendly graphics. Figure 6 illustrates exemplarily result diagrams generated by the planning tool.

Which products, in which tested condition are assigned to which planning period of disassembly, material recycling or other processes are presented in the result diagrams. E.g., products with a faultless visual test are assigned to the next test step, the mechanical test, or products with failures in the visual test are assigned to disassembly processes. The assignments can be graphically or numerically visualized for every decision made by means of the optimization model. They are provided online by the platform independent planning tool. Furthermore, diagrams are provided for the analysis of product specific costs considered in the planning approach, e.g. costs for testing and disassembly.

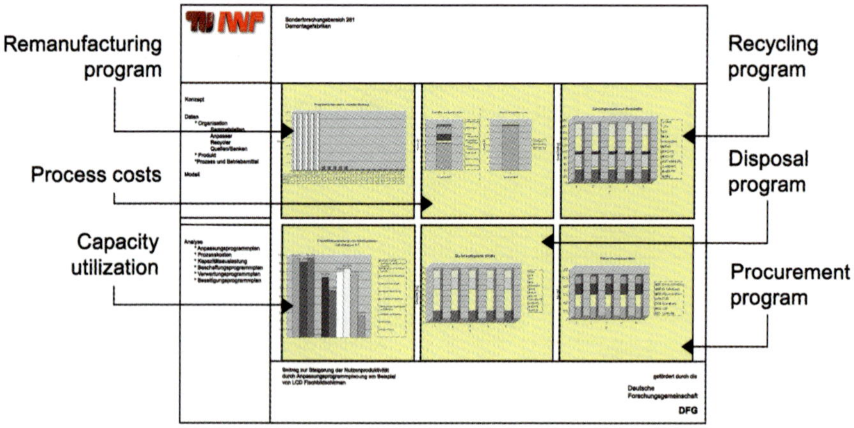

Fig. 6: Selected result diagrams

6.2.4 Conclusion

A concept for developing master production schedules for remanufacturing systems was presented. It considers organisational, product, process and equipment related frame conditions. The approach is based on the formulation of a program planning optimization model. The decision model was implemented as a web-based software tool and validated on the example of LCD Flat Screen Monitor remanufacturing. Based on a developed facility for the remanufacturing process, an optimal master production schedule considering capacity utilization, costs and revenues as well as the assignment of products and components to reuse or material recycling was generated.

References

1 Perlewitz, H.: Planung und marktorientierter Betrieb von Demontagefabriken. Dissertation, TU Berlin, 2000.

2 Franke, C.: Beitrag zur Steigerung der Nutzenproduktivität von Ressourcen durch Anpassungsprogrammplanung. Dissertation, TU Berlin, 2006.

3 Alexander, R.: Worldwide Monitor Market Tracker. iSupply, Stanford Resources 3. Quarter, 2004.

4 Franke, C.; Kernbaum, S.; Seliger, G.: Remanufacturing of Flat Screen Monitors. In: Proceedings of the 12th CIRP Seminar on Life Cycle Engineering (LCE). Grenoble, France, April 2005.

5 University of Tennessee 2001, University of Tennessee, US-Environmental Protection Agency, Desktop Computer Displays – A Life Cycle Assessment, 2003. Available at www.epa.gov (Access: November 2006).

6 Directive 2002/96/EC of the European Parliament and of the council on waste electrical and electronic equipment (WEEE) of 27 January 2003.

7 DisplaySearch 2004. Available at www.displaysearch.com (Access: November 2006).

8 Seliger, G.; Başdere, B.; Ciupek, M.; Franke, C.: Contribution to Efficient Cellular Phone Remanufacturing. In: Proceedings of the CIRP Seminar on Life Cycle Engineering. Copenhagen, Denmark, May 2003.

9 Franke, C.; Seliger G.; Yakut, E.: Mobile Phone Remanufacturing Program Planning in Networks. In: Proceedings of the Global Conference on Sustainable Design and Life Cycle Engineering. Berlin, Germany, 2004, pp. 191-198.

10 Franke, C.; Kernbaum, S.; Seliger, G.: Remanufacturing of Flat Screen Monitors – Economical and Technological Challenges until 2010 In: Proceedings of the 38th CIRP International Seminar on Manufacturing Systems (ISMS). Florianopolis, Brazil, May 2005.

11 Franke, C.; Kernbaum, S.; Seliger, G.: Remanufacturing of Flat Screen Monitors. In: Brissaud, D.; Tichkiewitch, S.; Zwolinski, P. (Ed.): Innovation in Life Cycle Engineering and Sustainable Development. Springer-Verlag, Berlin, Heidelberg 2006, pp. 139-152.

12 Kernbaum, S.; Franke, C.; Seliger, G.: Flat Screen Monitor Disassembly and Testing for Remanufacturing. In: Proceedings of the 13th CIRP Seminar on Life Cycle Engineering (LCE), Leuven, Belgium, May/June 2006.

13 Ciupek, M.: Beitrag zur simulationsgestützten Planung von Demontagefabriken für Elektro- und Elektronikaltgeräte. Dissertation, TU Berlin, 2006.

7 Enabling for Sustainability in Engineering

Application of technology and setting up business procedures and models in the framework of competition in market environments can considerably help promoting sustainability in engineering. New approaches in teaching and learning especially for open minded young students are a powerful leverage to overcome the forces of persistence in established societal life and immobile institutions.

Takata postulates the goal of manufacturing is no longer to produce products in an efficient way, but rather to provide the functions needed by society while minimizing material and energy consumption. Maintenance could affect business models in the sense of facilitate the transformation from product to service providers. A change of the paradigm of manufacturing is needed from "how to produce products most efficiently" into "how to avoid producing products while still maintaining customer satisfaction and corporate profits". With product life cycle management companies will sell utilization and customers will pay only for utilization. Lee and Ni state that maintenance considerably contributes to prolonging useful product and component application. In a "predict and prevent" paradigm a condition based maintenance scheme is under development. Considering current degradation and its evolution in different product areas is achieved by signal processing and feature extraction and extrapolating component behaviour. The Life Cycle Unit (LCU) concept developed in modular architecture of sensorial, actuating, information processing and communicating devices at the Technical University of Berlin enables for the business model of selling use instead of selling products. Miniaturised devices are presented.

Modularity contributes to ease the remanufacturing. A multicriteria methodology for modularization according to objectives of sustainability has been implemented applying integer linear programming. Finally competence management modelled in a market chain reference is identified as a significant innovative force for implementing sustainable processes and products in the future.

7.1 Maintenance Essential for Life Cycle Management

Shozo Takata, Tokyo, Japan

Everyone agrees that maintenance is a necessity. Nevertheless, maintenance has a negative image and is sometimes regarded as a necessary evil. But as the paradigm of manufacturing shift towards realizing a sustainable society, we should also begin to recognize the changing role of maintenance. The goal of manufacturing is no longer to produce products in an efficient way, but rather to provide the functions needed by society while minimizing material and energy consumption. Product life cycle management is becoming a crucial issue in order to achieve this goal. In this context, the role of maintenance should be redefined as an essential means for life cycle management. Maintenance is the most efficient way to keep the functional level of a product above the level required from the viewpoint of environmental impact.

Traditionally, the scope of maintenance activities has been limited to the operation phase. In considering the role of maintenance in life cycle management, however, we should be aware that there are close relationships between maintenance activities and those in other phases of product life cycle, such as the design, production, and end of life phases. These relations create the necessity for integration in terms of technologies as well as information throughout the product's life cycle to perform effective maintenance. Hence, we should investigate maintenance technologies more extensively in connection with, for example, design technologies such as product modeling and digital engineering.

Changing the role of maintenance could also affect the business models of manufacturing companies. Attention to maintenance could facilitate the transformation of businesses from product providers to service providers, with maintenance being a major service.

In this section, we first discuss life cycle management for closed-loop manufacturing, which could be an essential means for realizing a sustainable society, and the changing role of maintenance from the perspective of life cycle management. Then, we present a maintenance framework that shows management cycles of maintenance in the product life cycle. Finally, we point out importance of information systems for effective execution of maintenance activities ranging over the entire product life cycle, and show an example of computer support systems for predicting deterioration and failures of mechanical systems.

7.1.1 Life Cycle Perspective

Since the Industrial Revolution, people have been improving the quality of human life by increasing manufacturing capability. Mass production, however, has also brought about mass consumption of natural resources and energy, as well as mass disposal. The scale of our industrial activities has already extended beyond the limit. We cannot continue to consume resources and energy, and to dispose of waste without considering the impact of these activities on the environment. Therefore we need to change the paradigm of manufacturing from "How to produce products most efficiently" into "How to avoid producing products while still maintaining customer satisfaction and corporate profits". Closed-loop manufacturing has been proposed as a solution to this question [1].

The concept of closed-loop manufacturing can be expressed as "renewing functions while circulating material". There are, however, many ways to circulate material, which are called life cycle options, such as prolonged use by means of maintenance, product reuse, part reuse, recycling, and energy recovery. To realize closed-loop manufacturing, the product life cycle should be managed by selecting proper life cycle options.

In selecting life cycle options, we need to consider their environmental performance or eco-efficiency, which is defined as the ratio of provided value to environmental load. Therefore, we cannot always select the options with the lowest environmental load, because we need to consider the balance between environmental friendliness and customer satisfaction, as well as corporate profits. Nevertheless, the more the loop is inner, the smaller is the load applied on the environment. In this sense, maintenance is the most important means for life cycle management.

The Changing Role of Maintenance

In the past, maintenance was regarded as repair work. Machines were operated until they broke down, and there was no way to predict failures. With the development of reliability engineering in the 1950s, the concept of preventive maintenance was advocated, and Time Based Maintenance (TBM) was introduced. TBM was based on the so-called bathtub curve, which represents the increase in the failure rate of products after a certain period of operations. However, in many cases, product condition cannot be identified from the extent of the operational period, since the rate of deterioration depends not only on elapsed time but also on various other factors, including operational and environmental conditions. Therefore, TBM sometimes imposes unnecessary treatments, which often disrupt normal operations and induce malfunctions due to missed operations [2].

After the limitations of TBM as a means of preventive maintenance were recognized, the concept of Condition Based Maintenance (CBM) was proposed based on the development of machine diagnostic techniques in the 1970s. In the case of CBM, preventive actions are taken when symptoms of failures are recognized through monitoring or diagnosis. Therefore, CBM enables taking the proper actions at the right timing to prevent failures, if there is a proper diagnostic technique.

However, CBM is not always the best method of maintenance, especially from the perspective of cost effectiveness. When failures of machines or components are not critical, we can allow Breakdown Maintenance (BM), in which actions are taken after failures are detected. When the lives of machines or components can be estimated precisely, TBM is the most effective means of maintenance. Therefore, from the latter half of the 1980s, the importance of selecting proper maintenance strategies has been acknowledged in various areas. Reliability Centered Maintenance (RCM) [3] and Risk Based Inspection (RBI) or Risk Based Maintenance (RBM) [4] are the most well known methodologies for this purpose.

Although maintenance concepts and methodologies have advanced significantly over the past several decades, as explained above, maintenance still has a negative image because it is regarded as merely a measure against troubles. A maintenance department is usually regarded as a cost-centre, which does not create profits.

If we look at the role of maintenance from the perspective of life cycle management, however, we realize that the picture is completely different. The purpose of product life cycle management is to control the conditions of products to provide the functionality required by customers or by society, while keeping the environmental load at a minimum and maintaining appropriate corporate profits. There are two reasons why it is necessary to control the conditions of products. One is the change in product conditions due to deterioration. Another is the changing needs of customers or of society. The former is referred to as the product's physical life and the latter as its functional life. In both cases, the measure that should be considered first is maintenance including upgrade, because maintenance generates a lesser environmental load. If maintenance does not work well, other measures should then be considered, such as remanufacturing. Production of new products should be the last measure taken. In this context, the priority relationship between production and maintenance has been completely reversed. What we need to consider from the perspective of life cycle management is the notion of *"production free" rather than "maintenance free."*

The perspective of life cycle management for closed-loop manufacturing has brought about transformation of business models of the manufacturing companies from product providers to service providers [5]. Maintenance could be one of the major services associated with product life cycle management. If this business transformation goes further, companies will sell utilization and customers will pay only for utilization. In this context, achieving effective maintenance could be of benefit to companies, which can increase their profits by reducing maintenance costs, as well as to customers who can enjoy improved service quality [6].

7.1.2 Framework

The objective of maintenance is to preserve the condition of products to fulfill their required functions throughout their life cycle. Maintenance is an important part of life cycle management, whose main purpose is to enhance the eco-efficiency of the product life cycle as explained above. Therefore, we use the term "life cycle maintenance" to stress its role from the perspective of life cycle management.

As previously mentioned, there are two reasons why it is necessary to control the conditions of products: changes in the condition of products due to deterioration, and the changing needs of customers or of society. These changes generate gaps between the required function and the realized function. Maintenance is executed to compensate these gaps by means of treatment or upgrading, as shown in Figure 1. For this purpose, maintenance should involve the following activities.

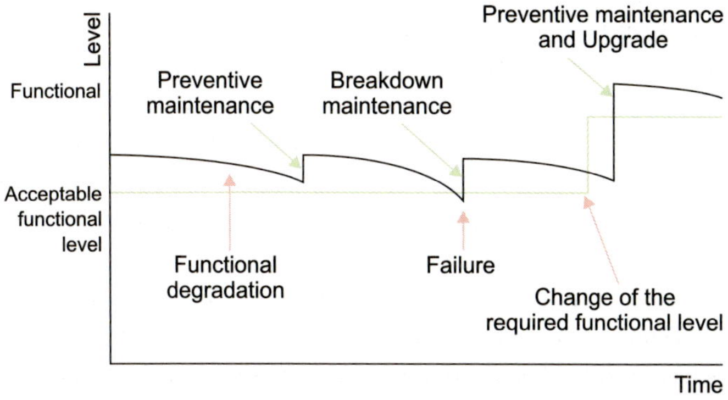

Fig. 1: Maintenance activities

- *Maintainability design:*
  Improving design based on evaluating maintainability in the product development phase and providing the design data for maintenance strategy planning and maintenance task control;
- *Maintenance strategy planning:*
  Selecting a maintenance strategy appropriate to each part of the product;
- *Maintenance task control:*
  Planning and executing the maintenance tasks based on the selected strategy;
- *Evaluation of maintenance results:*
  Evaluating the results of maintenance to determine whether the maintenance strategy planning and maintenance task control are appropriate;
- *Improvement of maintenance and products:*
  Improving maintenance task control, maintenance strategy planning, and even product design based on the evaluation of maintenance results;
- *Dismantling planning and execution:*
  Planning and execution of dismantling at the end of the product life cycle.

In life cycle maintenance, we have to manage the activities listed above in an effective way throughout the life cycle of the product. For this purpose, the following issues should be considered:

- *Adaptation to various changes during life cycle:*
  During the product life cycle, there could be various changes in the required functions, in the operating environment, in the operating conditions and in the product itself. Maintenance management should be flexible enough to adapt to these changes, because maintenance methods depend on these factors.
- *Continuous improvement of products:*
  In general, it is impossible to design a product perfectly. Therefore, maintenance should include a mechanism for continuous improvement of products based on experience and knowledge acquired during their life cycle. This mechanism is also effective for functional upgrade of products to cope with shortening the product life cycle because of rapid changes in users' needs and technology development.
- *Integration of maintenance information:*
  For effective maintenance management, all information associated with maintenance should be integrated in such a way that it is available from any phase of the life cycle. In the development phase, for example, it is essential to know the real operating situations and the problems encounte-

red during past operations. On the other hand, it is necessary to have exact design data for maintenance strategy planning and maintenance task control.

For fulfilling the requirements of life cycle maintenance described above, effective execution of a Plan-do-Check-Action (PDCA) cycle is essential. For this purpose, the framework for life cycle maintenance (Figure 2) has been proposed [7]. In this framework, maintenance strategy planning plays a key role. This planning involves selecting the strategy of maintenance among various options, such as BM, TBM, and CBM, based on the evaluation of potential problems that could occur during operation as well as evaluation of failure effects and effectiveness of maintenance technologies. Maintenance strategy planning serves as a bridge between the product development phase and the operation phase. It obtains design data and production records from the development phase, and determines the maintenance strategy for each component of the product.

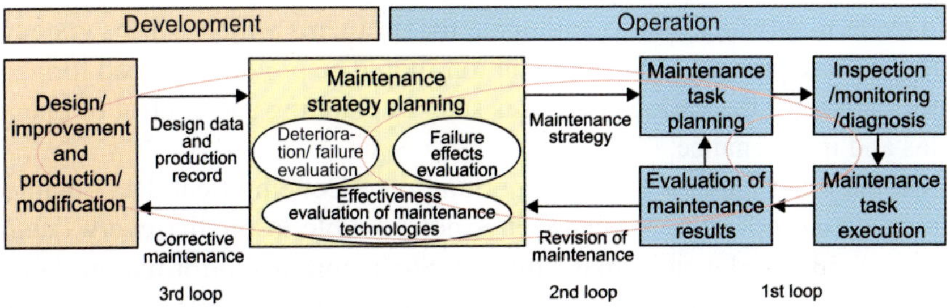

Fig. 2: Framework for life cycle maintenance

These strategies are passed on to the operation phase, where maintenance tasks are planned in terms of procedures and schedules based on them. After maintenance tasks, such as inspection, monitoring, diagnosis, and treatment, are executed, the results are evaluated by comparing the actual condition of the product to what it is anticipated when the maintenance strategy was selected. If there are discrepancies, the information is fed back to the maintenance strategy planning, where the maintenance strategies are revised based on re-evaluation of potential problems, taking the actual data into account. If corrective maintenance, i.e. design improvement, is needed, the information is further fed back to the development phase, where improvements and modifications of product design are performed.

As illustrated in Figure 2, there are three feedback loops. The first is the loop of operational phase maintenance task management, which consists of maintenance task planning, task execution, and assessment of maintenance results. This is the loop for controlling routine maintenance work. The second loop includes maintenance strategy planning. By means of this loop, the maintenance strategies can be improved based on the observation of actual phenomena and knowledge accumulated during the product life cycle. The third loop includes product development. This loop is essential for continuous improvement of the product during its life cycle.

These three loops provide effective mechanisms for adapting maintenance strategies to various changes, such as changes in operation conditions and environment, and also for continuously improving products.

7.1.3 Information Loop

For implementing the management loop shown in Figure 2, it is essential to facilitate information flow among each life cycle phase. There should be two types of information flow in the maintenance management as shown in Figure 3 [8]. First, it is important to evaluate various processes during the life cycle in advance and to anticipate the problems which will be encountered in these processes. Such prediction should be provided as feed forward information for the various purposes such as planning and control of operations and maintenance.

On the other hand, it is also important to collect empirical data obtained during actual operation, because it is not possible to predict every occurrence encountered in an actual situation. Such empirical information should be fed back to improve the operation and maintenance plan, and also to improve the product itself.

As pointed out above, predicting deterioration of the products is a key element in maintenance management. It is necessary at every phase of the product life cycle. While FMEA involves qualitative analysis, simulation of deterioration and failure enables quantitative analysis. Deterioration processes such as wear are simulated, and their effect on product behavior is evaluated based on the product model. Deterioration is induced by operational and environmental stresses. Since conditions of operation and environment are different from machine to machine and changeable over time, the progress of deterioration is not the same even in the case of the same type of machines. Therefore, we need to use a physical model in which specific conditions can be considered rather than a statistical model by which average values are discussed.

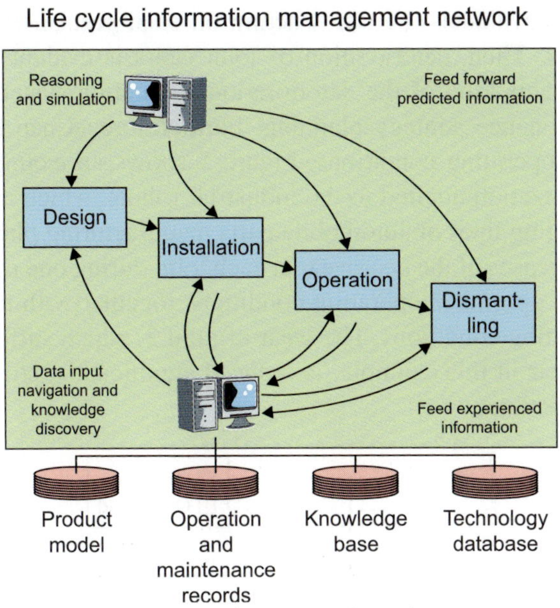

Fig. 3: Twofold data flow in maintenance data management

For evaluating deterioration under non-steady operating conditions, model-based deterioration simulation is effective [9, 10]. Figure 4 shows a procedure of deterioration simulation applied to joint gears of industrial robots.

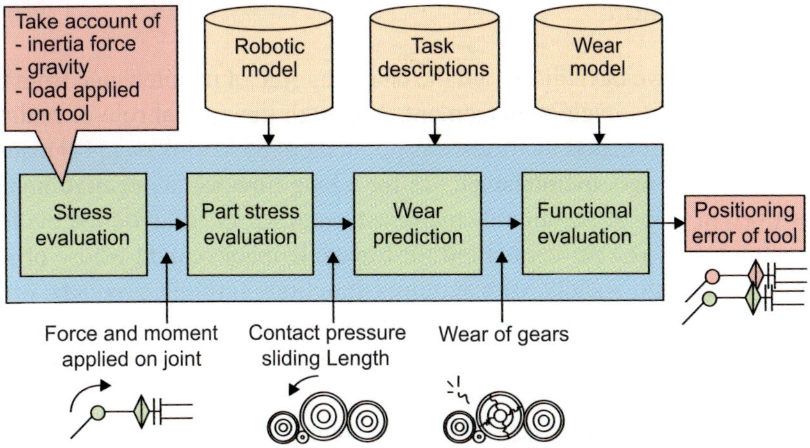

Fig. 4: Procedure of deterioration and failure simulation applied to joint gears of industrial robots

The system evaluates the stress acting on each part of the joints using a robot model. Then, deterioration of joint gears is evaluated by using a wear model. This kind of the deterioration simulation system is effective for the maintenance strategy planning. Furthermore, it can be applied to optimize the operating conditions. Figure 5 shows an example of results of such optimization applied to the industrial robots, which are actually in use in assembling lines of automobile parts manufacturing plant. The figure shows the increases of the gear wear at each joint during one cycle of operations with the optimized operating conditions together with those with the original operating conditions. The wear of joint 2, which suffers the largest amount of wear in this example, is reduced significantly, still maintaining the same cycle time.

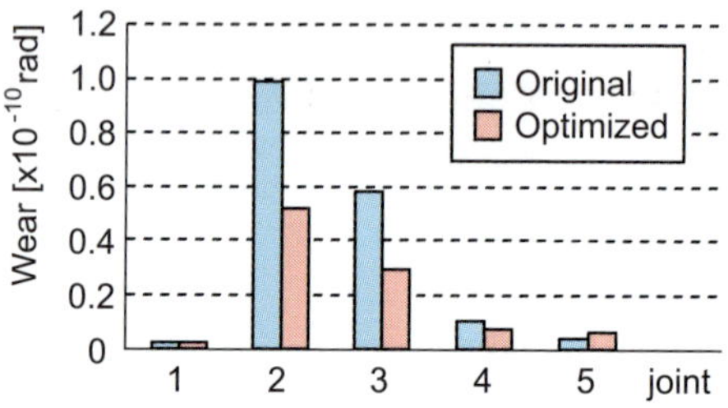

Fig. 5: Wear of each joint with original and optimized operating conditions

7.1.4 Conclusion

In this section, we have discussed the changing role of maintenance from the perspective of life cycle management. Although the critical role of maintenance within automated factories was pointed out by Yoshikawa [11] a quarter of a century ago, maintenance has for a long time had a negative image. However, in view of sustainable manufacturing, we should redefine the role of maintenance as a prime method for life cycle management whose objective is to provide society with required functions through products while minimizing material and energy consumption.

Though an enormous number of works have focused upon maintenance as a whole, they are dispersed across various areas and are not yet systematized. Therefore, we have proposed a maintenance framework that could help us to discuss maintenance technologies from various areas on the same table

Recent advancements in information and communication technologies could also improve maintenance effectiveness. There are many possibilities for making use of technologies as shown in the example of deterioration simulation of industrial robots.

References

1 Kimura, F.; Suzuki, H.: Product Life Cycle Modelling for Inverse Manufacturing, Life-Cycle Modelling for Innovative Products and Processes. Chapman & Hall, London 1995, pp. 80-89.

2 Takata, S.; Kimura, F.; van Houten, F.J.A.M; Westkamper, E.; Shpitalni, M.; Ceglarek, D.; Lee, J.: Maintenance: Changing Role in Life Cycle Management. In: Annals of the CIRP, Vol. 53/2, 2004, pp. 643-655.

3 Nowlan, F.S.; Heap, H.F.: Reliability-Centered Maintenance. In : Proceedings of the Annual Reliability and Maintainability Symposium, 1978, p. 38.

4 ASME: Risk-based Inspection – Development of Guidelines. Vol. 3, Fossil Fuel-fired Electric Power Generating Station Applications. ASME Research Report, CRTD-Vol. 20-3, 1994.

5 Arai, T.; Shimomura, Y.: Proposal of Service CAD System – A Tool for Service Engineering. In: Annals of the CIRP, Vol. 53/1, 2004, pp. 397-400.

6 Seliger, G.; Buchholz, A.; Grudzien, W.: Multiple Usage Phases by Component Adaptation. In: Proceedings of the 9th CIRP International Seminar on Life Cycle Engineering. Erlangen, Germany, 2002, pp. 47-54.

7 Takata, S.: Life Cycle Maintenance Management. In: Lee, J.; Wang, B. (Ed.): Computer-Aided Maintenance. Kluwer Academic Pub., 1999, pp. 209-230.

8 Takata, S.; Inoue, Y.; Kohda, T.; Hiraoka, H.; Asama, H.: Maintenance Data Management System. In: Annals of the CIRP, Vol. 48/1, 1999, pp. 389-392.

9 Takata, S.; Yamada, A.; Kohda, T.; Asama, H.: Life Cycle Simulation Applied to a Robot Manipulator – An Example of Aging Simulation of Manufacturing Facilities. In: Annals of the CIRP, Vol. 47/1, 1998, pp. 397-400.

10 Yamada, A.; Takata, S.: Reliability Improvement of Industrial Robots by Optimizing Operation Plans Based on Deterioration Evaluation. In: Annals of the CIRP, Vol. 51/1, 2002, pp. 319-322.

11 Yoshikawa, H.: A Japanese Project on Fully Automated Factory. In: Proceedings of the International Conference PROLAMAT '76, 1977, pp. 1-18.

7.2 Intelligent Maintenance Systems

Jay Lee, Haixia Wang, Cincinnati, USA
Jun Ni, Dragan Djurdjanovic, Ann Arbor, USA

Product and system maintenance plays an important role in sustainable development by prolonging the useful life of products and systems. It thus enhances the productivity of resources and reduces environmental burdens by avoiding unnecessary remanufacturing of products and systems. When applied to natural resources, maintenance helps to control pollution, keeping it within the predefined limits. For example, automotive maintenance can be conducted to control exhaust gas emission based on an OBD (On-board Diagnostics) system.

For years, maintenance has been seen as critical for maintaining productivity but has yet to be recognized as a key component of sustainable development. Currently, most field services depend on sensor-driven management systems that provide alerts, alarms and indicators of failure. However, the moment an alarm sounds, it is already too late to prevent the failure. Therefore, most maintenance operations today are either purely reactive (fixing or replacing equipment after it fails) or blindly proactive (assuming a certain level of performance degradation, with no input from the machinery itself, and servicing equipment on a routine schedule whether service is actually needed or not). Both of these scenarios could be extremely wasteful. Rather than reactive maintenance, "fail-and-fix," world-class companies are moving ahead towards a "predict-and-prevent" maintenance paradigm. A maintenance scheme, referred to as Condition Based Maintenance (CBM), was developed by considering current degradation and its evolution. It aims to provide warning of an impending failure on a particular piece of equipment, allowing that equipment to be maintained only when there is definitive evidence of an impending failure. CBM methods and practices have been continuously improved in the recent decades. Sensor fusion techniques are now commonly used due to their inherent superiority in taking advantage of information from multiple sensors [1, 2, 3, 4, 5, 6]. Existing prognostic approaches can be classified into three basic groups: model-based, data-driven, and hybrid approaches. Details can be found in references [2, 7]. However, the existing prognostic methods are application or equipment specific. For instance, a polynomial neural network was conducted for fault detection, isolation, and estimation in a helicopter transmission prognostic application. A stochastic model of fatigue crack dynamics in mechanical structures was built to predict remaining service time [8]. Fuzzy logic-based neural networks have been used to predict paper web breakage in a paper mill and the failure of a tensioned steel band with seeded crack growth [9]. Vachtsevanos and Wang [10] gave an overview of different CBM algorithms and suggested a method to compare their performance for a specific application.

In spite of the progresses in CBM, many fundamental issues still remain. The unmet needs in maintenance can be categorized into the following three groups (Figure 1):

- *Equipment Intelligence:*
  Intelligent monitoring, predict and prevent, compensation, and reconfiguration for guaranteed functionality and sustainability (self-maintenance);
- *Operations Intelligence:*
  Prioritize, optimize, and plan responsive maintenance scheduling for reconfiguration needs;
- *Synchronization Intelligence:*
  Autonomous information flow from market demand to factory asset utilization.

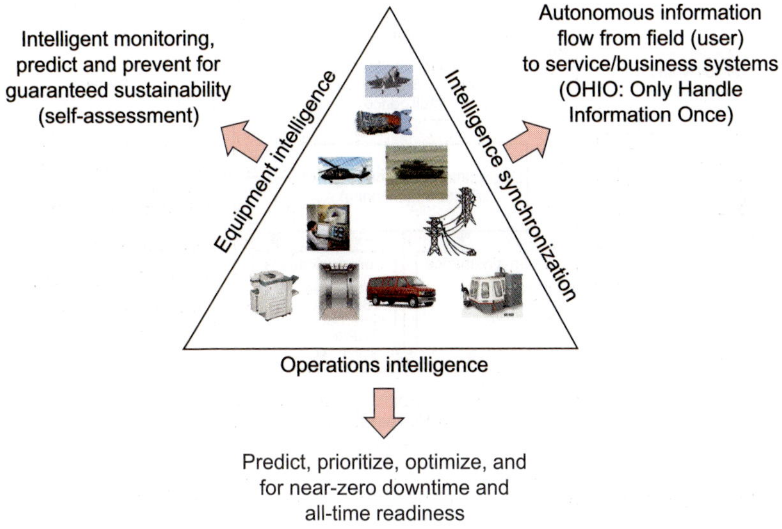

Fig. 1: Unmet needs in maintenance

To address these unmet needs, a National Science Foundation (NSF) Industry/University Cooperative Research Center on Intelligent Maintenance Systems (IMS) [23] has been established in 2001 with a vision to develop a systematic approach in advanced prognostics to enable products and systems to achieve near-zero breakdown reliability and performance and, ultimately, to realize closed-loop life cycle design for product reliability and system sustainability. As illustrated in Figure 2, the vision of the Intelligent Maintenance Systems mainly includes the following two loops:

- *Near-zero operation downtime to be achieved in the small loop:* Feedback of product health degradation information to production for responsive maintenance decision making. It is to realize equipment intelligence and operation intelligence.
- *Design for product and system reliability and sustainability to be achieved in the big loop consisting dotted lines:* Feedback of product health degradation and maintenance information to product redesign for sustainability. It is to realize synchronization intelligence.

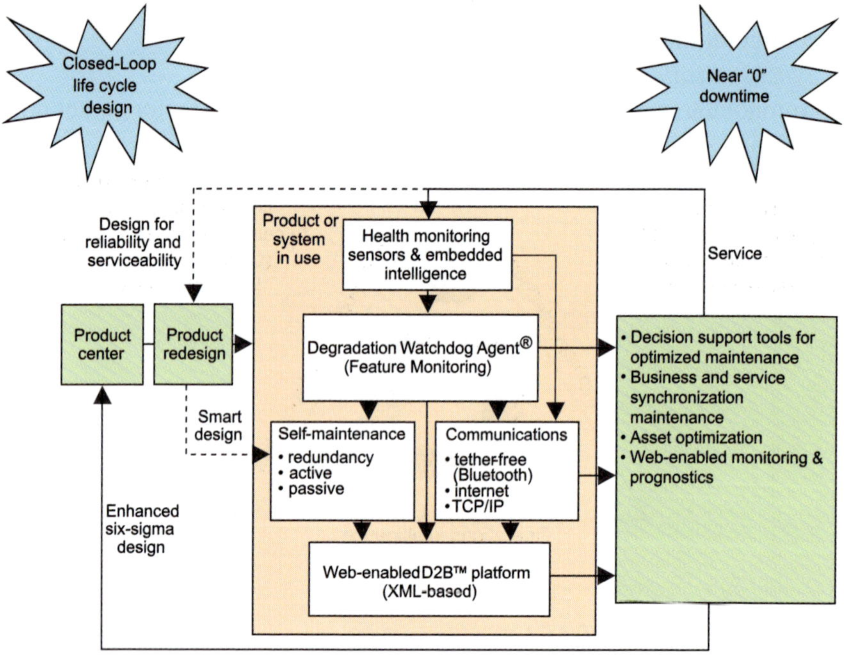

Fig. 2: Vision of Intelligent Maintenance Systems (IMS)

The near-zero breakdown performance can be achieved by intelligently monitoring, predicting, and optimizing product and system performance using the Watchdog Agent® (autonomic prognostics computing) and responsive maintenance decision making techniques.

7.2.1 Watchdog Agent® Technique

A "digital doctor" inspired by biological perceptual systems and machine psychology theory, the Watchdog Agent® consists of embedded computational prognostic algorithms and a software toolbox for predicting degradation of devices and systems. The current algorithms utilized by the Watchdog Agent® toolbox are grouped by their functions into: signal processing and feature extraction, performance assessment, diagnostics, and prognostics modules. The algorithms grouped in each module are summarized in Figure 3. Each of these modules is realized, in several different ways, to facilitate the use of the Watchdog Agent® in a wide variety of products and applications. The Watchdog Agent® is being developed to be extensible and adaptable to most real-world machine situations.

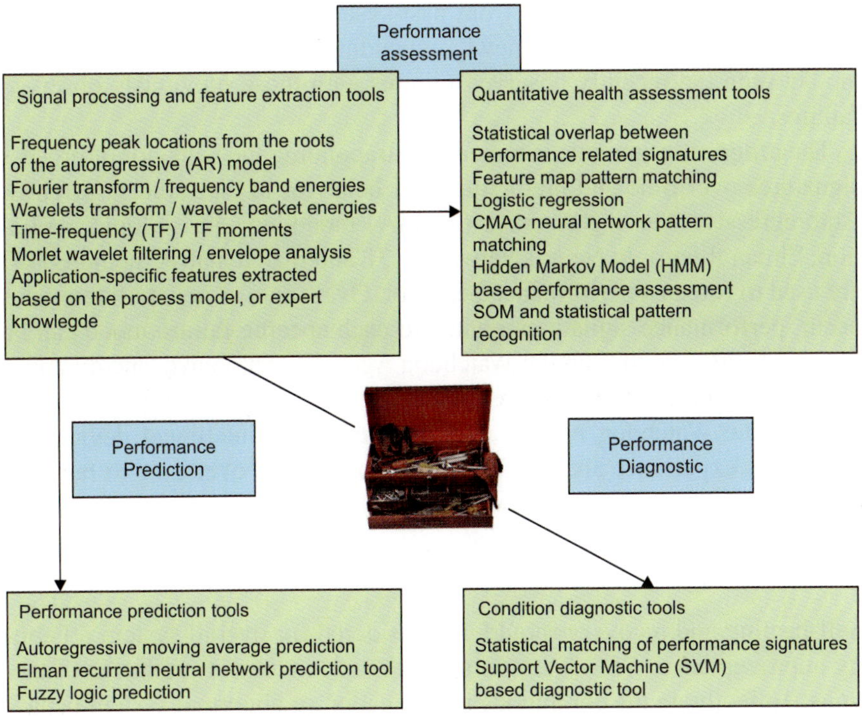

Fig. 3: Watchdog Agent® prognostics toolbox

The signal processing and feature extraction module transforms multiple sensor signals into features that are the most relevant to the description of a machine's health condition. The performance assessment module evaluates a machine's health condition based on the Confidence Values (CVs) of the extracted features. A CV is calculated based on the overlap between the most recently observed signatures and those observed during normal conditions, with a higher CV signifying a high overlap, and hence a health condition closer to normal [11, 12]. With multiple CVs, a machine's health condition can be represented by a radar chart in which each axis represents performance CVs related to key machine components, or to key individual features extracted from sensor readings.

The diagnostics module recognizes indications of failure in system behavior. Namely, if signatures describing system behavior in the presence of a given fault are available from past system operation, it is possible to identify the root causes of the newly impending fault by evaluating the overlap between the recently observed failures with those previously observed failures.

The prognostics module is aimed at extrapolating the behavior of process signatures over time and predicting their behavior in the future. Currently, Autoregressive Moving Average (ARMA) modeling, Fuzzy logic prediction, Elman Recurrent Neural Network [13], and match matrix methods [14] are used to forecast performance behavior. Over time, as new failure modes occur, performance signatures related to each specific failure mode can be collected and used to train the Watchdog Agent® to recognize, and diagnose those failure modes in the future.

Thus, the Watchdog Agent® is envisioned as an intelligent device that utilizes its experience and human supervisory inputs over time to build its own expandable and adaptable world model.

Self-healing Systems

Self-healing and recovery should be considered for overall system reliability improvement and for determining what to do in case of product malfunctions. To facilitate the continuous identification of abnormal system behavior, isolating the source of anomalous behavior and achieving the desired system performance in a complex system, a novel approach to realizing the immune system functionalities in an artificial system has been recently proposed in a manner analogous to ones in living organisms [15]. This anomaly detection capability should be performed without the need to a priori know the entire set of faults that the monitoring system is supposed to capture and mediate regardless of control inputs and external conditions. The proposed approach aims to enable machine self-healing by performing self-detection of anomalies, self-diagnosis, and control adaptation whenever a fault occurs, which corresponds to the set of generic Anomaly Detection Agents (ADAs), Diagnostic Agents (DAs) and Control Agents (CAs), as illustrated in Figure 4.

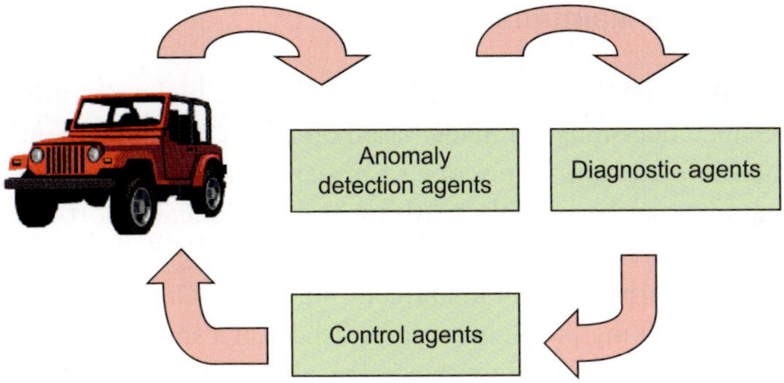

Fig. 4: Proposed immune system operation for an automotive system

The role of ADAs is:

- to identify the intrusion depicted in the degraded performance of the system. This task will be accomplished regardless of the character of control inputs and external disturbances through generic profiling indexed by clusters of system inputs and external conditions, following the ideas already developed [14, 16];
- to isolate the intrusion source by reconfiguring and reconnecting themselves until input/output sets with anomalous behavior are identified in as much detail as possible. The generic nature of the intrusion (anomaly) detection procedure will be crucial in enabling easy reconfiguration of the input/output sets on which ADAs would perform anomaly detection.

The role of DAs is:

- to recognize and to describe the character of the anomaly source, if such characterization is possible, that is, if input/output patterns from the anomalous subsystem have been observed in the past;
- in case input/output patterns characterizing the observed anomaly have not been seen in the past, to create a new DA that will be able to recognize the observed anomaly next time it occurs.

The role of CAs is:

- to utilize fault characterization from DAs in order to postulate control laws in such a way that desired performance of the anomalous subsystem is restored as much as possible. Hence, CAs would effectively yield the self-healing functionality to an autonomous system.

7.2.2 Responsive Maintenance

The full benefits of intelligent maintenance cannot be realized until responsive maintenance decisions are made based on predictive equipment degradation-related information. The scope of research in the maintenance decision-making area can be roughly partitioned into tactical (short-term) and strategic (long-term) maintenance decision-making.

Maintenance prioritization in manufacturing is a common tactical maintenance decision-making problem (accomplished on a shift-to-shift, or even hour to hour time-frame). Its importance has already been well-recognized in both the industrial and academic communities [17, 18]. However, online Work-in-Progress (WIP) and machine status information across a manufacturing system have not been well addressed in a maintenance prioritization study [19].

Factory-wide available online information about machine status and WIP has been used [20] to optimally prioritize a set of maintenance work-orders using discrete-event simulation of operations in a manufacturing process, coupled with a genetic algorithm based optimization procedure that maximized throughput associated with the execution of the prioritized maintenance work-orders. Figure 5 shows how potential benefits of employing the proposed systematic maintenance prioritization grow as the number of jobs for prioritization grows, as evaluated through a simulation study. Furthermore, the newly proposed prioritization method was tested in a major automotive plant, where it was demonstrated that the new maintenance prioritization methods yield a statistically significant increase in the expected system throughput. An increase of 3 % in the expected number of produced work-pieces was observed during one specific day of production, when four maintenance tasks were executed.

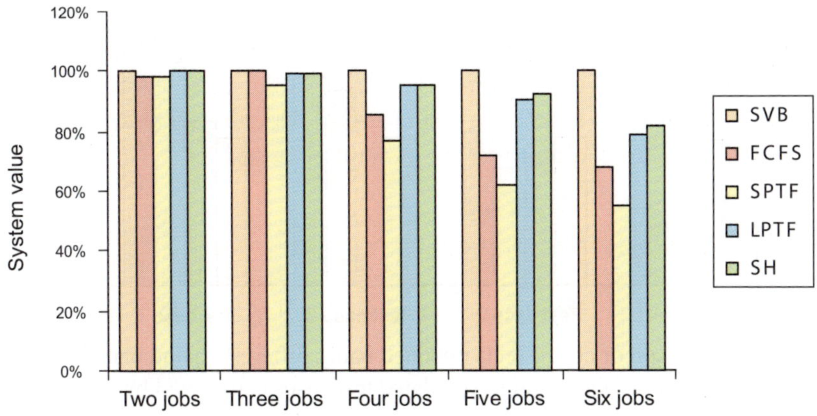

Fig. 5: Relative improvement of system productivity using SVB method proposed by [20], compared to more traditional policies: First-Come-First-Serve (FCFS), Shortest-Processing-Time-First (SPTF), Longest Processing-Time-First (LPTF) and a Static Heuristic (SH) method

By considering machine degradation as well as functional dependencies between machines, a strategic maintenance scheduling method has recently been proposed by [21], which utilizes discrete event simulation to evaluate effects of a maintenance schedule and a GA§ optimization to search for schedules that achieve a tradeoff between proactive maintenance that prevents failures from happening, and avoids maintenance-induced interruptions to normal production functions. Figure 6 shows the results of simulated comparisons of maintenance schedules based on the proposed predictive equipment condition information (Strategy D), as compared to the purely corrective maintenance strategy (Strategy A), the purely time-based preventive maintenance (Strategy B) and the current equipment condition-based information (Strategy C). In all test cases, the maintenance schedules obtained using Strategy D yielded more cost benefits by avoiding unscheduled downtime on production-critical machines. Only in Test Case 2, were the cost effects obtained using proposed Strategy D and purely reactive Strategy A statistically the same. This is understandable given that Test Case 2 was simulated under an extreme assumption that costs associated with scheduled and reactive (unscheduled) maintenance are the same (usually, costs of unscheduled maintenance are much higher than those associated with scheduled maintenance that takes place before a machine breaks down).

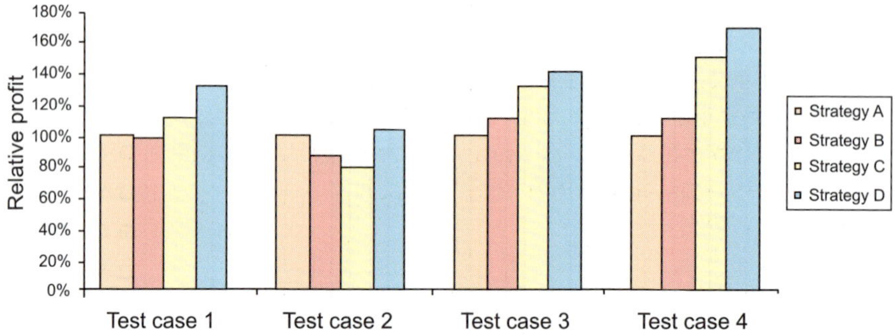

Fig. 6: Comparison of cost-effects obtained using different maintenance strategies in four different simulated test cases considered by [21]

7.2.3 Hierarchical Architecture

Intelligent maintenance systems are based on product degradation monitoring and assessment. In order to work toward the goals identified in the vision of IMS research is being conducted to develop methodologies for closed-loop life cycle design for product reliability and system sustainability. Two levels of research opportunities are identified as follows:

Research is planned to determine relations between product health management information and product parameters. These relations allow isolation of critical product parameters that cause failures, which is essential for the development of a methodology to revaluate product design in order to improve product robustness to failures.

Research is also planned to incorporate self-maintenance abilities into design based on the concept of functional maintenance [22]. Functional maintenance aims to recover the required functionality of a degradating machine by trading off functions, whereas traditional repair (physical maintenance) aims to recover the initial physical state by replacing faulty components, cleaning, etc. A product needs to be designed with function redundancy so that it can reconfigure its behavior to maintain the most important functions, while sacrificing less important functions, when a fault occurs. For example, when a car's engine does not work, in an emergency the car might operate for a while with its starting motor and battery. While the starting motor has an original function of starting the engine, it has a redundant function of operating the car for a while. This is reconfiguration of behavior, rather than reconfiguration of structure or adjustment of states.

These new frontier efforts will lead to a more fundamental understanding of reconfigurability and allow the closed-loop design of autonomously reconfigurable engineered systems integrating physical, information, and knowledge domains These autonomously reconfigurable engineered systems will be able to sense, perform self-prognosis, self-diagnose, and reconfigure the system to function uninterruptedly when subject to unplanned failure events.

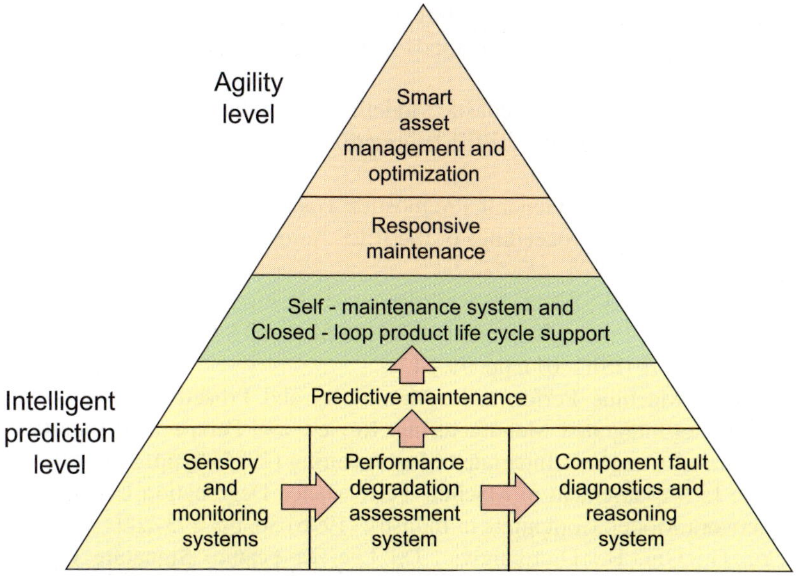

Fig. 7: Hierarchy of Intelligent Maintenance Systems (IMS)

References

1 Kemerait, R.: New Cepstral Approach for Prognostic Maintenance of Cyclic Machinery. IEEE Southeascon, 1987, pp. 256-262.

2 Hansen, R.; Hall, D.; Kurtz, S.: New Approach to the Challenge of Machinery Prognostics. In: Proceedings of the International Gas Turbine and Aeroengine Congress and Exposition, American Society of Mechanical Engineers, June 13-16, 1994, pp. 1-8.

3 Wilson, B. et al.: Development of a Modular In-situ Oil Analysis Prognostic System. International Society of Logistics (SOLE) 1999 Symposium, Nevada, Las Vegas, 30 August –2 September, 1999.

4 Reichard, K.; van Dyke, M.; Maynard, K.: Application of Sensor Fusion and Signal Classification Techniques in a Distributed Machinery Condition

Monitoring System. In: Proceedings of SPIE – The International Society for Optical Engineering, 2000, 4051:329-336.

5 Goodenow, T.; Hardman, W.; Karchnak, M.: Acoustic Emissions in Broadband Vibration as an Indicator of Bearing Stress. In: Proceedings of IEEE Aerospace Conference, 2000; 6: 95-122. Hall, L.D.; Llinas J. (Eds.): Handbook of Sensor Fusion, CRC Press, 2000.

6 Roemer, M.; Kacprzynski, G.; Orsagh, R.: Assessment of Data and Knowledge Fusion Strategies for Prognostics and Health Management. In: Proceedings of IEEE Aerospace Conference, 2001, pp. 62979-62988.

7 Bunday, B. D.: Statistical Methods in Reliability Theory and Practice. Ellis Horwood 1991.

8 Ray, A.; Tangirala, S.: Stochastic Modeling of Fatigue Crack Dynamic for On-Line Failure Prognostics. IEEE Transactions on Control Systems Technology 4 (1996) 4, pp. 443-449.

9 Swanson, D. C.: A General Prognostics Tracking Algorithm for Predictive Maintenance. In: Proceedings of the IEEE Aerospace Conference, 2001, Vol. 6, pp. 2971-2977.

10 Vachtsevanos, G.; Wang, P.: Fault Prognosis Using Dynamic Wavelet Neural Networks. In: Proceedings of the IEEE International Symposium on Intelligent Control 2001 (ISIC '01), pp. 79-84.

11 Lee, J.: Machine Performance Monitoring and Proactive Maintenance in Computer-integrated Manufacturing: Review and Perspective. International Journal of Computer Integrated Manufacturing (1995) 8, pp. 370-380.

12 Lee, J.: Measurement of Machine Performance Degradation Using a Neural Network Model. Computers in Industry (1996) 30, pp. 193-209.

13 Yu, G.; Qiu, H.; Djurdjanovic, D.; Lee, J.: Feature Signature Prediction of a Boring Process Using Neural Network Modeling with Confidence Bounds. Accepted for publication in the International Journal of Advanced Manufacturing Technology, Paper No. 2946, 2005.

14 Liu, J.; Djurdjanovic, D.; Marko, K.; Ni, J.: A Novel Method for Anomaly Detection, Localization and Fault Isolation for Dynamic Control Systems. Submitted to IEEE Transactions on Control Systems Technology, Paper No. 2005-114, 2005.

15 Djurdjanovic, D.; Marko, K.; Ni, J.: Immune System Engineering for Automotive Engine Systems. Project Sponsored by the National Science Foundation, Award No. CMS-0600200, 2006.

16 Liu, J.; Djurdjanovic, D.; Marko, K.; Ni, J.: Growing Structure Multiple Model System for Anomaly Detection and Fault Diagnosis. Submitted to ASME Trans., J. of Dynamic Systems, Measurements and Control, Paper No. DS-06-1152, 2006.

17 Mann, L. J.: Maintenance Management. Lexington. Lexington Books, Massachusetts 1983.

18 Levitt, J.: The Handbook of Maintenance Management. Industrial Press, New York 1997.

19 Koc, M.; Ni, J.; Lee, J.; Bandyopadhyay, P.: e-Manufacturing – Trends and Oppotunities. International Journal of Advanced Manufacturing Systems 6 (2003) 1, pp. 29-46.
20 Yang, Z.; Chang, Q.; Djurdjanovic, D.; Ni, J.; Lee, J.: Maintenance Priority Assignment On-Line Production Information. Accepted for publication in the Transaction of ASME, Journal of Manufacturing Science and Engineering, Paper No. MANU-05-1073, 2005.
21 Yang, Z.; Djurdjanovic, D.; Ni, J.: Maintenance Scheduling for a Manufacturing System of Machines with Adjustable Throughput. Submitted to IIE Transactions on Quality and Reliability Engineering, Paper No. UIIE-0447, 2005.
22 Umeda, Y.; Tomiyama, T.; Yoshikawa, H.: A Design Methodology for Self-maintenance Machines. ASME Journal of Mechanical Design 117, September 1995.
23 IMS: NSF I/UCRC Center for Intelligent Maintenance Systems. http://www.imscenter.net (Status: 2004).

7.3 Life Cycle Unit – Concept and Application

Günther Seliger, René Gegusch, Daniel Odry, Berlin, Germany

Less use of resources can be achieved by transforming cradle-to-grave-economies into sustainable cycle economies. One of the main objectives of this approach is to increase the use-productivity of resources.

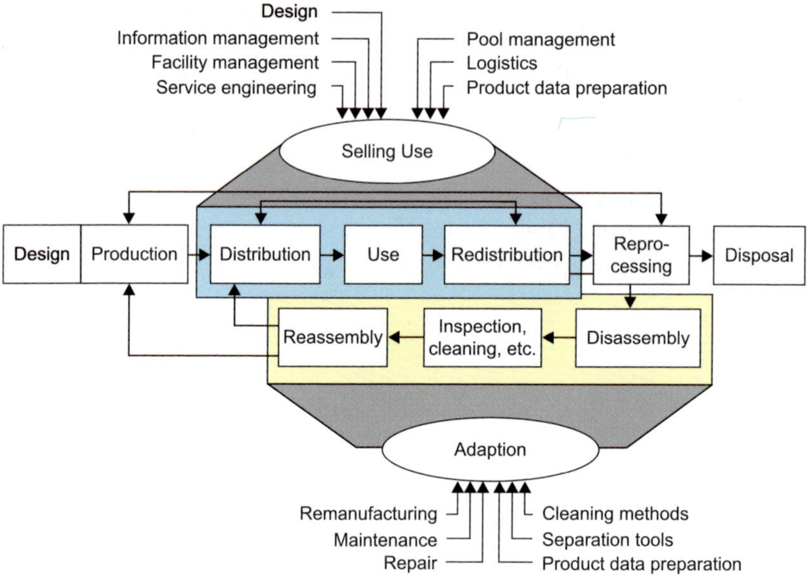

Fig. 1: Business model of a cycle economy

In order to successfully implement a cycle economy, the products and components involved must be designed considering their whole life cycle (Figure 1). This starts with their development, carries on for their use and ends up with their reuse or disposal [1]. A cycle economy can be fostered by introducing multiple usage phases of modular products and components with adaptation in and in between the usage phases. Adaptation may include maintenance, repair, up- and downgrading, enlargement, reduction, rearrangement, modernization and remanufacturing, which is associated with disassembly and reassembly in and between the usage phases. Innovative approaches for the mentioned processes facilitate selling use, e.g., condition based maintenance and repair can reduce downtime and increase the residual value of the products and components.

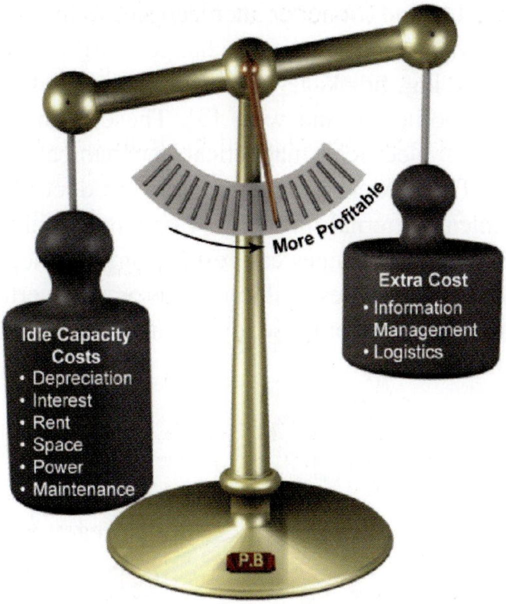

Fig. 2: Prerequisite for selling-use instead of selling products

For ease of adaptation, the products and components involved have to be modular, integrable, convertible, and diagnosable. The implementation of these properties in a high level increases the applicability and the availability of products and components in multiple usage phases [2].

A cycle economy is not only ecologically reasonable but is also a chance for new business. In the selling use approach the use provider offers a product's functionality in quality, time and location to the customer without passing the product out of his possession. Opposite to the traditional selling products approach, the use provider manages the costs of investment, operation, maintenance, and disposal of the product.

By selling functionality the provider is encouraged to maximise the utilization of his equipment. If the product is not in use, the use provider as owner has to bear the idle capacity costs (Figure 2). Consequently, the customer pays only for the use that he obtains by the product. This is profitable once the idle capacity costs of equipment are higher than the additional costs for logistics and information management facilitating access to functionality. Information and communication technology and logistics become technological enablers for the selling-use approach. Rental car companies are examples for the selling use approach.

Products and their components underlie deterioration by usage, which may limit their quality and sooner or later lead to a failure. Deterioration can be divided into physical and functional changes. Utilization leads to physical changes like ageing, breakage, corrosion, creep, deformation, fatigue, loss/displacement, pollution, and wear [3]. These changes are caused by biological, chemical, electrical, magnetical, mechanical, radiation or thermal mechanisms. Their extent is determined by product characteristics like material or treatment, external influences like temperature or dust and the type of usage. Functional changes can result from technical progress, legal amendments, a change in values, fashion trends or a converted usage and are attributed to parameters like purpose, duration, place or intensity of usage, policy, society and economy.

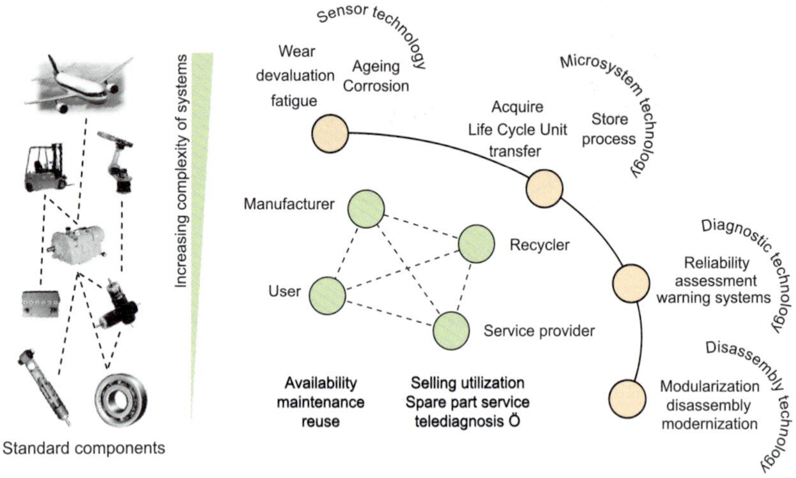

Fig. 3: Products, standard components, interest groups, and business areas

As manufacturers are confronted with increasing demands for product availability and reliability, the assessment and prediction of the product's behaviour is desirable. Knowledge about physical changes, components, paths of adaptation and other product features is also required for disassembly of deteriorated products, reassembly and distribution of adapted products for further usage phases. Due to the huge variety of existing products, this assessment is a difficult task that has to be solved individually for every product [4].

Product accompanying information systems – so called Life Cycle Units (LCUs) – integrated into products are capable of acquiring, processing, and communicating relevant product and process data for evaluation purposes during the entire product life span. In the proposed approach, the assessment focus is put on standard components like bearings, gears, compressors, pumps, dampers, filters, hose lines or pneumatic components (Figure 3). These standard components are integrated into various more complex products, e.g. assembly systems, ground conveyors or industrial robots. This way the assessment of a complex product is achieved by assessing its standard components.

By focusing on the assessment of standard components, the development effort for assessment of complex products is distributed technically and economically on many applications. Also, less overall expertise is needed in order to develop and debug a diagnostic system of a complex product or component, because each subsystem can be examined almost independently of the others. The solution space is significantly reduced for each LCU designer.

7.3.1 Architecture and Implementation

The LCU system is based on a modular architecture and can be built as any reasonable combination of the five modules marking, sensor, Life Cycle Board (LCB), actuator, and communication (Figure 4).

Markings are a cost-effective way to store static information. The function of this module is the storage of small data on a cost effective way. Single colour coding may provide e.g. information about the position of objects relevant for manufacturing processes. Further information can be encoded by colour and form of the marking. If more data shall be stored, for example detailed dis/assembly instructions, bar codes are better suitable.

Sensors are integrated into products and components to acquire relevant physical parameters, e.g. acceleration, force, temperature, or pressure. A direct registration of the product's and component's state, e.g. deterioration, with sensors is difficult. Further data processing is needed to provide this information, e.g. Fast Fourier Transformation (FFT) analysis.

Actuators may be integrated into the product to act against deterioration or to support processes like maintenance or disassembly. An actuator is an energy converter of chemical, electrical, magnetic or any type of radiation energy. For example, electromagnetic joining elements integrated into housings can be opened with electromagnetic signals controlled by the LCB to support maintenance and automated disassembly processes of the inner components of a product.

The processing and storing of the acquired data is done on the Life Cycle Board (LCB). The LCB is composed of one or more processors, memory devices and hardware interfaces for sensor, actuator, and communication modules. It controls the core functionality of the LCU and its components. The gathered sensorial data can be stored in memory and processed in order to generate relevant product and process information and knowledge, e.g. assess the product's condition.

The communication module allows the LCU to communicate with other LCUs and its environment. Wired and wireless technologies can be applied, e.g. CAN, Ethernet, RS485, Bluetooth, IrDA, or W-LAN, to transmit the gathered data and generated information and knowledge. While the wired communication technology is relatively inexpensive and reliable, wireless communication technology is gaining more and more importance because of its higher flexibility.

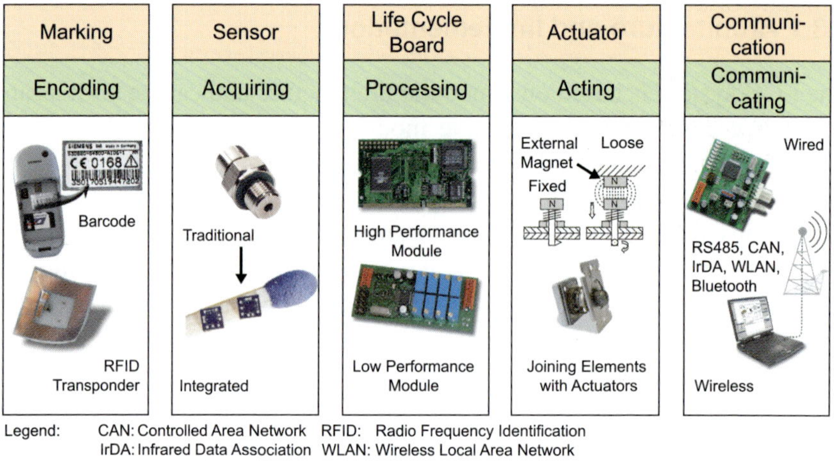

Fig. 4: Modular architecture of LCUs

IEEE1451 – Modular Communication Module for LCUs

The realization of Life Cycle Units (LCUs) for broad application areas requires a flexible and modular design. One of the biggest challenges lies in the realization of a modular hardware architecture of LCUs. This enables a flexible way of combining suitable LCU hardware. The producer of LCUs can then easily combine the requested modules.

Therefore, a modular design of the communication module is proposed. In the application of standard components equipped with LCUs, the communication of acquired, evaluated or stored data is important. However, the higher application determines what kind of communication standard can or shall be used, i.e. W-LAN, Ethernet, and RS485.

The IEEE1451 [5] concept offers such a modular design method for data acquisition with evaluation and a communication module. This facilitates to build small LCUs with modular communication interfaces by choosing the communication module with the according interface. Developers and manufacturers can select the required boards with sensor interfaces and processing power of the microcontroller and combine this with the needed communication module. As there are no standardized communication interfaces for LCU applications, the application itself also influences the selection, – i.e. high speed communication, low speed or wireless – this technique enables a high flexibility in the LCU configuration.

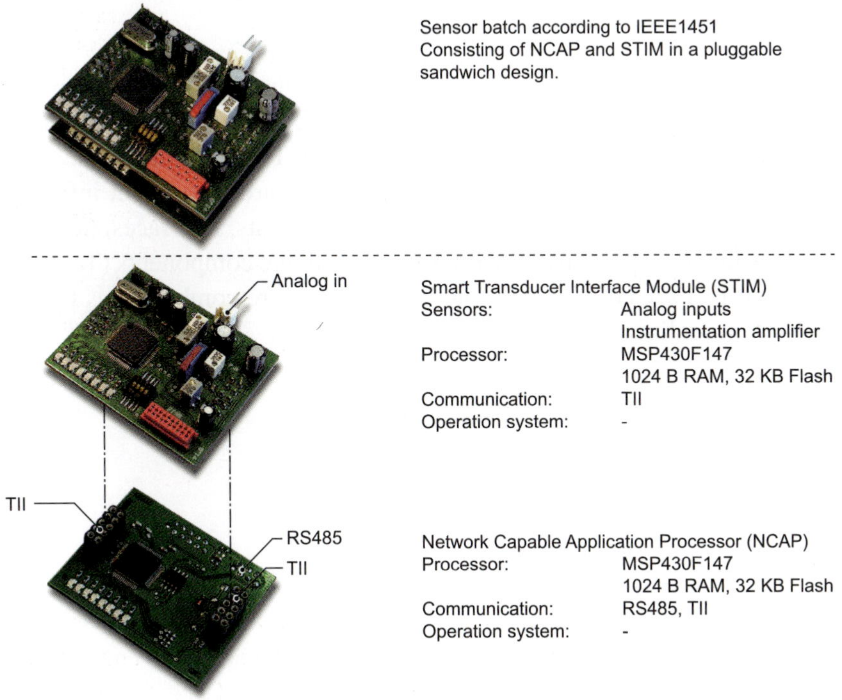

Fig. 5: Modular realizations basing on the IEEE1451 concept (NCAP/STIM) [6]

IEEE 1451 Concept

The Smart Transducer Interface Model (STIM) contains sensor elements, signal amplification and configuration, A/D converters, a processor for the implementation of the STIM functionality, and a Transducer Electronic Data Sheet (TEDS) [5]. The TEDS is stored in the processor's memory and facilitates automatic configuration of the sensor in a network. It contains data describing the STIM module and each sensor can be described in channel, sensor calibration data, and other data that can be sensor specific. The connection to the NCAP is implemented using a 10 wire Transducer Independent Interface (TII) for serial synchronous communication. The NCAP contains a processor and additional circuitry is needed for the connection to the TII and a specific network (Figure 5).

This way, STIMs can be developed regardless of the underlying network. For each network type, only one NCAP has to be developed.

LCU Configurator

The LCU Configurator [7] is a development tool that gives suggestions for the selection and combination of useable hardware for the realization of LCUs considering the application's requirements. This accelerates the decision-making process for the selection of suitable hardware. Input information, e.g. energy consumption, calculation speed, costs, interfaces, and sensor ranges, are being compared with data of available components (Table 1). The output is a selection of possible usable hardware combinations for the LCU system realization.

Tab. 1: Example of LCU configurator input criteria

Object	Function	Information input to LCU configurator
Life cycle microprocessor	Processing data	Computing performance Available interfaces Power consumption
Life cycle board	Hardware base for components	Max./Min. size of board Temperature ranges
Sensors	Acquiring data	Measuring range Tolerance Interface
Actuators	Acting	Tolerance Interface Power consumption
Communication	Communicating between microcontroller and sensor, actor and other LCUs	Type Bandwidth Range
Algorithms	Algorithms for diagnostic, prognostic and condition assessment	Expenditure evaluation Memory size requirements
Energy	Energy consumption level	Voltage Max. power consumption

7.3.2 Integration in Different Products

Washing Machine

To support the maintenance and disassembly of the washing machine, LCUs are integrated into standard components of the washing machine, i.e. electromotor, lye pump, or valve group on the washing machine. The data gathered by sensors are statistically analysed and classified by the LCU. The data can be accessed by connecting the LCU, i.e. using a PDA with InfraRed interface.

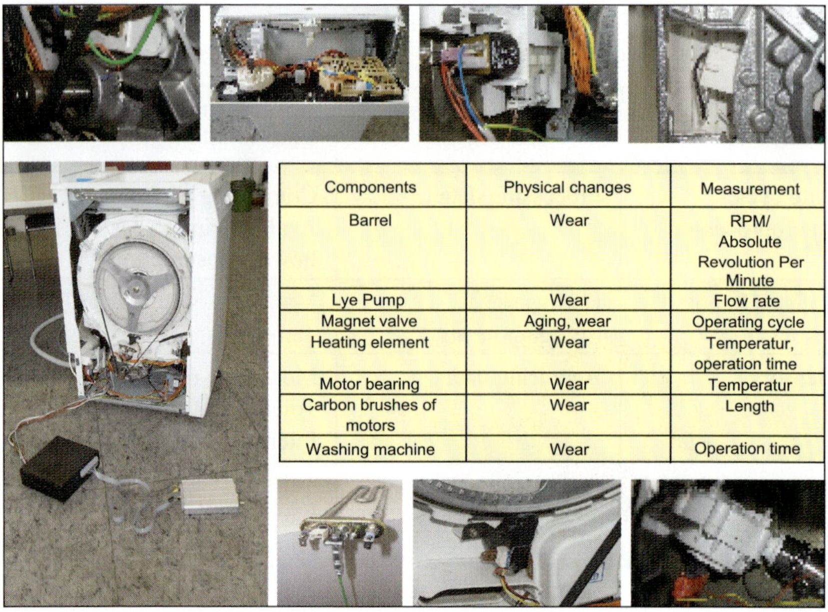

Components	Physical changes	Measurement
Barrel	Wear	RPM/ Absolute Revolution Per Minute
Lye Pump	Wear	Flow rate
Magnet valve	Aging, wear	Operating cycle
Heating element	Wear	Temperatur, operation time
Motor bearing	Wear	Temperatur
Carbon brushes of motors	Wear	Length
Washing machine	Wear	Operation time

Fig. 6: Washing machine LCU (aluminum-box) with sensors

A complete LCU was integrated into a washing machine for maintenance support and fault diagnosis. The LCU is equipped with a RS485 based modular sensors and gathers temperature and utilization times of the heating element, the motor, the lye pump, the water valve, and the machine itself. This way, information about the machine and its components (Figure 6) can be easily retrieved. The LCU evaluates the information. For the outside world, an interface based on infrared communication has been integrated into the LCU to allow easy information retrieval with small Personal Digital Assistants (PDA) or laptops. The PDA application shows information about the product and may give, i.e. instructions for maintenance to a service technician. All data and information can be used to focilitate the disassembly process, i.e. by providing disassembly plans of the machine and it's componemts to the hybrid disassembly system.

Disassembly Tool Kit

The Disassembly Tool Kit is another application example for LCUs integrated into standard components. A shear is a standard component; it is used for different applications in the industry, i.e. for cutting in the textile industry, or for destruction of flexible elements, i.e. tubes and cables in the disassembly process. An energy delivering module is another standard component; it supplies energy, in this case the energy module translatory drive delivery energy for the cut operations.

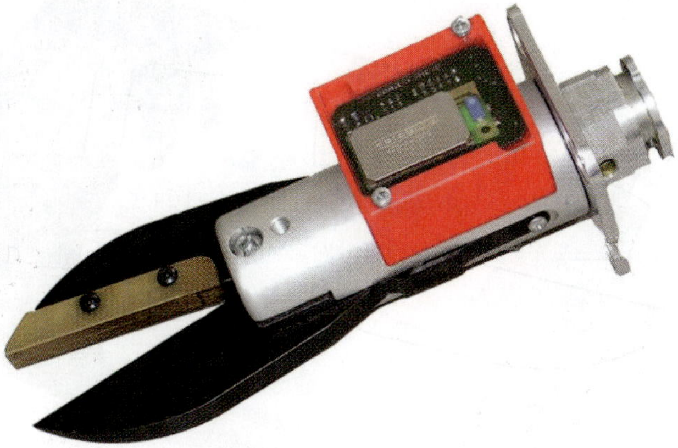

Fig. 7: Acting module shear with attached LCU (red box)

To define the condition of the tool system it is important to know if the shear works correctly or if the blades are blunt. This can be analyzed with the help of the following information:

- Was the shear completely closed?
  If a cut is successful, the shear must be completely closed.
- How much energy was needed to cut an object?
  A sharp shear needs less energy in cutting than a blunt one with the same physical parameters.

Combining these information enables the condition assessment of the whole tool. The detection if the shear is completely closed is done with the use of a miniature light reflection switch. Figure 7 shows the responsible LCU module, processing and communicating the light reflection data. The force information is handled by a LCU on the energy module translatory drive (Figure 8), which measures the level of applied power to push the blades together. For the energy measurement, strain gages are affixed to the piston rod. A digital amplifier is connected to the LCU microcontroller and amplifies data gathered from strain gages.

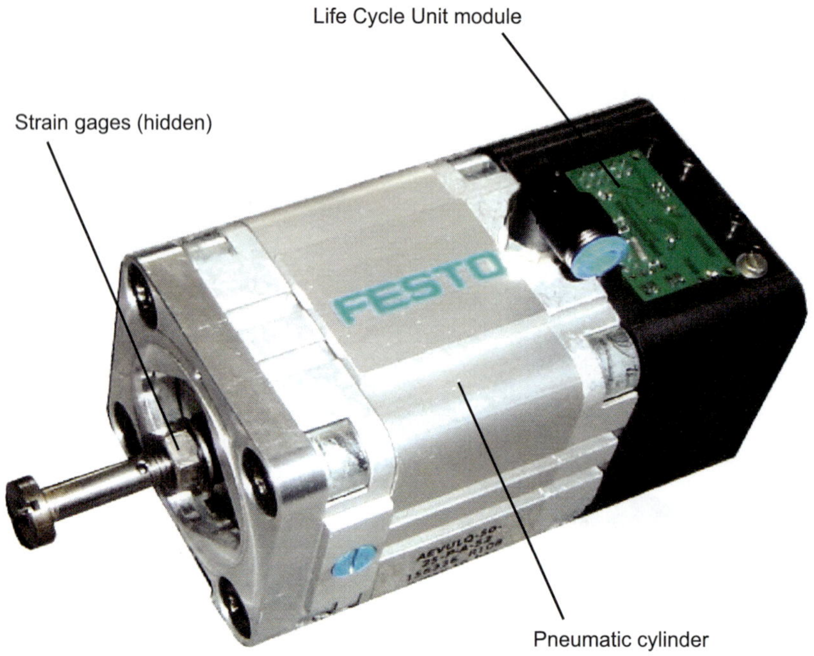

Fig. 8: Energy module translatory drive with attached LCU

An application running on a standard Microsoft Windows PC was implemented for the evaluation of the gathered data. On the timeline diagram shown on Figure 9 the signals of the LCUs are shown. The black line shows the measured level of energy being used for the cut operation form the LCU on the energy module translatory drive. The red bar shows the detection of the moment of the closed blades on the acting module shear. Each aspired cut operation can be seen by a rising of the measured energy. On the left diagram each rising is combined with a reported successful closing of the blades. This means successful cut operations all the time. On the right diagram this report is missing from the fifth energy incline. This means, that the cut operation was no more successful after the fourth task, beside there was more energy used for the cut operation. This indicates a failure operation and a possible need to repair the tool, i.e. by replacing the blades. Furthermore a constant incline of the applied energy can be used for the prognostic conclusion that the tool will fail, as a blunt shear requires a higher level of energy for a cut operation.

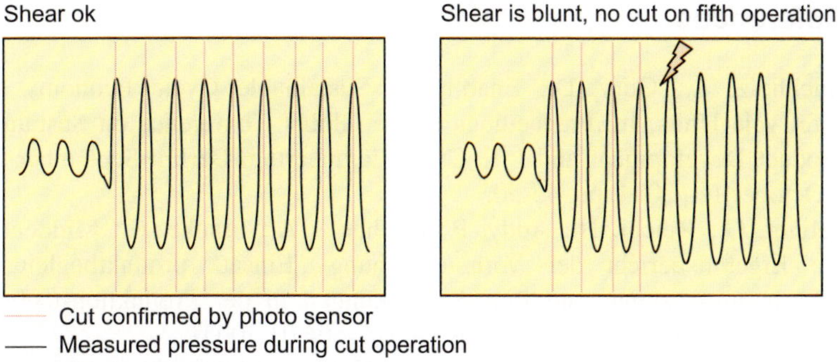

Fig. 9: Timeline diagram of level of applied energy with signals (red bar) from acting shear module LCU, OK-case (left) and Error-Case (right)

7.3.3 Conclusion

Information and communication technology – also in form of LCUs – can significantly contribute to save resources. It is an essential part of the required infrastructure which enables the provision of functionality in quality, time and location as desired by the customer within the selling-use approach. System complexity is one major challenge that has to be overcome. For example, today more than 50 % of all defects in automobiles are resulting from electronic system failures. Innovative solutions to cope with this challenge have to be found and implemented. Promising approaches may include further modularization and standardization of the systems and of their development process, e.g. software rapid prototyping and hardware in the loop simulation.

References

1 Buchholz, A.; Odry, D.: Enabling a Sustainable Cycle Economy with LifeCycle Units. In: Proceedings of the Global Conference on Sustainable Product Development and Life Cycle Engineering, Berlin, September 29th – October 1st, 2004, pp. 45-50.

2 Seliger, G.; Reichl, H.; Addy, P.; Buchholz, A.; Franke, C.; Middendorf, A.: Ergebnisbericht der Voruntersuchung „Einsatz produktbegleitender Informationssysteme und ihre Auswirkungen in der Produktionstechnik". Berlin; http://www.epi.tu-berlin.de (Access: 04.01.2005).

3 Grudzien, W.: Beitrag zur Steigerung der Nutzenproduktivität von Ressourcen durch eine Life Cycle Unit. Dissertation TU Berlin, Berlin, 2002.

4 Buchholz, A.; Franke, C.: Assessment of Standard Components for Extended Utilization. In: Proceedings of the Colloquium "e-ecological Manufacturing", Berlin, March 27-28, 2003, pp. 39-42.

5 IEEE 1451.3: Standard for a Smart Transducer Interface for Sensors and Actuators – Digital Communication and Transducer Electronic Data Sheet (TEDS) Formats for Distributed Multidrop Systems. IEEE, New York 2003.

6 Buchholz, A.: Zustandsorientierte Instandhaltung von Standardkomponenten mit Life Cycle Units. Dissertation, TU Berlin, 2005

7 http://demonty.ipk.fhg.de/index.php/Disassembly_Knowledge_Platform (Access: November 2006).

7.4 Miniaturized Life Cycle Units

Herbert Reichl, Andreas Middendorf, Berlin, Germany

7.4.1 Sustainment of Microsystems

The elements of sustainment include reliability, testability, diagnosibility, reparability, maintainability, warranty/guarantee, upgradeability, obsolescence management, technology infusion and insertion, qualification and certification, configuration control, regression testing, and cross-platform applicability. Sustainment costs can be a significant contributor to the total life cycle cost of the system.

Reliability is one of the key elements for sustainability [1]. Reliability is simply spoken the ability of a product to fulfil the required functions over a certain time while experiencing usage loads. Reliability influences the time and cost of sustainment activities such as testing, diagnosis, repair, servicing, scheduled and unscheduled maintenance etc. Reliability also plays an important role in supporting product reuse decisions, which is also a major aspect of sustainability.

The World Business Council for Sustainable Development (WBCSD) defines: "Eco-efficiency is achieved by the delivery of competitively-priced goods and services that satisfy human needs and bring quality of life, while progressively reducing environmental impacts and resource intensity throughout the life-cycle to a level at least in line with the earth's estimated carrying capacity" [2]. In short, it concerns about creating more value with less impact.

The three main objectives of eco-efficiency are:

- *Reduction of the consumption of resources:* This includes the minimized use of energy, materials, water and land. Useful strategies according to system reliability are the enhancement of recycling and of product durability.
- *Reduction of the impact on nature:* This includes the minimizing of air emissions, water discharges, waste disposal and the dispersion of toxic substances. System reliability helps to prevent unexpected averages in this context
- *Increasing product or service value:* This means providing more benefits to customers through product functionality, flexibility and modularity.

The first two objectives aim at the environmental impact. The achievement meaning product and service value is the target of the third objective. Reliability helps to longer the lifetime of products and the knowledge of reliability attributes is a fundamental basis for risk management and protection.

Condition monitoring is a main prerequisite to increase the reliability of technical systems. The LCU-concept is one suitable solution for condition monitoring. The following sections describe the concept and the realization of miniaturized LCUs.

7.4.2 Modular Microsystems

Miniaturisation and integration of electronic systems results to an amazing increase of reliability in the past and the process is still going on. Less components with more functionality and complexity are more reliable as more components with a lower functionality. Therefore an ongoing miniaturisation of the LCU is crucial.

The concept of miniaturized LCU (μLCU) is a sophisticated kind of condition monitoring because the hardware in combination with the software not only collects, stores, or communicates data from the use phase, it also computes the implemented models simultaneously. These are important differences to data loggers which are also available for a wide range of applications. The LCU concept is mainly helpful in the use phase for predictive maintenance, but offers also additional and important information for the design and manufacturing phase. Another aspect is the possibility to interpret the data if the appliance has a second lifetime via ReUse. The appropriate circuits for shock detection have been realized in a separate TB-BGA package. In the configuration the sensors are used for shock detection only to reduce the power consumption. Figure 1 shows a prototype of the μLCU with acceleration, humidity and temperature sensors already included. The larger 3-layer block on top of the stacks, where the acceleration sensors for the 3 axis are placed, could lapse if only temperature changes are needed for the lifetime modelling. Also the small FR4 carrier board is only needed for programming and testing.

The suggested use scenarios have different requirements to an appropriate μLCU. In general the appropriate device or functional module requires sensors for environmental or performance parameters, a processing capability, memory and an interface for programming and readout. The intended purpose determines the hardware of a μLCU and the appropriate analysis algorithms.

Fig. 1: μLCU demonstrator in match X technology

Because of increasing demands on reliability aspects for more and more complex systems and a longer expected lifetime of electronics, e.g. for automotive applications, the combination of classic reliability prediction with physics-of-failure and statistical methods with condition monitoring approaches make sense from an economical and environmental point of view as well as from the scientific perspective. Especially the transfer from lead to lead-free with the change of technologies and questions on the availability of components shows the demand on advanced test strategies where the μLCU concept is able to give support.

The functionality of the μLCU can be either integrated using system resources or an external add-on to the monitored device. Here an add-on system was taken to have an independent system for many applications. This enables also the possibility to use a higher reliable technology for the μLCU in comparison to the application.

Concept of the μLCU Design

In order to meet the requirements of various possible applications on the one hand and to keep development and manufacturing cost low on the other hand, a modular concept was chosen.

To realize such a modular concept with the required functionality and a very small package size, the TB-BGA-package (top-bottom-ball-grid-array, developed by the Fraunhofer IZM [3]) is used. This package system enables three-dimensional micro systems with very high pin-counts at reduced requirements on space. To meet various reliability specifications, a variety of basic materials like ceramics or FR4 is available.

For the prototype of the μLCU a TB-BGA package with an area of 17.5 x 17.5 mm^2 with 79 pins is used. Four packages are stacked, the top cover is populated with the external sensors (temperature), leading to an overall height of about 12 mm. The stack is soldered directly on a printed wiring board which carries the energy supply and the interface connectors. The pin assignment is identical for all modules enabling an arbitrary order.

A 5 V power supply with a maximum power output of 0.5 W is needed. Power-saving-algorithms are used to keep the power consumption as low as possible. The test equipment for μLCUs is shown in Figure 2.

Sensors

The μLCU can be equipped with up to eight sensors. In the current setup two temperature sensors and three acceleration sensors (one for each axis) were integrated. The power on/off sensor as well as additional sensors can be added on the PWB carrying the stack. A humidity sensor is applied in one prototype μLCU only. Due to the bad long-term stability, humidity sensor are at this time not applicable for longer measuring.

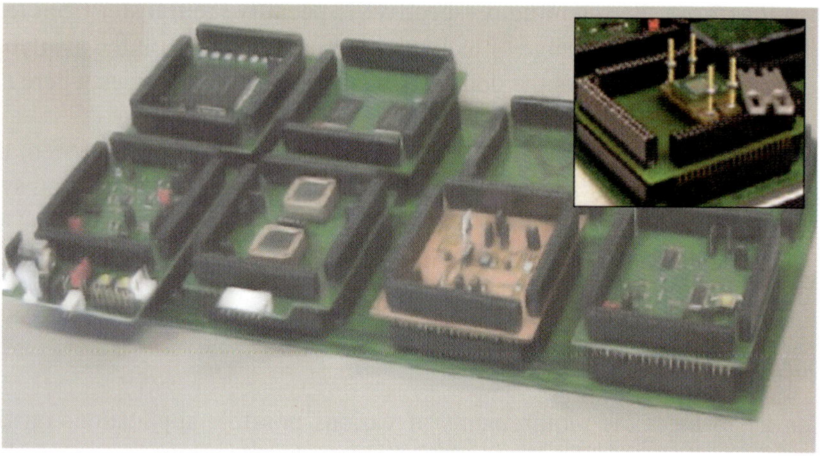

Fig. 2: Test and development equipment for different technologies

One temperature sensor is positioned right beside the micro controller for self-monitoring. The second temperature sensor and the humidity sensor (only one prototype) are positioned on the top cover. It is a monolithic temperature sensor with on-chip signal conditioning and can be operated over the temperature range of -50 °C to +150 °C.

7.4.3 The eGrain Concept

In the field of microelectronics, further reductions in size of semiconductor-chip structures and greater chip complexity are predicted. Future microelectronic applications require significantly more complex devices: Besides the trend towards higher integration density, there is also a demand for more functionality and increased performance. The mainstream planar technology is marked by physical and technological limitations which have a severe impact on system characteristics. Performance, multi-functionality and reliability of microelectronic systems will be mainly limited by the wiring between the subsystems (so-called "wiring crisis"). This leads to a critical performance bottleneck. Conventional fabrication is based on embedded technologies but there are serious disadvantages:

- Resulting large chip areas cause yield problems.
- The chip partition with the highest complexity determines the process technology.
- Embedded technologies are cost intensive.

3D System Integration creates a basis to overcome these drawbacks. 3D integrated Systems-on-Chips (3D-SOC) show reduced chip areas and enable optimized partitioning (Figure 3).

Fig. 3: Chip-on-chip technology (flip-chip interconnects) [3]

Both decrease the fabrication costs. An additional benefit of 3D-SOC's is a minimal interconnection length and the elimination of speed limiting inter-chip interconnects.

A new conception of system integration is based on so-called eGrains ("Electronic Grains") [3]. eGrains are tiny, autonomous and functional units with flexible communication possibilities. The vision of eGrains represents a new approach to systems integration that will help to develop complex, flexible and cost-efficient integrated systems based on ultra-small subcomponents. To reduce lateral space, new integration technologies using the third dimension are necessary. There is a large spectrum of 3D integration technologies which can be classified in three categories:

- stacking of packages,
- stacking of chips, and
- vertical system integration.

Stacked packages can be realized with the so-called chip-in-polymer technology based on embedding of thin chips into build-up layers of PCBs. The resulting Chip-Size-Packages (CSP's) with copper-filled vias through the PCB are then stacked and electrically interconnected by solder balls. A stacking-of-packages technology based on top-bottom ball grid arrays is available by the Match-X association. State-of-the-art for stacking of chips is mainly chip-on-chip system technology based on flip-chip interconnects. Vertical System Integration – VSI® is based on thinning, adjusted bonding and vertical metallization by inter-chip vias placed at arbitrary locations. Thin layers have to be stacked and connected. The whole system of an eGrain can be divided into four basic components:

- Sensors and Actuators are the interface between the environment and the eGrain. According to the application, the sensors are able to measure temperature, pressure, humidity, light, sound, acceleration, magnetic fields etc.
- Wireless Communication is the basic ability of an eGrain. The distances between two nodes can differ from a few meters to kilometres.
- Computation is required to handle communication and the data processing of the sensors. Depending on the necessary speed and on the complexity of the algorithms, energy-saving 8 bit micro controllers or powerful 32 bit microprocessors can be used.
- The Power unit provides the energy for the various parts of the eGrain. The functionality envisioned for eGrains can be achieved only if the total power consumption is limited to the microwatt level and if efficient power management strategies are applied.

3D System Integration supports technologies to create eGrain applications. An eGrain works as a μLCU, if sophisticated algorithms to represent the life cycle and/or to predict the life time are implemented.

7.4.4 Monitoring Electronic Assemblies

The most interesting application of μLCU is the determination of the state of the electronic assembly after a certain time in use. The used state depends on the initial state and the experienced loads. The initial state is determined by the material parameters, geometry and manufacturing variability [4].

Now a simple way to determine these conditions is to measure a set of parameters for an electronic assembly degrading under load. This could be e.g. the resistance of a solder joint going to infinity with the rupture of the joint. Unfortunately the resistance changes significantly only at time of rupture and the early warning time are rather short.

The longer way to the system state is the application of failure models. The according models are called "failure physics models". Input parameters are material parameters (influenced by manufacturing variability) and the relevant loads. Much research has been undertaken in this field. The main tasks are:

- methods for the fast determination of the input parameters (especially material parameters),
- validation of the models for the application with life cycle information units, and
- accuracy of the calculated system condition.

Many of the current models were developed for single failure mechanisms only and require extensive calculation. They are inadequate for the utilization to a complete and thereby complex electronic assembly.

Nevertheless, the application of failure models to electronic assemblies has the opportunity to estimate the remaining lifetime. Having a tool which determines the actual condition using the experienced loads, the remaining lifetime can be determined with the assumption of the future use.

Condition Monitoring for Insulated Gate Bipolar Transistor

Figure 4 shows one typical example from a wide range of power modules which are used for supply and control of electrical motors. Several IGBT chips are connected in parallel in order to achieve a desired current capability. They are indispensable and will become more importance for energy efficient systems in the future. It is a key technology which enables sustainability for and by electronic and electrical appliances.

Figure 5 shows the typical technology for the interconnection of these components in detail. On a copper base plate, which is often part of the housing, a DCB (Direct Copper Bonded) ceramic (mostly Al_2O_3) has the function to fullfil the electrical and thermal interconnection as well as the mechanical carrying. One electrode of the power semiconductor has a solder connection (backside bulk contact) and the others are wire bonded.

Fig. 4: Assembled power module with IGBT and diodes

The geometry of the bond wire interconnection is susceptible to the formation of cracks (Figure 5). The ends of the wedge with the sharp angle promote the start of the crack. Figure 6 shows a crack which has reached the complete area. As one consequence the current density and the local temperature is increasing.

The effect is expedited. The wedge will be damaged and the parallel and redundant interconnections have to carry the additional current. Each temperature range has an impact to the mechanical stress and results to thermomechanical fatigue of the wire bond interconnection. A module with IGBTs damaged in this kind will fail in a foreseeable time.

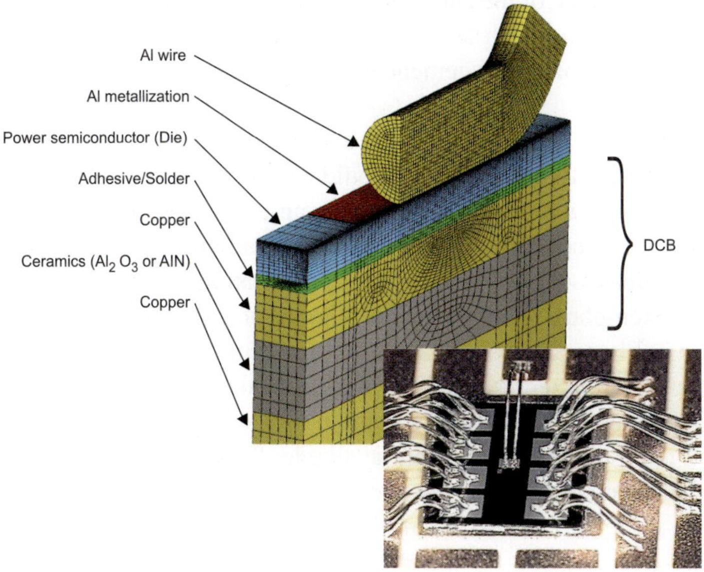

Fig. 5: Typical arrangement for power modules (left), IGBT mounted on a DCB-structure (right)

Failure Mechanism and Modelling

The thermal mismatch results in a strong dependence between the temperature swings affecting the module and the number of temperature changes until failure occurs [5].

Fig. 6: Fatigue of a bonded interconnection

The characteristic of the curves suggests the use of a LCU concept, which detects in a first approach temperature changes. The mechanical construction and the thermal management have to guarantee a small temperature swing.

Notice that the data are statistical and a logarithmic scale is used. This means that an individual product could react with great differences. The approach is that each μLCU collects data and transfers the data to a collective system. The collective system makes the additional statistical calculation and feedbacks the perhaps necessary adaptation to the individual μLCU [16]. The exemplary result of the calculation of one single μLCU is shown in Figure 7. The temperature classes are transformed to a "Wöhlerkurve"[1].

For the different temperature classes, an attainable lifetime is predefined in the program. To reduce the computing expenditure for the calculation, different interpolation routines are available. In this application, the μLCU collects the temperature ranges and calculates the cumulative lifetime. A marker is set if the cumulative lifetime reaches, e.g. 70 % of the attainable lifetime.

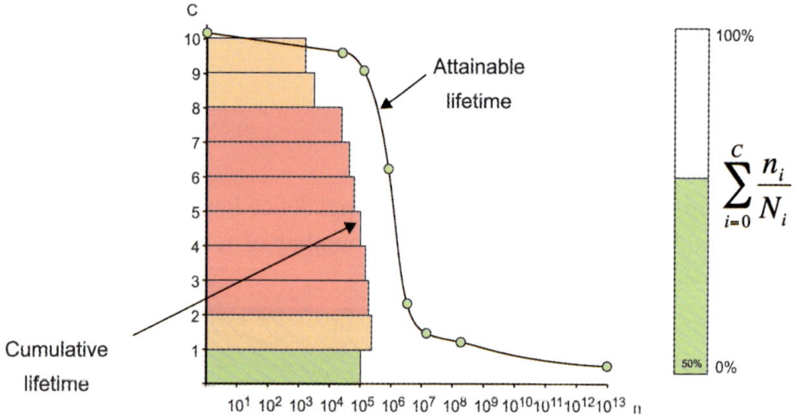

Fig. 7: Calculation of the μLCU

Additional temperature dependent failures are the crack of the solder interconnection between the DCB and the chip and intrinsic shorts or interruptions of the power semiconductor. In principle for both failure mechanisms models are available and needed for the holistic approach to give support on the lifetime estimation of the complete module.

[1] In fatigue testing, a specimen is subjected to periodically varying constant-amplitude stresses by means of mechanical or magnetic devices. The diagram plotted is called the stress-cycle diagram or S-N diagram. The stress-cycle diagram is often called ''Wöhlerkurve'' in German.

The failure accumulation model basis in this case is the Palmgren-Miner hypothesis. If the indicator column on the right side in Figure 7 reaches the 70 – 100 % level the failure probability of the system is increasing significantly. A transformation from temperature changes to shear stresses has to be defined in the experiments. This transformation could be improved, if experiences with field data respectively results from other μLCUs make it necessary.

References

1 Middendorf A. et al.: Reliability of Electronic Systems as an Indispensable Condition for Sustainable Development. The World of Electronic Packaging and System Integration, ddp goldenbogen 2004, pp. 367-374.
2 Eco-efficiency – Creating More Value with Less Impact. World Business Council for Sustainable Development, http: www.wbcsd.org (Access: October 2000).
3 Fraunhofer IZM: Berlin – Marketing – MatchX, eGrain, CoC.
4 Middendorf, A., et al.: Embedded Life-Cycle Information Module for Monitoring and Identification of Product Use Conditions. In: Proceedings of the EcoDesign 2003, Tokyo, Japan, December 8-11, 2003.
5 Poech, M.-H. et al.: Investigations on the Damage Mechanism of Aluminum Wire Bonds Used for High-power Applications. In: Proceedings of the EUPAC, 1996, DVS 173, pp. 128-131.

7.5 Modularity for Ease of Remanufacturing

Günther Seliger, Marco Zettl, Berlin, Germany

Increasing the use productivity of resources is a major approach in the framework of sustainability in manufacturing. Physical products are means of satisfying human needs. For example, a mobile phone is a mean to satisfy the need for communication, and an automobile for mobility. The design of products has a significant impact on resource consumption along the product life cycle (Figure 1).

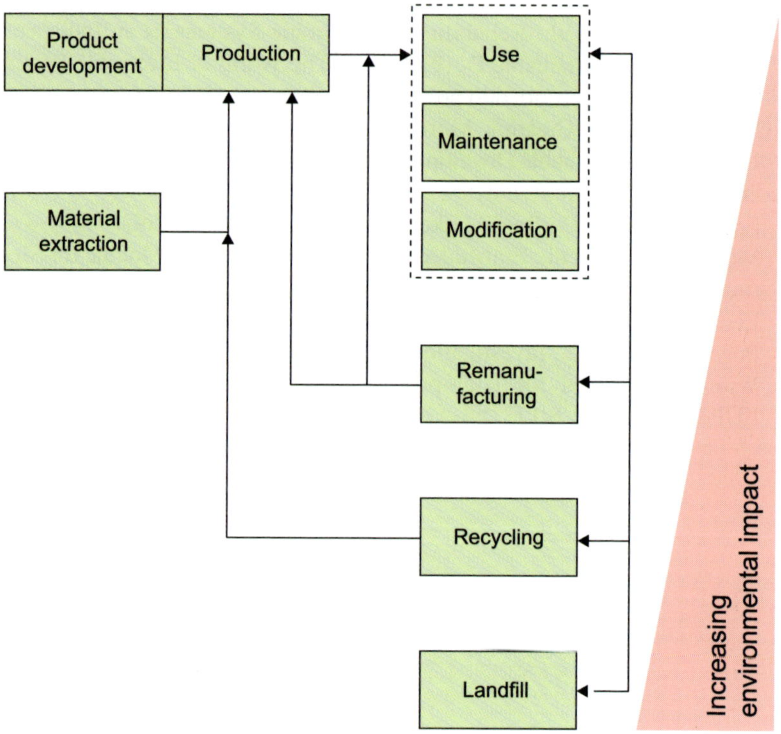

Fig. 1: Life cycle phases of a product based on [1]

As already discussed expanding the use phase of products by maintenance [2] and modification is a significant contribution to increase the use productivity of resources. Remanufacturing between use phases, allowing the reuse of products, modules, and components in multiple use phases is also a significant contribution to achieve this goal. NASR defines remanufacturing as reviving a product to a like-new condition in terms of performance and durability [3]. Remanufacturing can also be interpreted as the adaptation of a used product to a new use phase characterized by changed physical and functional requirements [4]. Remanufacturing processes are sorting, cleaning, testing, non-destructive disassembly, and reassembly. With reference to the product aspect remanufacturing in this book is often addressed also as adaptation. Thereby kinds of adaptation are repair, up- and downgrading, enlargement and reduction as well as rearrangement and modernization [1]. However, current product designs often do not support efficient and thus profitable maintenance, modification, and remanufacturing processes.

Modularity has been identified as an enabler to improve product design regarding to maintenance, modification, and remanufacturing. This has been demonstrated by the realized disassembly tool kit [5, 6]. Further advantages of modularity can be identified along the product life cycle through the economy of scale. Therefore various criteria along the life cycle have to be considered for the development of modular product architectures. The importance of these criteria depends on, e.g. the business model, product class, and reuse possibility after a use phase. A modularization methodology supporting the development of use-oriented modules has been realized prototypically as a web-based module configurator. Use-oriented modules are incorporating at least one function and are optimized for the reuse in its original and in other applications. The procedure of applying the modularization methodology of the module configurator is described and discussed exemplarily in the case study "Development of a Modular Mobile Phone Kit".

7.5.1 Remanufacturing

Currently, products are disposed and recycled at the end of their use phase. The end of a use phase is reached if a product does no longer satisfy customer demands. Physical changes, e.g. aging, and deformation, as well as functional changes, e.g. change of fashion, laws, and technology, are causing the end of a use phase. Related to the component level these kinds of wear appear in different quantities and values. Consequently, at the end of a use phase not all components of a product are worn [7].

Closing the loop to transform the cradle-to-grave economy in a cradle-to-cradle economy is a significant contribution to increase the use productivity of resources. Remanufacturing as a mean for allowing multiple use phases is an important circle element from the ecological point of view. A recent study comparing remanufacturing, recycling, and land-filling for two life cycles of a mobile phone is indicating that major savings on energy consumption and green house gas emission can be estimated by remanufacturing. Nevertheless, remanufacturing can also be a profitable business field. In fact, companies in Europe and North America are already making significant profits by selling remanufactured products and components. However, the potential of remanufacturing is not fully exploited yet. E.g. complex and manual processes, various product models, high spare-part costs, quality problems as well as technological and stylistic obsolescence are making remanufacturing of many products often unfeasible and unprofitable. An analysis of cost structures in the remanufacturing industry revealed the major cost drivers: acquisition of cosmetic parts, non-destructive disassembly and reassembly, and manual testing of functionality [7, 8].

Modular designed products support an increased efficiency and thus profitability of remanufacturing by non-destructive disassembly of worn modules and simple reassembly of repaired or new modules. This assumption is illustrated exemplarily on a modular product, a Personal Computer (PC) and on a non-modular product, a mobile phone. The Flection GmbH, a European-wide acting company is making significant profits by remanufacturing and recycling of Information Technology (IT) hardware, in particular PCs [8]. The remanufacturing process comprises manual sorting and identification, manual cleaning, testing, certified data deletion, manual non-destructive disassembly, adaptation, reassembly, and packaging. The applied kinds of adaptation are in particular up- and downgrading, enlargement, and reduction. The adapted PCs are sold in markets of industrialized and emerging countries in Europe and North Africa. Modular design and standardized modules, e.g. graphic cards and hard drives, contribute to a significant cost reduction of remanufacturing.

Recellular Inc., a US American company is making profits by remarketing mobile phones mainly in low-income countries, e.g. in South America and Africa. Hereby, the activities are predominantly limited to refurbishment. The applied processes are manual sorting, cleaning, data deletion and firmware upgrade, and packaging. Disassembly, reassembly, and adaptation are carried out for only 10% of the mobile phones. The applied kinds of adaptation are the repair of electronic components and the modernization of housing components. Consequently, the remanufacturing of many mobile phones with, e.g. broken housing components, is not profitable. These mobile phones are usually disposed and recycled. Especially the low grade of modularity and standardization contributes to less profitability in remanufacturing of mobile phones. However, it has been shown that hybrid non-destructive disassembly of mobile phones is feasible and contributes to cost reduction [6, 7, 8, 10]. Further cost reductions are possible by improving the mobile phone design as discussed in the case study "Modular Mobile Phone Kit".

Moreover, modular designed products, e.g. a PC, can also be easily maintained and modified during the use phase contributing to expand the use phase. Many customers use this opportunity to save costs instead of purchasing a new product. The effort and thus the costs for maintenance and modification of products with a low grade of modularity, e.g. mobile phone, are high. As a consequence, in Germany mobile phones are used 18 months in average [7].

Modularity is a powerful enabler to increase the use productivity of resources by expanding a use phase and allowing multiple use phases. In order to increase the reuse potential of modules in their original and other applications, the composition of functions incorporated in modules have to be optimized.

7.5.2 Modularity

In a common understanding, modularity is an approach, which supports the handling of complex systems by structuring them into several independent sub-systems [11, 12, 13, 14]. Complex systems include products and technical systems as well as processes and organizations. The grade of the product modularity increases by structuring a system in a set of sub-systems or so-called modules [15]. In context of technical products modules are assembly groups which are from a functional, logistical, and production technological point of view considered as sufficient units [16]. The effort for a pre-assembly of assembly groups can be much higher than the assembly of modules [17].

Product architectures are called modular in case they are composed of functional and physical independent assembly groups thus modules [18]. Functional independency is given if a module fulfills its designated function autonomously not limited to specific products or product variants. Modules interact with other modules by exchanging energy, signal and material flows by means of module interfaces. Physical independency is given if the module interfaces allow ease of assembly, non-destructive disassembly, and reassembly.

Kinds of Modularity

Kinds of modularity describe how modular product architectures can be structured. Kusiak and Piller identified six kinds of modularity: Component Swapping Modularity [19], Bus Modularity [14], Individual Modularity [20], Fabricate to Fit Modularity [20], Component Sharing Modularity [19], and Open Modularity [14]. These kinds of modularity have in common that a set of standardized modules, often manufactured with different properties, are assembled in different products or product variants to fulfill their designated function. The modules are equipped with standardized module interfaces, supporting ease of disassembly and (re)assembly. They are assembled mostly on a core module or so-called product platform. A product platform comprises as a set of standardized functional carriers forming a joint structure on which different product models and product classes can be developed and produced [12, 21, 22, 23, 24]. Hereby, functional carriers are defined as components or assembly groups fulfilling a specific function. Moreover, platforms can also be based on the application of standardized processes and technologies, core competencies and knowledge, and fixed organizational and logistical operations [15, 20, 25]. The functional carriers assembled on a platform have small or no influence on properties directly perceived by the customer. They are not responsible for the product's brand (Figure 2).

Fig. 2: Audi, Bentley and Volkswagen automobiles based on the platform PL62 of Volkswagen [26, 27, 28, 29]

Module Interfaces

The physical independency of modules is achieved due to a disassembly and (re)assembly suitable design of module interfaces. Module interfaces can be defined as technical interfaces describing the relations between modules [30]. According to Pahl [13], possible relations of interfaces are limited to an energy, material, and signal flow. Pimmler [31] adds geometric relations.

(Re)assembly and disassembly suitable designed module interfaces allow a postponement of the final assembly step and increase the efficiency of manufacturing, maintenance, modification, and remanufacturing. Module interfaces should be standardized [18] to allow an efficient adaptability of a product, e.g. upgrade and enlargement, as well as a module reuse in different products, product variants and generations. In order to assure the functionality of module interfaces the early definition in the design process is crucial. Module interfaces have to be designed simple and in a way that they safely allow the transfer of energy, signal, and material flows [32]. Thereby, the disassembly and reassembly frequency has to be considered.

Potentials and Risks

Companies implementing the concept of modularity aim on meeting diversified customer requirements and product costs, both main elements of Mass Customization [33]. Product variations can be realized by offering products with varying modules in amount and characteristics. The lower product costs result from lower development costs and costs for manufacturing due to bigger lot sizes and economies of scale in purchase [34]. In addition the reduction of product and process complexity contributes to further cost savings. As discussed before, modularity also contributes to the establishment of profitable business in service and remanufacturing.

Nevertheless, modularity implies also economical and ecological risks along the product life cycle. Economical risks are in particular high development and changing costs of modular product architectures, low product differentiation, and increased competition and imitation. Environmental risks are an increased material and energy consumption thus increased resource consumption during production, maintenance, modification, remanufacturing, and recycling. Reasons for increased resource consumption are in particular: additional components for the housing of modules and for complex module interfaces supporting non-destructive disassembly and reassembly, frequently replacing and disposing modules because of failure and technological fresh up, and disposing worn and not worn functionality incorporated in same modules.

7.5.3 Module Configurator

Reasons for forming modular product architectures by grouping functional carriers to modules are identified along the product life cycle. These reasons are also called module drivers [11]. The complexity of developing use-oriented modules is increasing with the number of functional carriers of a product and thus with the number of all feasible modular product architecture combinations as well as with the number of relevant modularization criteria. Thereby, the possible compositions of modules have varying ecological and economical impacts. Consequently, the designer has to be supported by a software tool to handle this complexity.

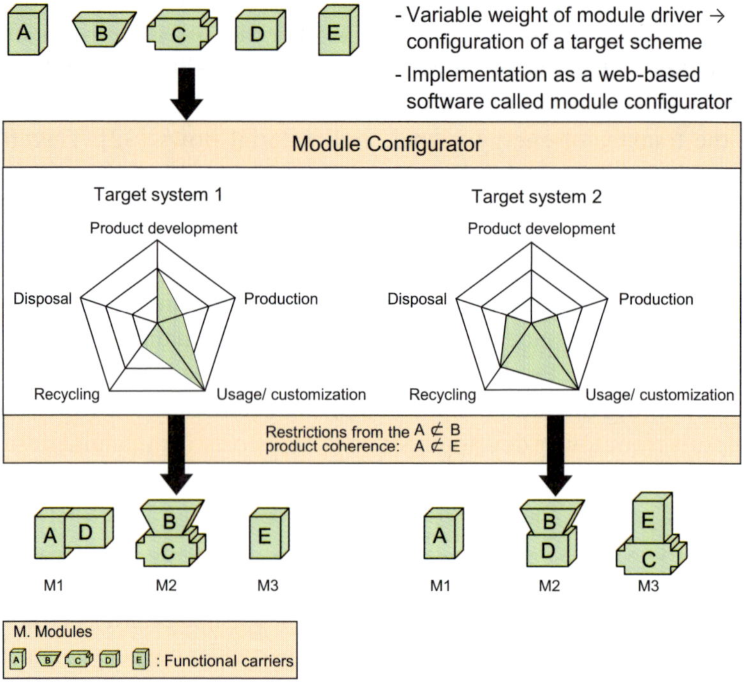

Fig. 3: Module configured by life-cycle oriented module drivers

A multi-criterion modularization methodology based on module drivers has been developed and prototypical implemented as the module configurator. The principal idea of the module configurator is based on the evaluation of module configurations by module drivers and the definition of a target scheme by a variable weighting of module drivers which are assigned to the product life cycle phases. Changing the target scheme effects a change of the module configuration. That means modular product architectures can be optimized to ease remanufacturing after the first use phase as well as to ease production (Figure 3). The steps of the development are described in the following.

Analysis of Existing Methodologies

Some existing methodologies are focusing on the evaluation of relations between functional carriers, e.g. energy and material flow, such as METUS from Göpfert [18], the heuristic approach from Stone, Wood and Crawford [35] and the Design Structure Matrix from Pimmler and Eppinger [31]. Modular product architectures developed with the help of these methodologies are characterized by strong dependencies of parts and components within modules and weak dependencies between modules. The advantage of this approach is that based on the products' function structure, composed of functions, sub-functions and energy, material, and signal flows, the modular product architecture can be defined quickly. Due to weak dependencies between modules the complexity of module interfaces is low compared to the interfaces within a module. However, these kinds of methodologies do not lead to an optimal solution since the life cycle phases of a product are not considered.

The methodology from Kusiak and Huang offers the possibility to group functional carriers to modules considering aspects of production [19]. This methodology points in the right direction but is focused on production.

Erixon uses module drivers, which are partly defined along the product life cycle, and the Module Indication Matrix (MIM) to identify modular product architectures [11]. The approach of module drivers is appropriate for the development of the module configurator. Gu presents an integrated modular design methodology for life cycle engineering. Modular product architectures are randomly generated and evaluated according to life-cycle criteria to find a good solution. The implemented algorithm is fast but the result may not be the optimal solution [36].

Development of the Module Configurator

The principal idea from Gu of generating and evaluating modular product architectures using a mathematical algorithm as well as the module driver approach from Erixon form the basis for the development of the module configurator. However, in order to find the optimal solution, the benefit of all feasible modular product architectures shall be evaluated. The module drivers are identified by analyzing and comparing modular and non-modular products according to the life cycle phases. Hereby, potentials and risks of modularity are considered. In addition relevant criteria and module drivers of the analyzed methodologies are used to complement the list of module drivers as illustrated in Table 1.

The module driver Time-to-market is aiming on the optimization of the product development process to reduce the time to market. The specification of the module driver is the Development Time, which is the time span from the idea to the start of mass production of a functional carrier. If two or more functional carriers have a similar development time then the benefit of grouping them to modules is high.

The second module driver Assembly/ Configuration is aiming on the reduction of the module interface complexity. The specification of this module driver is the energy, material, and information flow intensity. Module interfaces with high flow intensity are more complex, which affects the development time and the effort in production. Therefore, the benefit of combining functional carriers with high flow intensity to a module is high.

Tab. 1: List of module drivers

Product life cycle phase	No.	Module driver	Specification for mathematical description
Product development	1	Time-to-market	Development time
Product development and production	2	Assembly/ Configuration (complexity of module interfaces)	Energy flow intensity
			Material flow intensity
			Signal flow intensity
	3	Core compentences and supply chain	Manufacturing location
Production, use and remanufacturing	4	Reuse of functional carrier	Other product class
			Other product model
			Other product generation (carry over)
Use and remanufacturing	5	Use of functional carrier	Use ratio
	6	Product innovation	Innovation cycle
	7	Maintenance	Inspection cycle
			Maintenance cycle
			Repair cycle
	8	Adaption/ Modification	Substitute function
Remanufacturing	9	Remanufacturing after the first use phase	Remanufacturing possibility
Recycling	10	Recycling after the first use phase	Recycling possibility
Landfilling	11	Landfilling after the first use phase	Landfilling possibility

The module driver Core Competences and Supply Chain is aiming on grouping functional carriers to modules, which are or can be manufactured by the same supplier. This module driver is affecting the product development and production phase. Combinations of functional carriers, which can be purchased or manufactured by the same supplier, are evaluated with a high benefit.

The fourth module driver Reuse of Functional Carrier is aiming on the development of standardized modules and shall optimize the life cycle phases production, use and remanufacturing. Standardized modules reduce the effort of production and can be reused in different product models, classes, and generations. Maintenance, modification, and remanufacturing are optimized by reducing costs for spare parts, disassembly and reassembly processes. Consequently, the grouping of functional carriers with the same specification contains a high benefit.

The module driver Use of Functional Carrier is aiming on the optimization of the use phase by grouping functional carriers to modules which have a similar use ratio. E.g. rarely used modules can be detached from the product to save energy. The benefit of combining functional carriers with a similar use ratio is high.

The sixth module driver Product Innovation is assigned to the life cycles use, modification and remanufacturing in particular. The specification of this module driver is the innovation cycle. That means, if two functional carriers have a similar innovation cycle than the grouping to modules has a high benefit in product development, use, and remanufacturing.

The module driver Maintenance is aiming on the optimization of the use and remanufacturing phase. The specifications of this module driver are inspection, maintenance, and repair cycles. The grouping of functional carriers with similar specifications has a high benefit since the modules can be easily tested, maintained, changed, and repaired.

The eighth module driver Adaptation/ Modification is also assigned to the life cycle phases use and remanufacturing. Hereby, the modification and remanufacturing is optimized by grouping functional carriers to modules, which are planned to be substituted by another functionality during and after the use phase. Thus combinations of functional carriers, which are substituted to a similar date, have a high benefit.

The module drivers Remanufacturing, Recycling, and Landfilling After the First Use Phase are aiming on the optimization of the product's treatment after the first use phase. Combinations of functional carriers to modules with the same designated purposes after the first use phase are evaluated with a high benefit. This module driver contributes to efficient non-destructive disassembly of modules and fosters the reuse of modules.

As mentioned before, the described module drivers shall be used to evaluate all feasible modular product architectures. The number of modular product architectures is increasing with a growing number of functional carriers of a product. This number is reduced by defining two constraints. The constraint Predefinition of Functional Carrier Combinations is aiming on excluding combinations of functional carriers to modules thus limiting the number of possible modular product architectures. The constraint Pre-clustering of Functional Carriers is aiming on the grouping of functional carriers according to standardized, market specific, and additional functionality. The configuration of modules is carried out within these groups.

The generation and evaluation of all feasible modular product architectures is supported by the developed module configurator. A C++ program generates all feasible modular product architectures and formulates the Integer Linear Program (ILP). The ILP is solved by a CPLEX solver. The product related data and the weighting of the module drivers are entered in a MySQL data base via a website based on HTML and PHP. The result, the optimal modular product architecture, is stored on the database and is illustrated graphically on the website (Figure 4).

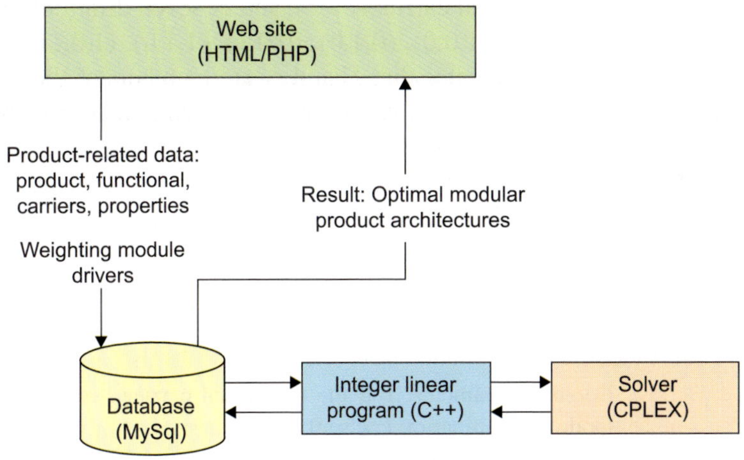

Fig. 4: Principal structure of the module configurator

The ILP is a special form of a Linear Program (LP), which is a main part of operations research. Thereby, an optimal solution for a linear objective function is calculated. The value of the objective function is restricted to equations and inequations, so-called constraints of the LP. LPs are strong tools and do not require a specific development of an algorithm. In the present case the variables have to be integers. Therefore an ILP was chosen.

A product or sub-group is made of i different functional carriers. Each functional carrier can only be assembled in one module. Some of the functional carriers are not allowed to be in the same module. So these combinations should be excluded.

In the following the ILP with indices and sets, parameters, decision variables, the objective function, and constraints are described.

Short Cut	***Explanation***
$i \in I$	Set of functional carriers
$k \in K$	Set of module combinations $\lvert K\rvert=2^I-1$
$i \in I_K$	Set of functional carriers in module combination k
$n_k=\lvert I_k\rvert$	Number of functional carriers in module combination k
$s \in S$	Set of module driver
$(c, d) \in B$	Set of incompatible functional carriers ($c, d \in I$)

Each combination of functional carriers i to a module is described as a vector, in which the component i indicates whether the functional carrier is a part of the respective module (if component i is 1) or not (if component i is 0). E.g., for three functional carriers c_1, c_2, and c_3 seven different module combinations are possible (1,0,0), (0,1,0), (0,0,1), (1,1,0), (1,0,1), (0,1,1), (1,1,1), whereas (1,1,0) indicates that c_1 and c_2 are members of the module and c_3 not. Consequently $\lvert K\rvert=2^I-1$ different module combinations are possible.

Parameters:

Short Cut	***Explanation***
a_{ik}	Boolean parameter: 1 if the functional carrier i is contained in module combination k, else 0
c_{is}	Boolean parameter: 1 if the functional carrier fulfils the function of the module driver s, else 0
E_i	Number of input spots for functional carrier i
F_k	Number of input spots for module combination k
M_{is}	Set of functional carriers i, which have a specific attribute for module driver s
N_{sk}	Benefit of module combination k for module s
N_{si}	Benefit of functional carrier i for module s
O_i	Number of output spots for functional carrier i

P_k	Number of output spots for module combination *k*
W_s	Weight of module *s*
a_{ik}	Boolean parameter: 1 if functional carrier *i* is contained in module combination *k*, else 0

Decision variables:

Short Cut	***Explanation***
X_k	Binary variable: 1 if module combination *k* is chosen, else 0.

Objective function:

$$max \sum_S \sum_K W_s * N_{sk} * X_k$$

The value of the objective function is maximal for the optimal modular product architecture. The benefit of the module combinations N_{sk} and the module drivers weight W_s are parameters and X_k is the decision variable, which determines the chosen module combination. Consequently, the optimal modular product structure depends on the weighting of the module drivers, because the benefit of the module combinations is based on the product related data. The product related data is the input for the optimization problem.

Definition of the mathematical constraints for the ILP:

$$\sum_K \left(a_{ik} * X_k\right) = 1 \ \ \forall i \tag{1}$$

Each functional carrier has to be in one module combination. Otherwise this constraint is violated.

$$\sum_S W_s = 1$$

The sum of all module driver weights W_s is one.

$$\sum_B a_{ck} * X_k + a_{dk} * X_k \leq 1 \qquad \forall (c,d) \in B \tag{2}$$

Two incompatible functional carriers *(c, d)* have to be in different modules.

In the following the mathematical terms of the module drivers are described. Every module driver can be put in one of five classes (S_1, S_2, S_3, S_4, and S_5) and for each class the benefit N_{sk} is calculated differently.

$$N_{sk} = \left(1 - \frac{\max\limits_{i \in I_k}(N_{si}) - \min\limits_{i \in I_k}(N_{si})}{\max\limits_{i \in I}(N_{si}) - \min\limits_{i \in I}(N_{si})} \right) * \frac{b_k - 1}{n_k} \qquad \forall s \in S_1; k \in K \tag{1}$$

In the first class the benefit is calculated by multiplication of two terms. The first term calculates the difference between the maximum and minimum benefit of a functional carrier *i* for a module driver *s* and correlates it to the maximum overall difference. For example, the module driver *s* is defined as product innovation. In the numerator the difference of the longest and shortest innovation cycle of all functional carriers *i* in one specific module combination *k* is calculated. In the denominator the greatest possible difference of all functional carriers *i* is calculated. The resulting fraction correlates the time intervals. The second term evaluates the number of functional carriers in a module combination and awards a greater number. If there is only one functional carrier in a module combination then the benefit is *0*. Members of this class are the module drivers: product innovation, time-to-market, and maintenance.

$$N_{sk} = \begin{Bmatrix} 1, \text{ if } N_{si} = N_{sj} \forall i, j \in I_k; i \neq j \\ 0, \text{ else} \end{Bmatrix} * \frac{b_k - 1}{n_k} \qquad \forall s \in S_2; k \in K \tag{2}$$

The first term is *1*, if all functional carriers in module combination *k* fulfill the condition of module driver *s,* else it is *0*. For example, *s* is the module driver recycling after the first use phase. If the functional carrier *i* shall be recycled after the use phase then the benefit of the functional carrier N_{si} is *1*, else it is *0*. And if the benefit of all functional carriers N_{si} in module combination *k* is equal then the benefit N_{sk} of the module combination *k* is *1*, else it is *0*. The second term has got the same influence as in class one. Members of this class are the module drivers: Modification and remanufacturing, recycling, and land-filling after the first use phase. The constraint pre-clustering of functional carriers is member of this class as well.

$$N_{sk} = \begin{cases} 1, \text{ if } \; 0 < \dfrac{F_k + P_k}{\sum\limits_{i \in I_k} (E_i + O_i)} \leq 1 \\ 0, \text{ else} \end{cases} \quad \forall s \in S_2; k \in K \tag{3}$$

In the third class the benefit is calculated in one term. If the quotient of the sum of input and output spots after the modularization and the sum of input and output spots before the modularization is in range of 0 and 1, then the benefit N_{sk} is 1, else it is 0. Member of this class is the module driver assembly/ configuration.

$$N_{sk} = \begin{cases} 1, \text{ if } \; \bigcap\limits_{i=1}^{I_k} M_{is} \neq \varnothing \\ 0, \text{ else} \end{cases} * \frac{b_k - 1}{n} \quad \forall s \in S_4; k \in K \tag{4}$$

The first term of the benefit is *1* if the intersection of all functional carriers, which are in module combination k, is not empty, else it is *0*. For example, the set M_{is} are all locations, in which the functional carrier *i* can be manufactured. If the intersection of all functional carriers, which are in module combination *k* is not empty, then the benefit is *1*, else it is *0*. Members of this class are the module drivers: Core competences and supply chain, and reuse of functional carrier.

$$N_{sk} = \begin{cases} 1, \text{ if } \; c_{ij} = 1 \forall i, j \in I_k; i \neq j \\ -\infty, \text{ else} \end{cases} \quad \forall s \in S_5; k \in K \tag{5}$$

In the last class the benefit is *1*, if all functional carriers in module combination *k* are compatible, else it is $-\infty$. In this way it will be prevented that a module combination will be chosen, which has got two incompatible functional carriers, because the benefit is $-\infty$. Member of this class is the constraint pre-definition of functional carrier combinations.

Modularization Procedure with Module Configurator

In the following the handling of the modularization process supported by the module configurator is described. According to Pahl and Beitz [13] and to VDI 2221 [37] the modularization of products is located in the embodiment design phase.

In the first step the complexity of the modularization task may be reduced by clustering a complex product, e.g. automobile, in appropriate sub-groups including 20 to 30 functional carriers.

Afterwards, the product-related data have to be entered in the MySQL database of the module configurator. The product-related data are characterized by: the identifier of a functional carrier and the relations and properties according to the selected module driver specification. The user of the web-based software is guided by an intuitive website layout. Prerequisite for this step is the developed function structure and principal solution as well as the knowledge about the properties of the functional carriers.

In a third step the constraints of the product structure have to be set. Hereby, not reasonable combinations of functional carriers can be eliminated. Moreover a pre-grouping of functional carriers can be carried out in order to reduce the complexity of the optimization problem and thus the calculation time. This constraint is reasonable in case the product structure exceeds 20 functional carriers.

In a fourth step the module drivers and specifications are weighted according to their importance for the respective optimization problem. The weighting is mainly influenced by strategic decisions made in the product planning phase.

In the fifth step, the calculation of the optimal modular product structures is carried out. The modular product architecture with the highest benefit for all possible sets of modules is illustrated in Figure 5. The sets of modules are characterized by a different number of modules.

Finally, the user can choose the most appropriate modular product structure suggested by the software. The result depends very much on the weighting of the module drivers as well as on the product related data. Consequently, step two and four have to be carried out carefully.

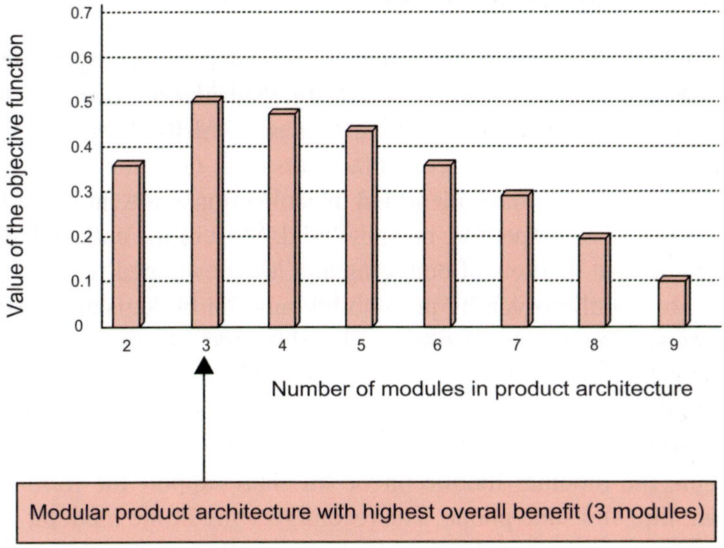

Fig. 5: Illustration of results on website

Limitations

The developed module configurator is capable to calculate the modular product architecture with the highest benefit for a pre-defined fixed number of modules. The optimal modular product architecture for every possible number of modules will be calculated and visualized. Limitations of the current module configurator version are the following:

- Depending on the amount of selected module drivers, many product specific data have to be entered in the database. However, this step can be skipped, if the functionality is improved with a software interface allowing the import of data from other applications, e.g. a product data management system.
- For the calculation of the optimal modular product architecture the module configurator is limited to products or sub-groups with a total number of 20 to 30 functional carriers. However the pre-clustering of functional carriers is a powerful mean to reduce the calculation time.

7.5.4 Case Study – Modular Mobile Phone Kit

Background of the case study is the dramatically increasing number of mobile phone users worldwide, with a estimated number over 1.3 billion in 2003 [38], combined with a short use phase, which is in average 18 to 24 months [39]. Third party remanufacturers, e.g. Greener Solutions and Shields Environmental, have identified the collection and treatment of used mobile phones as a competitive business field. Most of the collected mobile phones are sold in markets of emerging and less developed regions. These activities are mainly driven by profitability and not by legislation, such as the European directive on Waste of Electrical and Electronic Equipment (WEEE) [40].

Goal of the case study is the development of a modular mobile phone kit suitable for efficient maintenance, modification, and remanufacturing. In addition, the modular mobile phone kit shall support the form-factors candy-bar, flip, and slide phone. The reconfiguration of the modular mobile phone shall be carried out by simple disassembly and reassembly operations. The case study modular mobile phone kit was an integrated part of an industry project in cooperation with the Original Equipment Manufacturer (OEM) Motorola Germany.

The development of the modular mobile phone was carried out according to the design methodology from Pahl and Beitz [13]. Firstly, the product structure and the functional properties of mobile phones with the form factors candy-bar, flip and slide from one OEM have been analyzed. Weak points of the existing mobile phone design have been identified with respect to the goal of the case study. Based on this analysis product requirements have been formulated. Further product requirements of customers, the OEM and net providers have been gathered and grouped respectively. Moreover, scenario management has been carried out to identify future robust product requirements for the mobile phone design. On the basis of the complete list of requirements, a functional product structure and a principal solution for a modular mobile phone have been developed.

The development of the modular mobile phone kit has been carried out in two steps. First, the modular architecture of the mobile phone has been developed without considering the form-factors. This is reasonable, since mobile phones consist of similar functional carriers. Combinations of functional carriers were excluded and pre-grouped according to standard, market-specific, and optional functionality. The groupings of functional carriers have been carried out within these groups supported by selected weighted module drivers. The selected module drivers are:

- reuse of functional carrier,
- product innovation,
- core competencies and supply chain, and
- maintenance.

The module drivers Reuse of Functional Carrier and Product Innovation gained the highest weight. The first step is finished once the best modular mobile phone architecture with the respective module configurations has been selected. In a second step, the most appropriate arrangement of the configured modules has been identified. Hereby, the different form-factors of the mobile phones have been considered. The arrangement of the modules is characterized by a platform, the printed circuit board, and several attached modules [10].

Finally, the best modular product architecture with its modules and interfaces are designed. The Embodiment Design Phase ends with the virtual and physical prototype of the modular mobile phone. The virtual prototype is illustrated in Figure 6.

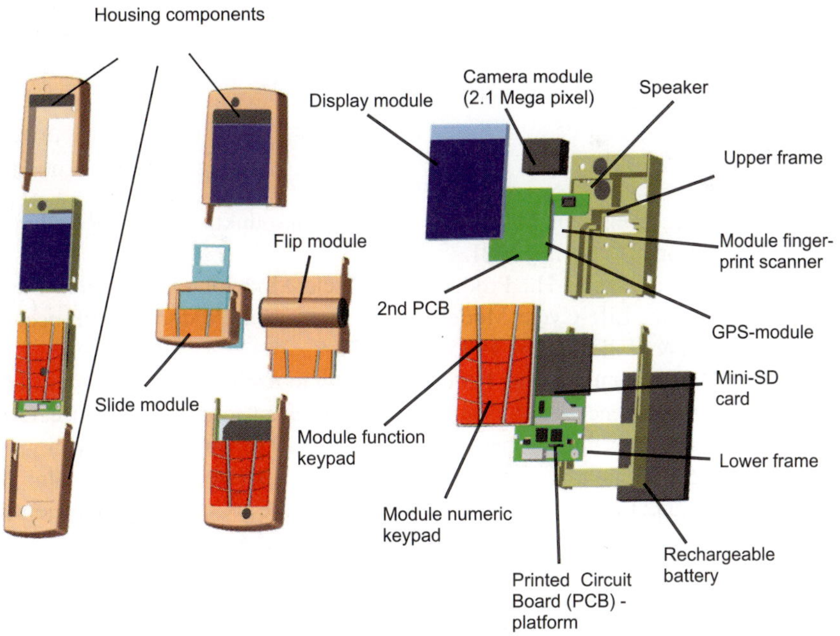

Fig. 6: Virtual prototype of the modular mobile phone kit [9]

The case study was carried out in parallel to the development of the module configurator. Consequently, this study had a significant impact on the development of constraints, module drivers, and the evaluation methodology of the module configurator.

Perspectives in the application of the module configurator are the development of manufacturing equipment, supporting efficient adaptation to changing functional requirements in multiple use phases. Manufacturing equipment in industrialized countries may have other functional requirements on accuracy, productivity, grade of automation, and maintainability, than for use in emerging and developing countries. Reasons are different conditions related to the skills of employees to operate the manufacturing equipment, the availability and the affordability of spare parts and auxiliary supplies, and the production volume. Not only in manufacturing but also in health care, mobility, agriculture, infrastructure for information and communication modular design might provide ease of access for communities in less developed regions with limited buying power. Thus modular design could contribute to wealth generation, better standard of living, development of skills, education, and entrepreneurial competencies. In this sense, modularity is an engineering tool for increasing the equity factor in global wealth distribution.

References

1 Müller, K.: Wege zur Steigerung der Nutzenproduktivität von Ressourcen. Dissertation, TU Berlin, 2001.

2 Kimura, F.; Suzuki, H.: Product Life Cycle Modelling for Inverse Manufacturing. In: Life-Cycle Modelling for Innovative Products and Processes. Chapman & Hall, London 1995, pp. 80-89.

3 Nasr, N.: Remanufacturing from Technology to Applications. In: Proceedings Global Conference on Sustainable Product Development and Life Cycle Engineering, Berlin, Germany, 29.09. – 01.10.2004, pp. 25-28.

4 Takata, S.; Kimura, F.; van Houten, F.J.A.M; Westkämper, E.; Shpitalni, M.; Ceglarek, D.; Lee, J.: Maintenance: Changing Role in Life Cycle Management. In: Annals of the CIRP, Vol. 53/2/2004, pp. 643-655.

5 Rebafka, U.: Beitrag zur Entwicklung modularer Demontagewerkzeuge. Dissertation TU Berlin, 2003.

6 Seliger, G.; Consiglio, S.; Odry, D.; Zettl, M.: Development of Intelligent Modular Tools for Disassembly. In: Proceedings of 15th International CIRP Design Seminar 2005, Shanghai, China, 22.05. – 26.05.2005, S. 182 – 187 (CD).

7 Basdere, B.: Beitrag zur Steigerung der Nutzenproduktivität von Ressourcen durch Anpassen von Mobiltelefonen. Dissertation, TU Berlin, 2004.

8 Seliger, G.; Skerlos, S.J.; Basdere, B.; Zettl, M.: Design of a Modular Housing Platform for Remanufacturing of Multiple Cellular Phone Models, Eco-Design 2003, Tokio, 2003.
9 http://www.flection.de (Access: October 2006).
10 Seliger, G.; Kernbaum, S.; Zettl, M.: Approaches for Sustainable Manufacturing. In: Proceedings of the 7th International Conference on Frontiers of Design and Manufacturing, June 19th to 22nd 2006, Guangzhou, China.
11 Erixon, G.: Modular Function Deployment – A Method for Product Modularization. Doctoral Thesis, Royal Insititute of Technology, Stockholm 1998.
12 Ulrich, K.; Tung, K.: Fundamentals of Product Modularity. In: Issues in Design Manufacture/Integration, DE-Vol. 39. ASME, MIT Cambridge, Massachusetts, 1991.
13 Pahl, G.; Beitz, W.: Konstruktionslehre – Grundlagen erfolgreicher Produktentwicklung, Methoden und Anwendung. Springer-Verlag, Berlin, Heidelberg, New York 2005.
14 Piller, F.-T.: Kundenindividuelle Massenproduktion. Carl Hanser Verlag, München, Wien 1998.
15 Schmidt, H.: Beitrag zum Variantenmanagement und zur Prozessoptimierung im Wagenkastenbau von Schienenfahrzeugen. Dissertation, TU Berlin, 2002.
16 Weißner, R.: Vernetzte Produktentstehungsprozesse, Tagungsband zum Dresdner Produktionstechnik Kolloquium, Dresden, 1999.
17 Eversheim, W: Komplexität im Unternehmen verringern und beherrschen – Optimale Gestaltung von Produkten und Produktionssystemen. In Adam: Komplexitätsmanagement, Schriften zur Unternehmensführung, Band 61, Wiesbaden, 1998.
18 Göpfert, J.: Modulare Produktentwicklung – Zur gemeinsamen Gestaltung von Technik und Organisation. Gabler Verlag/Deutscher Universitätsverlag, Wiesbaden 1998.
19 Kusiak, A.; Huang, C.-C.: Design of Modular Digital Circuits for Testability. IEEE Transactions on Components, Packaging and Manufacturing – Part C 20 (1997) 1, pp. 48-57.
20 Piller, F. T.; Waringer, D.: Modularisierung in der Automobilindustrie - neue Formen und Prinzipien. Shaker Verlag, Aachen 1999.
21 Meyer, M.H.; Lehnerd, A.P.: The Power of Product Platforms. The Free Press, New York 1997.
22 Robertson, D.; Ulrich, K.: Planning for Product Platforms. Sloan Management Review, Summer 1998, pp.19-31.
23 Baldwin, C. Y.; Clark, K. B.: Modularisierung – Ein Konzept wird universell. Harvard Business Manager 20 (1998) 2, S. 39-48.
24 Otto, K.; Wood, K.L.: Product Design. Preentice Hall, Upper Saddle River, New Jersey 2001.
25 Schmidt, H.; Zimmer, H.; Seliger, G.; Kross, U.: Concepts for Product Platforms – Applying a Successful Concept from Elsewhere to the Construction of Rail Vehicles. RTR (2001) 2/3, pp. 4-11.

26 http://www.audi.de (Access: August 2006).
27 http://www.volkswagen.de (Access: August 2006).
28 http://www.bentleymotors.com (Access: August 2006).
29 http://www.wikipedia.org (Access: August 2006).
30 Sanchez, R.; Mahoney, J.T.: Modularity, Flexibility, and Knowledge Management in Product and Organization Design. In: Research Working Paper Number 95-0121. University of Illinois at Urbana-Champaign, Illinois 1995.
31 Pimmler, T. U.; Eppinger, S. D.: Integration Analysis if Product Decompositions. In: Proceedings of the ASME 6th International Conference on Design Theory und Methodology, Minneapolis, 1994.
32 Blackenfelt, M.: Managing Complexity by Product Modularisation – Balancing the Aspects of Technology and Business During the Design Process. Doctoral Thesis, Royal Insititute of Technology, Stockholm 2001.
33 Tseng, M.: Design for Mass Customization. In: 46th CIRP General Assembly, 26.-31.8.1996 Como, Italy, Annals of the CIRP, Vol.45/1/1996, S. 153-156.
34 Piech, F.: Darstellung der Strategie von Volkswagen. Association of European Automotive Analysts, Meeting, Wolfsburg, 2000.
35 Stone, R. B.; Wood, K. L.; Crawford, R. H.: A Heuistic Method to Identify Modules from a Functional Description of a Product. In: Proceedings of Design Engineering Technical Conferences, Atlanta 1998.
36 Gu, P.; Solale, S.: Product Modularization for Life Cycle Engineering. Robotics and Computer Integrated Manufacturing (1999) 15, pp. 387-401.
37 N.N.: VDI-Richtlinie 2221 – Methodik zum Entwickeln und Konstruieren technischer Systeme und Produkte. Verein Deutscher Ingenieure, Beuth Verlag, Berlin 1993.
38 Marcussen, C.H.: Mobile Phones, WAP and the Internet – The European Market and Usage Rates in a Global Perspective 2000-2003, 2003.
39 Wiechens-Schwake, R.: Vodafone D2 GmbH, Juni 2003.
40 The European Parliament and the Council: Directive on Waste from Electrical and Electronic Equipment (WEEE), Brussels, 2002.

7.6 Competence Management in Education for Sustainability

Günther Seliger, Carsten Reise, Berlin, Germany

For managers in globally operating companies being under permanent pressure of innovation in competitive processes and products, abstract goals of sustainability are often beyond reach in their daily business. Permanently, they have to perform by meeting short term requirements of customers, shareholders, employees, suppliers and societal environment. Universities provide a powerful means of societal innovation by highly motivated students and creative teaching staff for developing creative future scenarios for sustainability and for prototypical implementation. The modern teaching and learning arena is coined by ubiquitous access on knowledge domains provided from continuously improved information and communication tools. Knowledge is a resource without wear but even expanding by application. Consequently, knowledge provided to more than five billion global population presently still without internet access might enable for creating wealth on own initiative.

The once utopian dream of total eradication of poverty from the world has created the powerful implementation of micro financed predominantly women entrepreneurial initiative by 2006 noble price winner Muhammed Yunus for more than two million families in rural Bangladesh [1].

Engineering students from developed countries are already active as agents of change in this environment. New type of courses with engineering student teams in global cooperation developing engineering artefacts are under development [2, 3, 4]. An enormous market potential is hidden beyond the poverty of four fifths of global population. However, due to ecological limitations of our planet, this market potential cannot be exploited based on the presently applied technology of developed countries. Food, clothing, housing, transport and education provided in developed countries' way for global population of presently 6.5 billion people would exceed globally available resources. Consequently, innovative changes in technology as tools for sustainable development in societal, ecological and economical framework are required.

Educational institutions nowadays have the unique opportunity to cope with this challenge. A competence market chain reference model is presented indicating the chances for educational institutions in a global competition on markets for education. Within this framework, sustainability in engineering by the leverage of concerned and motivated creative students can be developed.

7.6.1 Competence Market Chain

Based on a market-oriented approach for professional education starting latest from university studies and continuing in further professional training, the idea of competence management is developed. The market mechanism of demand and supply seems appropriate to understand the cause-effect-relationship between teachers, learners, employers, and business where business comprises not only profit-oriented entrepreneurs but also non-profit and public services. These relationships are described in the competence market chain model (Figure 1).

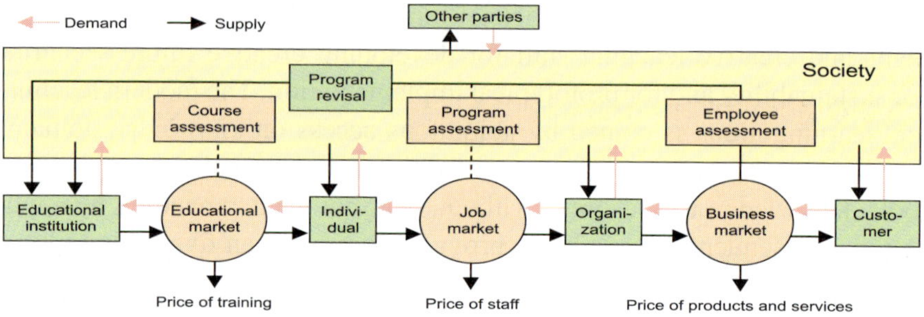

Fig. 1: Competence market chain reference model [5]

The demand from customers and the supply from organizations determine the price of rapidly changing products and services. From this business market, the organizations have to derive their demand for competencies dynamically. This demand is served by appropriately trained individuals. An individual can derive his personal demand for competencies from the labour market in order to arrange a supply of courses on the education market from educational institutions. As suppliers for the next instance and customers of the one before, organizations as well as individuals have to manage their competence portfolio to succeed in their business respectively job enterprise [5].

The competence market chain's key element is a portfolio that represents the potential of an individual for the performance to fulfil relevant functions or tasks. The representations of competencies are the medium to communicate and match demand and supply. Therefore, the development of a competence portfolio for the education as well as the labour market is becoming more fundamental. On the one hand, the taxonomy of the portfolio has to be standardized to fulfil the general interests on the individual mobility and compatibility. In order to make the portfolio comparable between learners, teachers, and employers, a transferable competence description and competence structure is required. The challenge is to find a common description of competencies, directed to the indicated sustainable educational approach. The threat of environmental damage on global living conditions will increase the younger generation's awareness of how to cope with this challenge. Course contents under the condition of the competence market chain will be adapted to the requirements of developing tools for sustainability in engineering. Pilot projects in educational environment can stimulate creative imagination about sustainable products and processes. Current approaches in developing a competence portfolio for paradigm change to sustainability may reveal a sufficient description base to start from. It is better to start with imperfection than not to start because of missing perfection. The competence model can be improved based on in-process experiences [5].

7.6.2 Outcome Orientation

Education in schools, universities as well as institutions for postgraduate professional education in Germany is traditionally input oriented (Figure 2). This means that quality of the graduates is controlled by process parameters i.e. the contents the teacher presents to the students. The Accreditation Board of Engineering and Technology (ABET) changed the accreditation mode of U.S. study programs from input to outcome in their Engineering Criteria 2000. In outcome driven studies, the quality of the graduates is controlled by product properties. One of the criteria mentioned is: "Engineering programs must demonstrate that their graduates have the ability to design a system, component or process to meet desired needs" [6].

The outcome orientation is based on the ability to create artefacts instead to reproduce knowledge. In reference to the agreement signed in 1989 by the Washington accord, a multinational network of accreditation agencies, the outcome orientation will gain more and more importance in next years [7]. Active and communicative forms of study and labour will ascend in engineering education and the contents of teaching by presentation of expert knowledge will decline.

Nowadays, the experience on how to assess initiative and creative abilities of a student instead of his memorized information is very low. Tools and test methods have to be developed to verify and certify competence sets of individuals. The introduction of the outcome-oriented EC 2000 by ABET has spawned much research on this question. If we succeed in certifying competencies in a sufficiently objective way, it will be possible to combine courses from various sources and locations freely. In the extreme, this could mean that there will be hardly home universities but true mobility among acknowledged universities [8].

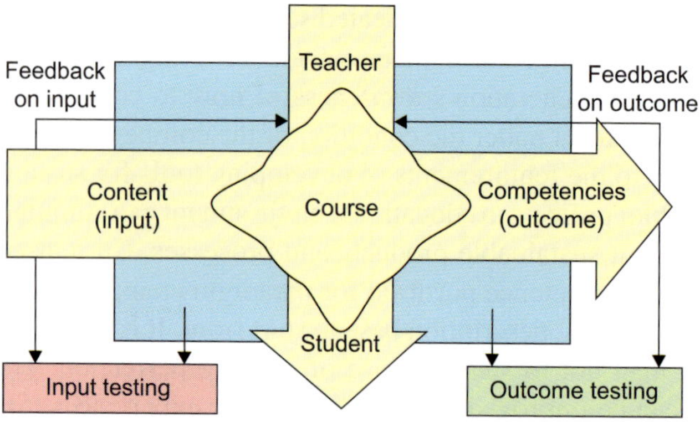

Fig. 2: Input- and outcome-orientation of courses [5]

Competencies differ in how well they can be identified and assessed. The more generic a competence is and the wider it can be applied, the harder it will be to assess, describe, and certify it. Some competencies can hardly be assessed in a rule based system of tests, e.g. personality related competencies.

7.6.3 Reference Model

A reference model for integrative competence management, which is based on markets and competition between educational institutions, students and employing industry has been presented (Figure 3).

The competence portfolio lies in the centre of the reference model with the intention of dynamically specifying and combining competencies. This is subject to the management circle of setting goals, plan, decide, act and control. It is interlinked with the players of the engineering education field, technology based organisations, learning engineers and technology oriented universities, which are connected together within the socio-political framework.

The players are connected by the markets they act on. On the labour market, the competence portfolio is the main criterion to match industrial demand for workforce with the available students. On the educational market, students cover their demand to acquire a competence portfolio that enables them for positioning themselves on the labour market. By applying the mechanisms of free market trade on competence portfolios students become managers of their own competence portfolio.

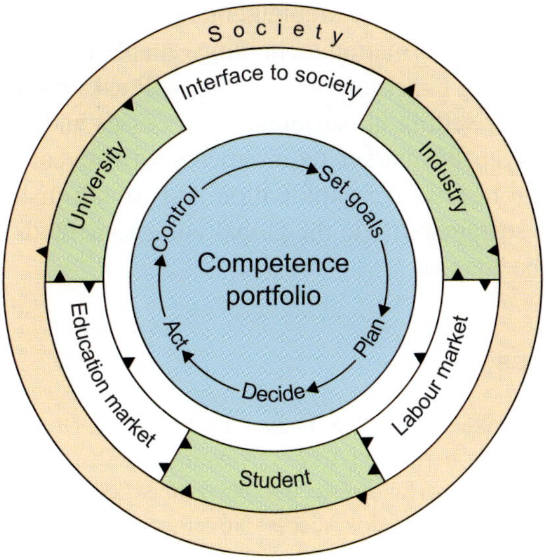

Fig. 3: Reference Model for Integrative Competence Management [9]

The competence management system is built on extensive information flow, to which organizational and analytical information technological tools are essential. The competence portfolio with its objects and relations as well as direction to act can be described as a relational data base consisting of linked tables. Therefore, competencies can be read, deleted or registered in the portfolio. Scenario development can support the planning of objectives and the simulation based decision making for study module, curricula, and human resources [10]. The applicant's study progress or structural changes of an organization can be observed in the competence portfolio. Furthermore, the planning in case of divergence can be adjusted. Experts can facilitate the assortment and conformal-to-standards formulation of competencies, e.g. during the analysis of organizational processes, description of teaching materials or the competence assessment. With the enrolment in study modules, it can be verified with information technology if the participants fulfil the requirements [9].

7.6.4 Conclusion

The transparency of competencies in portfolios is a market driven important step to advance the principle of competition in an educational environment. Changes in the higher education system to outcome orientation and new examination types support the implementation of competence portfolios. Communication and information technology enable for global exchange of teaching and learning experiences. Students will ask for acquiring competencies relevant for setting up sustainable processes and products. Finally, an education driven by market requirements will reveal the global challenges of change in resource exploitation and wealth distribution. Driven by demands for sustainability in the global village, methods and contents of education will be adapted.

References

1 Yunus, M.: Banker to the Poor. University Press Ltd, Dhaka, 1998.
2 http://www.get.tu-berlin.de (Access: November 2006).
3 http://www.gpd.tu-berlin.de (Access: November 2006).
4 http://www.gpe.tu-berlin.de (Access: November 2006).
5 Meyer, M.: A Vision of Sustainable Education with Competence Management. In: Seliger, G.; Nasr, N.; Bras, B.; Alting, L. [editor]: Proceedings of the Global Conference on Sustainable Product Development and Life Cycle Engineering. uni edition GmbH, Berlin 2004, pp. 319-325.
6 The Accreditation Board for Engineering and Technology: Engineering Criteria 2000. 3rd Edition, Baltimore, MD. 1999-2000.
7 Washington accord – A multinational agreement signed in 1989. http://www.washingtonaccord.org (Access: November 2006).
8 Heitmann, G.: Reformen der Ingenieurausbildung in internationaler Perspetive. Arbeitspapier der Netzwerk AG Internationalisierung der Technischen Universität Berlin. http://www.tu-berlin.de/zek/leit/aginter.html (Access: November 1999).
9 Meyer, M.: Management von Ingenieurkompetenzen im Spannungsfeld beruflicher Arbeitsteilung. Dissertation, TU Berlin, 2006.
10 Gausemeier, J.: Szenario-Management: Planen und Führen mit Szenarien. Carl Hanser Verlag, München, Wien1996.

8 Roadmap

Günther Seliger, Berlin, Germany

A considerable gap remains between the increasing public awareness about social, ecological and economical challenges and the implementation of sustainability in societal life in different global regions. Every point on earth can be reached physically in less than one day by modern traffic means and immediately by internet communication. So in principle, everyone has access to everyone in our global community. Globalization has become an inevitable condition of life, confronting mankind with respective complexity in how to cope with the intertwined social, ecological, economical and technological challenges. Apparently, industrialized as well as emerging countries have to face these challenges together in order to conserve the resources and the ecosystem of the planet for future generations. Mankind as a community under the framework of globalization can only survive if the distribution of wealth is shifted to less concentration, i.e. increasing the equity, and if more use is provided by fewer resources, i.e. increasing the use productivity of resources.

Are societal institutions, existing legislations, rules and habits, communication and education, processes and products adequately set? Due to an ever increasing speed of innovation, education - learning and teaching - becomes a coining element of leadership and teamwork in the management of societal institutions. Modern information and communication technology gives opportunities for an immediate exchange of documents and real-time communication across the globe. Internet-based learning and teaching in global teams in cooperation with partner universities has been implemented since a few years. The framework of sustainability in its global thinking / local acting scheme offers attractive contents in global engineering education.

The challenge is set. Figures 1 and 2 derived from [1] and Mathias Wackernagel [2] specify relevant aspects of how in the second half of the 20th century parameters, all determined by human activity, have developed. About one quarter of earth's surface accounting for 11.3 billion hectares can be considered as biologically productive area contributing to the regeneration of resources. The average amount of biocapacity per capita on earth in 2001 is calculated, by dividing the productive area by 6.15 billion people and results in 1.8 global hectares biocapacity per capita. The list of countries in Figure 1 shows how the ecological footprint mostly exceeds the biological capacity. The diagram curves in Figure 1 show humanity's total ecological footprint and the respective CO_2 portion of it from 1961 to 2001. Since 1985, resource consumption on a global level is higher than the ecological capacity.

	Population [mio.]	Ecological Footprint [global ha/cap]	Biological Capacity [global ha/cap]	Ecological Deficit (-) or Reverse (+) [global ha/cap]
World	6148.1	2.2	1.8	-0.4
Brazil	174.0	2.2	10.2	8.0
China	1292.6	1.5	0.8	-0.8
Germany	82.3	4.8	1.9	-2.9
India	1033.4	0.8	0.4	-0.4
Japan	127.3	4.3	0.8	-3.6
Korea Rep.	47.1	3.4	0.6	-2.8
Russia	144.9	4.4	6.9	2.6
UK	59.1	5.4	1.5	-3.9
USA	288.0	9.5	4.9	-4.7

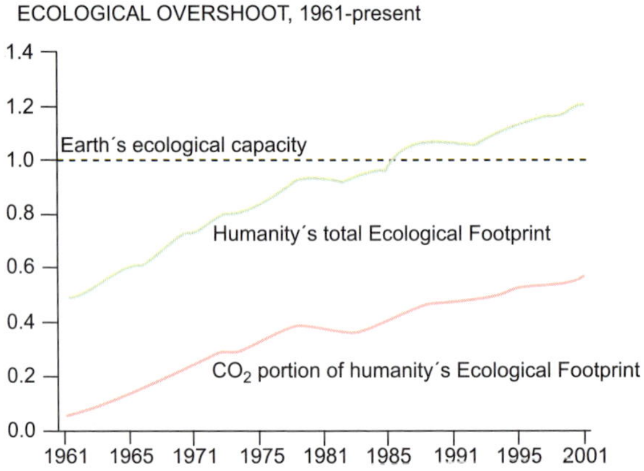

- 11.3 billion ha biologically productive area, about one quarter of earth´s surface; consisting of 2.3 billion ha of water, 9 billion ha of land
- 11.3 billion divided by 6.15 billion people gives average amount of biocapacity on planet per person: 1.8 global ha/cap

Fig. 1: Ecological footprint [1]

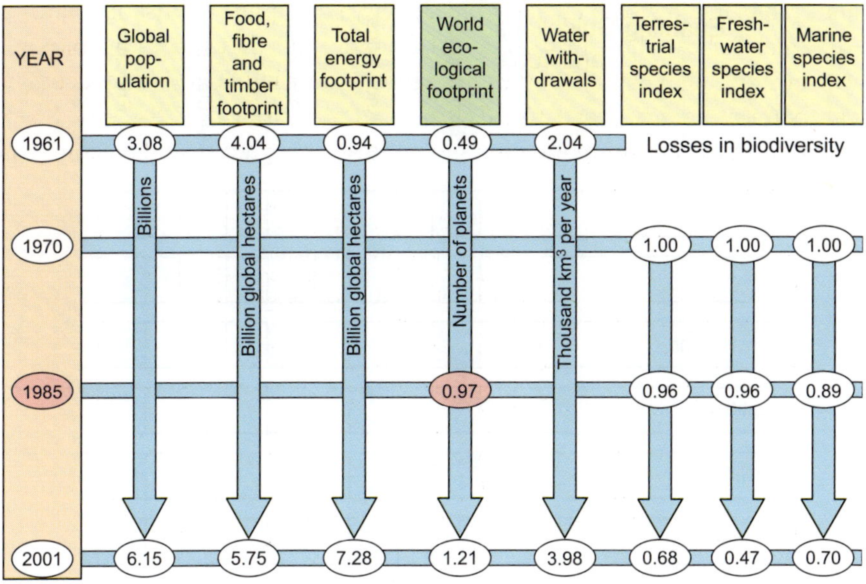

Fig. 2: The living planet through time

Figure 2 indicates that global population has increased from 3.08 billion in 1961 to 6.15 billion in 2001. Water withdrawals in the same time have increased from 2.04 to 3.98 thousand km^3 per year. Total energy consumption in 2001 is more than seven times the amount of 1961. Remarkable losses have occurred in biodiversity, where the indices since 1970 show an exponential decrease.

Engineering in a broader perspective of potentials and applications must investigate how to cope with the challenge by increasing the use productivity of resources. Researchers from engineering science, e.g. manufacturing, medical, transportation, design, information, process, electrical, and civil engineering, integrate their domain-specific knowledge and experiences thus developing methods and tools in management and technology for useful applications in selected processes and products according to criteria of sustainability.

An initiative on Sustainability in Engineering could be structured as shown in Figure 3.

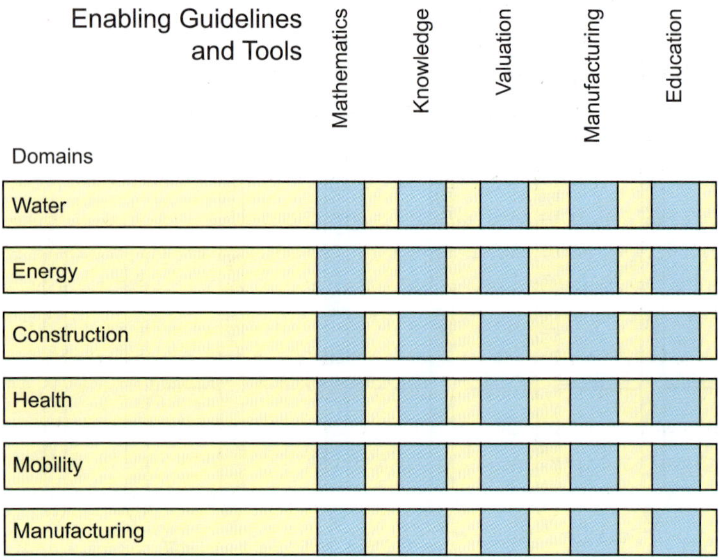

Fig. 3: Research areas

Water, energy, construction, health, mobility and manufacturing are domains of engineering activities to be directed along the guidelines of sustainability. Mathematics and knowledge creation by information science provide tools for modelling solutions without expensive realizations. Manufacturing gives methods for realizing products in processes. Valuation helps considing the manifold of sustainability criteria by creating physical artefacts and related services. Education enables for convincing and instructing people about the advantages and methods of sustainability in engineering.

The domains and enabling guidelines and tools could be covered in selected aspects of expertise in research and development by the partners of the initiative. Research clusters are described by how partners coming from their own areas of competence, by interdisciplinary approaches, identify areas of mutual interest and contribute in systemic integration to cope with the challenge of sustainability in engineering. It is expected that crossing disciplinary borders and referring to multiple criteria helps improving the design and valuation of processes and products.

A common understanding beyond disciplinary borders shall be gained by developing indicators of sustainability. Wealth is created by growth based on environmentally friendly management and technology. Processes and products are adapted to economical, environmental and social conditions in different regions of the globe, thus creating sustainability according to the regional and local conditions in the global network of demand and supply. Representing imagination and experience, the partners apply their common understanding by exemplarily developing physical artefacts and related services of sustainability in and across their respective scientific technical community. Extraction of freshwater and processing wastewater, energy transformation and storage, energy and resource saving mobility and construction, adaptable equipment for multiple usage phases in agriculture, health, mobility, achieved by respective processes of maintenance and (re)manufacturing represent means of fulfilling elementary human needs by resource saving technologies. Modeling techniques from mathematics and information science, valuation according to sustainability indicators giving guidelines and tools for design of processes and products, dissemination of knowledge and experiences on sustainable processes and products by innovative tools of education enable for the development of sustainable and innovative means fulfilling human needs.

References

1 http://www.footprintnetwork.org (Access: November 2006).

2 Wackernagel, M.: Ecological Footprint Accounting. In: Keiner, M. (Ed.): The Future of Sustainability. Springer-Verlag, Berlin, Heidelberg 2006.